# Lecture Notes in Physics



149

# Disordered Systems and Localization

## Proceedings of the Conference Held in Rome, May 1981

Edited by C. Castellani, C. Di Castro and L. Peliti

Springer-Verlag Berlin Heidelberg GmbH 1981

**Editors**

Claudio Castellani
Istituto di Fisica, Università dell'Aquila
l'Aquila, Italy

Carlo Di Castro
Luca Peliti
Università di Roma, Istituto di Fisica, "G. Marconi"
Piazzale Aldo Moro 2, 00185 Roma, Italy

ISBN 978-3-540-11163-4 ISBN 978-3-540-38636-0 (eBook)
DOI 10.1007/978-3-540-38636-0


Originally published by Springer-Verlag Berlin Heidelberg New York in 1981

2153/3140-543210

# INTRODUCTION

The 1970s have witnessed the success of a paradigm(+), the renormalization group, introduced in condensed matter physics to face a serious challenge to statistical mechanics: the understanding of critical phenomena. The paradigm was rapidly applied to an increasingly wide spectrum of problems, some of which at first sight seemed quite far from ordinary phase transitions.

At the turn of the decade, the attention of condensed matter physicists has been drawn to those problems which refuse to comply with the prevailing views: transitions which lack an evident order parameter or a reliable mean field theory as a starting point.

Some of these problems, like defect unbounding transitions, are now quite well understood, whereas others, like the spin glass transition, have shown unexpected difficulties.

In most of these still open problems the role of disorder is essential. Some transitions, however, in which disorder plays no role have not yet been driven into the general framework: this is the case of the correlation-induced metal-insulator transition. The interplay between disorder and correlation also appears as one of the most promising fields of research.

The renormalization group as such is not challenged in these problems. What is at stake is the procedure which makes its application reliable : e.g. the identification of the order parameter. It is at this stage that new concepts will be probably needed to proceed further.

Only three years ago P.W.Anderson was introducing his lectures on amorphous systems at Les Houches Summer School(++) with the statement: "If I had been giving a set of lectures on the problem of disorder a few years ago, I should almost certainly have devoted the majority of my time to a set of methods and a subject which, for practical purposes, has been ignored and will be ignored here at Les Houches; one type or another of multiple scattering theory".

(+) T.S.Kuhn - The Structure of Scientific Revolutions - The University of Chicago Press - Chicago-London, (1968).

(++) P.W.Anderson, in: R.Balian, R.Maynard and G.Toulouse (Eds.): Ill - Condensed Matter, North Holland (Amsterdam, New York, Oxford) (1979).

This last method would in fact appear as the most direct approach to disordered systems within a traditional scheme of mean field theory: to substitute to the disordered medium an ordered one with the same average properties. In this way the problem of disorder is essentially bypassed, and the machinery proper to ordered systems may be applied.

He then continued: "Multiple scattering is a paradigm (in the Kuhn sense) of the old attitude; localization and percolation are the corresponding paradigms of the new. These are phenomena which are specific to disordered and random systems; and they require a finite randomness before they become manifest."

"Non-ergodicity" was finally added by him as a closely related concept to complete the picture of the new scheme of interpretation associated with disorder.

When we organized this Conference on "Disordered Systems and Localization" it seemed to us that the time was ripe for testing these proposed paradigms and verify how far the "redefined group of practitioners" had gone in solving "all sorts of problems" left open after the main initial achievements.

For this reason we decided to combine three subjects, which seemed, at first sight, rather far from one another: percolation, disordered magnetic systems and metal-insulator transitions. We also scheduled the program in such a way that each session had almost the same number of contributions on each of the three subjects.

For the reader's convenience, however, the proceedings are divided in three chapters, each one devoted to one of the main subjects. In this way the present state of art is immediately apparent, and only a second thought shows the links and the perspectives which became evident in the discussions during the Conference.

Simple percolation theory has clarified the basic geometrical aspects, like connectivity properties, related to disorder. The combination of geometrically and thermally driven transitions has led to a variety of interesting results. The still unclear aspects of disorder, e.g. spin-glasses or metal-insulator transitions, seem however more strictly related to the peculiar dynamics of most disordered systems. Percolation therefore will hardly have a further impact in explaining them. Its most promising perspectives are instead in the direction of refined extensions of simple percolation theory suitable for applications to systems like water, gels, suspensions, and so on. Nevertheless the knowledge of the structure of the incipient infinite cluster below $p_c$ and of the infinite cluster above $p_c$ appears relevant for a more complete understanding of both theory and experiments.

The nature of the spin-glass state and the existence of the freezing transition are still strongly debated. However, these systems are beginning to comply with the paradigm of ordinary critical phenomena since a mean field theory, albeit with a complicated order parameter, is now available with a reasonable confidence. The presence of remanence effects characterizes the spin-glass state. The onset of irreversible behaviour, which was identified as one of the key mechanisms for disordered systems, is fully characterized by means of the order parameter related to the spontaneous breaking of replica symmetry. In this way the new paradigm has not only found its mathematical formulation, but has also enlarged its perspectives to include frustration.

In disorder-induced metal-insulator transitions one has reached a rather good understanding of localization. The lower critical dimensionality for the mobility edge has been located and thus the critical properties near it have been studied by means of the renormalization group approach. This has also been used more phenomenologically to obtain general information on localization at various space dimensions ($d = 2$ is one of the still controversial cases).

The introduction of correlation effects changes, however, the nature of the order parameter. Localization by itself seems not to be the complete paradigm for the metal-insulator transition. The order parameter associated to the correlation induced metal-insulator transition, if it exists, is far from being characterized.

The Conference has been the first of a series of meetings the Gruppo Nazionale di Struttura della Materia del Consiglio Nazionale delle Ricerche plans to devote each year to a different topic within the field of condensed matter physics. It was made possible by a grant of the Consiglio Nazionale delle Ricerche and by the sponsorship of the Gruppo Nazionale di Struttura della Materia and the Faculty of Sciences of the University of Rome. The Accademia dei Lincei, via its Centro Linceo Interdisciplinare di Scienze Matematiche e Loro Applicazioni, gave the Conference its beautiful setting in the gardens of Villa Farnesina. Carla Carbone, Chiara Prandi and Lucia Pratolini gave us assistance during the preparation of the Conference and its development. The help of Gigliola Gori made editing a pleasant task.

Immediately following the Table of Contents, the reader will find the list of contributed papers to the Conference that have not been published in these Proceedings for reasons of space. The abstracts of most of them have been published in the Abstract Booklet, which was distributed to the participants at the Conference. The contributions to the Conference follow, arranged in sections depending on the topic treated. The list of participants closes the volume.

While the Conference was being prepared, the news of the untimely death of John Hubbard reached us. His work had been of the greatest importance for many of the problems touched by the Conference and we had been honoured by his participation to the International Advisory Committee. As an homage to John Hubbard's memory, a commemorating address was delivered by T.M.Rice during the Conference. A summary of this talk as well as a list of John Hubbard's scientific contributions opens these Proceedings.

The Editors

## TABLE OF CONTENTS

# LIST OF CONTRIBUTED PAPERS NOT PUBLISHED IN THIS VOLUME

DECIMATION METHOD FOR LOCALIZATION PROBLEM WITH APPLICATIONS TO NON-SIMPLE SYSTEMS
*H.Aoki*

CONDUCTIVITY OF A 1-D SYSTEM OF INTERACTING FERMIONS IN A RANDOM POTENTIAL
*W.Apel*

ANALYSIS OF THE LOCALIZATION PROCESS IN AN EXACTLY SOLVABLE QUASI-PERIODIC POTENTIAL
*S.Aubry*

A TOPOLOGICAL THEORY OF CRYSTAL, INCOMMENSURATE AND AMORPHOUS STRUCTURES
*S.Aubry*

PHYSICAL PROPERTIES OF INHOMOGENEOUS MIXTURES OF ISING AND HEISENBERG SUBSTANCES
*M.Ausloos, P.Clippe, J.M.Kowalski, A.Pekalski and J.C.Van Hay*

MAGNETIC AND CRYSTALLOGRAPHIC DISORDER IN THE FRUSTRATED SYSTEM $CsMnFeF_6$
*L.Bevaart, H.A.Groenendijk, A.J.Van Duyneveldt and M.Steiner*

ELECTRON CORRELATIONS AND LOCAL CHARGE FLUCTUATIONS AT THE METAL-INSULATOR TRANSITION (MIT) IN $VO_2$
*A.Bianconi and S.Stizza*

METASTABLE STATES, INTERNAL FIELD DISTRIBUTIONS AND MAGNETIC EXCITATIONS OF SPIN GLASSES
*A.J.Bray and M.A.Moore*

THEORY OF METAL-NONMETAL TRANSITION IN LIQUID METAL ALLOYS
*F.Brouers, J.D.Franz and C.Holzhey*

ON THE THEORY OF LOCALIZATION IN DISORDERED ALLOYS
*F.Brouers and J.D.Franz*

MONTECARLO APPROACH TO THE HUBBARD MODEL
*C.Castellani, C.Di Castro, F.Fucito, E.Marinari, G.Parisi and L.Peliti*

DECIMATION IN QUANTUM SYSTEMS
*C.Castellani, C.Di Castro, J.Ranninger*

SUPERCONDUCTIVITY NEAR THE METAL-INSULATOR TRANSITION
*T.Chui, G.Deutscher, P.Lindenfeld, W.L.Mac Lean, K.Mu*

MEAN-FIELD THEORY OF AN ISING SPIN GLASS
*G.Corbelli, G.Lo Vecchio and G.Morandi*

ANISOTROPY FIELDS IN TRANSITION METAL SPIN GLASS ALLOYS
*A.Fert and P.M.Levi*

PERCOLATION IN INSULATING SPINELS: B-SUBLATTICE ($Zr\,Cr_x\,Ga_{2-x}\,O_4$)
*D.Fiorani, S.Viticoli, J.L.Dormann, M.Nogues, A.P.Murani, J.L.Murani, J.L.Tholence and J.Hamman*

TRANSPORT PROPERTIES OF SPIN GLASSES
*K.H.Fischer*

LOCAL ORDER AND DYNAMICS IN LIQUID ELECTROLYTES: SMALL ANGLE NEUTRON SCATTERING
*M.P.Fontana, G.Maisano, P.Migliardo, F.Wanderlingh, M.C.Bellissent and M.Roth*

# COMMEMORATION OF JOHN HUBBARD (1931-1980)

BY

T.M.RICE

The world of physics has suffered a great loss with the recent untimely death of John Hubbard. This is particularly true of those of us in the field of condensed matter physics and especially those who are concerned with the topics of localization and disorder that are the subject of this conference. It is very appropriate that at this conference we commemorate John Hubbard's passing by recalling some of his numerous important contributions to condensed matter physics and to the problems under discussion here.

When John Hubbard started his career in theoretical physics it was known that a surprisingly good understanding of the electron gas in metals could be obtained by ignoring the Coulomb interaction among the electrons but there was no understanding of how to develop a consistent way to treat these interactions. John Hubbard's doctoral thesis developed the dielectric approach to the problem and it was soon followed by a series of papers which are at the heart of the modern many body theory of the electron gas. J.R. Schrieffer has said of these papers, which had wide impact on the physics of metals, "His early work on the theory of exchange and correlation in the electron gas remains a classic".

Shortly thereafter, John Hubbard turned his attention to a different way of treating the many body problem. The result was a short but very influential paper which developed the method of functional integration. Over the years since his original contribution, many others have applied this method to a variety of problems, especially to the problems of electron localization. Indeed in his last major contribution John Hubbard came back to this method and applied it to the development of a first principles theory of the magnetism of iron and other transition metals. This work resolved the difficult theoretical problem of reconciling the simultaneous localized and itinerant behaviour of the magnetic electrons in 3d-metals and yielded a single model which gives reasonable values of both the magnetic moment and Curie temperature.

John Hubbard however is best known for the classic series of papers that treated electron correlations in narrow band materials. While the importance of correlation in causing the breakdown of band theory and insulating character of magnetic insulators was known from the work of Mott, Peierls, Van Vleck and Anderson, it was John Hubbard who put the problem on a firm foundation. The famous Hubbard Hamiltonian for electron correlation is as crucial and fundamental as the Ising and Heisenberg Hamiltonian for localized spins and by now has spawned almost as much work. However the large literature on the Hubbard Hamiltonian that now exists also serves to emphasize the importance of his original contribution and the depth of his understanding. W.Kohn has described his contribution as "the basis of much of our present thinking about the electronic structure of large classes of metals and insulators". It is also the basis of much of what we are discussing at this conference.

John Hubbard studied at Imperial College, University of London, receiving B.Sc. and Ph.D. degrees in 1955 and 1958 respectively. Most of his scientific career was spent as Head of the Solid State Theory Group at the Atomic Energy Research Establishment in Harwell, England. He visited a number of institutions in the U.S. at various times in his career and in 1976 he joined the staff of the IBM Research Laboratory at San Jose, Ca., a position he held at his death.

John Hubbard's work was characterized by great originality and by an uncommon ability to obtain elegant mathematical formulations and solutions of very difficult and fundamental problems. His passing leaves a void in the theoretical physics community which will not be filled.

# SCIENTIFIC PAPERS BY JOHN HUBBARD


1. Plasma Oscillations in a Periodic Potential: The One-Zone Theory. Proc. Phys. Soc. A67, 1058-1068 (1954).

2. On the Interaction of Electrons in Metals. Proc. Phys. Soc. A68, 441-443 (1955).

3. The Dielectric Theory of Electron Interactions in Solids. Proc. Phys. Soc. A68, 976-986 (1955).

4. The Description of Collective Motion in Terms of Many-Body Perturbation Theory. Proc. Roy. Soc. A240, 539-560 (1957). (Reprinted in The Many Body Problem, ed. D. Pines, Benjamin, N.Y., 1961).

5. The Description of Collective Motion in Terms of Many-Body Perturbation Theory II. The Correlation Energy of a Free Electron Gas. Proc. Roy. Soc. A243, 336-352 (1957). (Reprinted in The Many Body Problem, ed. D. Pines, Benjamin, N.Y., 1961).

6. The Description of Collective Motion in Terms of Many-Body Perturbation Theory III. The Extension to a Non-Uniform Gas. Proc. Roy. Soc. A244, 199-211 (1958).

7. The Instabilities of Cylindrical Gas Discharges with Field Penetration. AERE-T/R2668 (1958).

8. Theoretical Problems Suggested by Zeta (with W.B. Thompson, S.F. Edwards and S.J. Roberts). Second U.U. International Conference on the Peaceful Uses of Atomic Energy 32, 65-71 (1958).

9. The Calculation of Partition Functions. Phys. Rev. Lett. 3, 77-78 (1959).

10. Long - Range Forces and the Diffusion Coefficients of a Plasma, (with W.B. Thompson).Rev. Mod. Phys. 32, 714-717 (1960).

11. The Friction and Diffusion Coefficients of the Fokker-Planck Equation in a Plasma. Proc. Roy. Soc. A260, 114-126 (1961).

12. The Friction and Diffusion Coefficients of the Fokker-Planck Equation in a Plasma II. Proc. Roy. Soc. A260, 371-387 (1961).

13. Electron Correlations in Narrow Energy Bands. Proc. Roy. Soc. A276, 238-257 (1963).

14. Electron Correlations in Narrow Energy Bands II. The Degenerate Band Case. Proc. Roy. Soc. A277, 237-259 (1964).

15. Electron Correlation in Narrow Energy Bands III. An Improved Solution. Proc. Roy. Soc. A281, 401-419 (1964).

16. Exchange Splitting in Ferromagnetic Nickel. Proc. Phys. Soc. A84, 455-464 (1964).

17. Correlations in Partly-Filled Narrow Energy Bands. Proc. Bull. Int. Conf. on Materials. (Gordon and Breach, N.Y., 1965).

18. Electron Correlations in Narrow Energy Bands IV. The Atomic Representation. Proc. Roy. Soc. A285, 542-560 (1965).

19. Covalency Effects in Neutron Diffraction from Ferromagnetic and Anti-Ferromagnetic Salts (with Dr. W. Marshall). Proc.Phys. Soc. 86, 561-572 (1965).

20. Weak-Covalency in Transition Metal Salts (with D.E. Rimmer and F.R. Hopgood). Proc. Phys. Soc. 88, 13-36 (1966).

21. Electron Correlations in Narrow Energy Bands V. A Perturbation Expansion about the Atomic Limit. Proc. Roy. Soc. A296, 82-99 (1966).

22. Electron Correlations in Narrow Energy Bands VI. The Connection with Many-Body Perturbation Theory. Proc. Roy. Soc. A296, 100-112 (1966).

23. The Approximate Calculation of Electronic Band Structures. Proc. Phys. Soc. 92, 921-937 (1967).

24. Electron Correlations at Metallic Densities. Phys. Letts. 25A, 709-710 (1967).

25. The Approximate Calculation of Electronic Band Structures II. Application to Copper and Iron (with N.W. Dalton). J. Phys. C. (Proc. Phys. Soc. 2) 1, 1637-1649 (1968).

26. Generalized Spin Suscepibility in the Correlated Narrow-Energy-Band Model (with K.P. Jain). J. Phys. C. (Proc. Phys. Soc. 2) 1, 1650-1657 (1968).

27. Spin-Waves in the Paramagnetic Phase (with J.L. Beeby). J. Phys. C. (2) 2, 376-377 (1968).

28. Collective Motion in Liquids (with J.L. Beeby). J. Phys. C. (2) 2, 556-571 (1969).

29. The Approximate Calculation of Electronic Band Structures III. J. Phys. C. (2) 2, 1222-1229 ( 1969).

30. Spin Correlation Functions at High Temperature (with M. Blume). Phys. Rev. 81, 3815-3830 (1970).

31. Spin Correlation Functions in the Paramagnetic Phase of a Heisenberg Ferromagnet. J. Phys. C 4, 53-70 (1971).

32. Spin Correlations in the Paramagnetic Phase. J. Appl. Phys. 42, 1390 (1971).

33. Critical Behaviour of the Ising Model. Phys. Letts. 39A,365 (1971).

34. Scaling Relations in the Wilson Theory. Phys. Letts. 40A, 111 (1972).

35. Wilson Theory of a Liquid-Vapour Critical Point (with P. Schofield). Phys. Letts. 40A, 245 (1972).

36. The Approximate Calculation of Electronic Band Structures V. Wave Functions (with P.E. Mijnarends). J. Phys. C5, 2323 (1972).

37. A Perturbation-Theoretic Derivation of Wilson Theory. J. Phys. C6, 2765 (1973).

38. On the Perturbation Theory of Critical Phenomena. Phys. Letts. 45A, 349 (1973).

39. The Critical Correlation Function of the Ising Model in a Magnetic Field and on the Co-Existence Curve. J. Phys. C7, L216 (1974).

40. Generalized Wigner Lattices in One Dimension and Some Applications to TCNQ Salts. Phys. Rev. B17, 494-505 (1978).

41. Electronic Structure of One-Dimensional Alloys. Phys. Rev. B19, 1828-1839 (1979).

42. The Magnetism of Iron. Phys. Rev. B19, 2616-2636 (1979).

43. Magnetism of Iron II. Phys. Rev. B20, 4584-4595 (1979).

44. Many-Body Theory. Contribution to the Theoretical Physics Division (AERE Harwell, U.K.) 15th Anniversary Progress Report (1980).

45. Calculation of the Magnetic Properties of Iron and Nickel by the Functional Integral Method. To be published in the Proceedings for the Symposium on Electron Correlation and Magnetism in Narrow Bands, November 1980, Susono, Japan.

46. The Magnetism of Nickel. To be published in Phys. Rev. B , (1981).

47. The Magnetism of Iron and Nickel. To be published in Journal of Applied Physics (the Proceedings of the 26th Annual Conf. on Magnetism and Magnetic Material, Dallas, Texas, November 1980).

48. On the Resolved Power of Time-Reversed Wavefront Imaging Devices. Accepted for publication in the Journal of the Optical Society of America (1981).

49. On the Neutral-Ionic Phase Transformation (with J. Torrance), to be published, Phys. Rev. B (1981).

50. On the Magnetic Hysteresis of an Assembly of Small Particles. IBM Internal Publication (1981).

# PERCOLATION AND RELATED TOPICS

# SCALING PROPERTIES OF PERCOLATION CLUSTERS

D.Stauffer
Institut für Theoretische Physik, Universität, 5000 Köln 41, W.Germany

Abstract:

Some of the progress since 1979 is reviewed for methods and results in simple percolation and lattice animal theory. The figure below gives the number of publications with "percolation" etc in the title, as a function of time. Obviously the field is not dead.

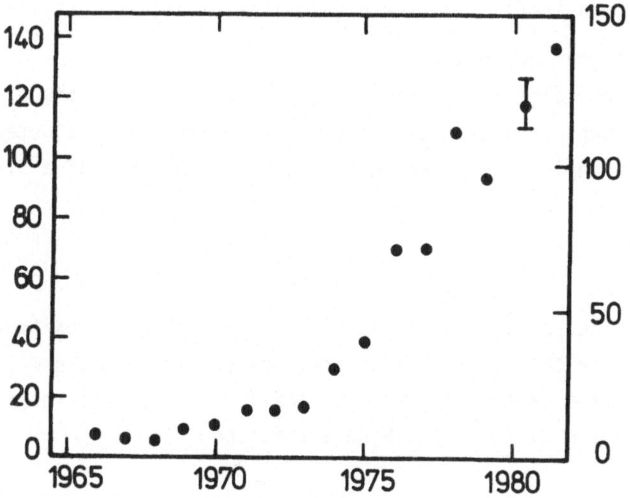

## I. INTRODUCTION

Percolation is a still growing field of research. Recent reviews[1-5] summarize the various aspects of percolation theory; the present article thus merely reviews some of the more recent progress. As for many other fields of research, one may either look at the numerous more or less speculative and fascinating applications and modifications of the theory; or one restricts oneself to the details of the basic, most simple, model. At this conference on disordered systems and localization, Stanley will emphasize (I hope) the first approach, and thus I take the second. In this sense "percolation" refers here only to random site percolation (in a few cases bond percolation) on a periodic lattice in d dimensions, $1 < d < 6$. First we look at new methods, and then at new results, for percolation problems, including lattice animals. The field is too large to cover all important publications; the present selection is based subjectively on the author's experience (and lack of it, respectively).

Let us shortly repeat what percolation is:
Imagine each site of a periodic lattice to be occupied with probability p and empty with probability 1-p, independent of the status of its neighbors. A cluster then is a group of neighboring occupied sites, as in the following example of a single cluster on a square lattice:

If p increases, then at the percolation threshold $p = p_c$ a phase transition occurs in the sense that at $p_c$ for the first time an infinite cluster appears which percolates from top to bottom, from left to right, etc. Scaling theory concerns itself mainly with critical phenomena close to this phase transition.

## II. METHODS (mostly Monte Carlo)

What new improvements exist for Monte Carlo simulation of percolation, for precise determination of critical exponents, or for relations with other phase transitions ?

### a) Growth of one cluster

For Monte Carlo simulations, Leath's method[6] of letting a single cluster grow has been improved recently[7,8]. One starts with a single occupied site and then lets the cluster grow by adding more neighbors to it, with probability p for each place. The computer saves a lot of effort if at each step of adding new neighbors only those neighbor sites of occupied places are filled (with probability p) which have not been investigated at previous stages of the growth process. Then the total number of sites to be investigated (i.e. either to be filled or to be left empty forever) is proportional to the number s of sites in the cluster, and not (as in the original method[6]) to the larger cluster volume ( $\propto s^{1+1/\delta}$). As a result, ref.8 seems to be the most comprehensive investigation of cluster properties in a single original paper. For example this Monte Carlo simulation[8] confirmed reasonably a prediction of de Gennes[9] for the fraction of cluster sites which can be removed without splitting the cluster.

b) Iteration for $p_c$

In the more usual Monte Carlo simulation, all lattice sites are investigated in a regular manner, and each site investigated is either filled or left empty. Here the percolation threshold $p_c$ can be found with good accuracy,even after only a few samples of the same lattice have been investigated, by the following iteration method. This iteration is not really new[10,11] but seems not yet described in the open literature.

Let $p_{co}$ be a rough estimate for the position of the threshold. First one fills the lattice with probability $p_{co}$ for each site. Then one checks if a cluster percolates from top to bottom. If yes (no), we shift our estimate to $p_{c1} = p_{co} - \Delta$ ($p_{c1} = p_{co} + \Delta$). Then, using the same sequence of random numbers as before, we fill the lattice with this new probability $p_{c1}$ and check if it percolates. This process is repeated until for the first time the last estimate, $p_{ci}$, gives a nonpercolating sample whereas the previous one, $p_{c,i-1} = p_{ci} + \Delta$, gave a cluster percolating from top to bottom.(In the case of no percolation at $p_{co}$, we need percolation at $p_{ci}$.) Now we know that the true threshold for this sequence of random numbers lies between $p_{ci}$ and $p_{c,i-1}$. From now on, by repeated dichotomy with $p_{c,i+1} = p_{ci} \pm \Delta/2$, $p_{c,i+2} = p_{c,i+1} \pm \Delta/4$, etc, we can estimate the threshold with exponentially increasing accuracy, using five to ten such iterations. (For two dimensions, $\Delta \simeq \frac{1}{4} L^{-3/4}$ is reasonable if $p_c$ is known already quite accurately.[12] An additional simplification for square bond percolation is described in ref. 13.) For all these iterations, which give just one estimate for $p_c$, we use the same random numbers; thus we employ the fact that a computer is not really random (if functioning properly) but produces even these (pseudo-)random numbers according to a well-defined sequence if one puts the computer's random number generator always back to the same initial status (same seed).

Repeating this process many times, with different sequences of random numbers, we get as many different estimates for $p_c$ (apart from "accidental" coincidences due to finite accuracy). Their average $\langle p_c \rangle$ is our final estimate for the percolation threshold. Its "error bar" determines the correlation length exponent $\nu$ due to finite size scaling[10,11,12] in systems of size $L^d$:

$$(\langle p_c^2 \rangle - \langle p_c \rangle^2)^{1/2} \propto L^{-1/\nu} \quad (L \to \infty) \tag{1}$$

(note $\xi \propto |p-p_c|^{-\nu}$ for the correlation length). Eq(1) can also be regarded as a large-cell real-space renormalization group result, in the limit of large L. Lattices of up to hundred million sites have been used[12] for this purpose. How can these sites be stored ?

c) A Fortran Program

Table I: Part of Fortran subroutine to check if system percolates from top (k=1) to bottom (k=L) in simple cubic lattice. For empty sites, Level(i,j) = Max, for occupied sites in the top-most plane, Level(i,j) = 1. For the lower planes k = 2,3,... the function LASS and the array Level indicate the grouping into separate clusters. Ranf is the random number generator, with $0 \le Ranf < 1$; MinØ gives the smallest of its three arguments.

```
        Lp1 = L+1
        Index = 1
        N(1) = 1
        Do 2 k = 2,L
        Iconn = Ø
C       Iconn equals 1 if one site in plane k is connected to top level
        Do 3 i = 2,Lp1
        Do 4 j = 2,Lp1
          If (Ranf(j).GT.p) Goto 7
C         If yes, new site is empty
          Mold = LASS(Level(i,j))
          M1   = LASS(Level(i-1,j))
          M2   =      Level(i,j-1)
          MTR  = MinØ(Mold, M1, M2)
          If(MTR.eq.Max) Goto 5
C         If yes, all three previously investigated neighbors are empty
          If(MTR.eq.1) Iconn = 1
C         If yes, one of the three neighbors was connected to top level
          If(Mold.LT.Max) N(Mold) = MTR
          If( M1 .LT.Max) N( M1 ) = MTR
          If( M2 .LT.Max) N( M2 ) = MTR
          Goto 8
5         Index = Index + 1
          MTR = Index
8         Level(i,j) = N(MTR) = MTR
          Goto 4
7         Level(i,j) = Max
4       Continue
3       Continue
        If(Iconn.eq.Ø) Goto 6
2       Continue
        Ispan = 1
        Return
6       Ispan = -1
        Return
        End
```

In such a Monte Carlo simulation on big computers it is not necessary to store the whole lattice with all its $L^d$ sites: In three dimensions only one Level(i,j) with i and j = 1,2,⋯,L+1 needs to be stored, and in two dimensions only one row[14]. (It is practical to regard the boundary sites Level(1,j) and Level(i,1) always as empty and to avoid periodic boundary conditions[15].) Thus for 10 000 × 10 000 lattices in two dimensions[12] only one array of size 10 001 is necessary to store the occupation status, and simulations of larger systems are possible[14]. Much larger are usually the storage requirements for the array N above, which stores cluster labels.

To simplify future work along these lines we gave on the previous page the main part of a Fortran program (J.Kertesz, priv.comm.; P.J.Reynolds, Thesis, MIT 1979) based on the Hoshen-Kopelman technique[14]. It calculates Ispan = 1 if a cluster extends from top (k = 1) to bottom (k=L) of a $L \times L \times L$ simple cubic lattice; otherwise Ispan = -1. In the first part of the program (not shown) one sets Level(i,j) = 1 for all occupied sites in the top level (k=1), and Level(i,j) = Max otherwise, including Level(1,j) = Level(j,1) = Max. Here Max is an integer larger by unity than the largest Index needed in the program for the array N. (Typically, Max = $0.2\ L^d$ is sufficient; recycling of unused labels is possible to save memory space for N.) Thus N has the dimension Max - 1. The function LASS is a simplified version of the Classify subroutine of ref.14 and looks for the root of the cluster in the label tree. Modifications are necessary if one also wants to count the number of finite clusters in the sample.

Table II. Function LASS for the main program of table I. Max is set equal to 25001 in this example; one should first check that in the other program Index and MTR never become larger than 25000 in this case.

```
      Function LASS(M)
      Dimension N(25000)
      Common N, Max
      If(M.LE.0 .or. M.GT.Max) Stop 2
C     Omit this last line after program has been tested sucessfully
      If(M.NE.Max) Goto 1
      LASS = Max
      Return
C     The site was empty
1     MØ = M
      M = N(M)
      If(MØ.NE.M) Goto 1
      LASS = M
      Return
      End
```

This type of program is also useful if one makes a Monte Carlo simulation of one single cluster with fixed size s and fluctuating shape[2]. For then at each exchange of sites one has to check if the exchange would not split the cluster.

### d) Nightingale renormalization

A numerical method, different from Monte Carlo simulation, is called Nightingale renormalization, phenomenological renormalization or transfer matrix approach[16,17] and is also based on finite size scaling. Ref.16 calculates exactly the correlation length in an infinite strip of width n by a transfer matrix. As in renormalization theory[2,3,1], a similarity assumption relates the two correlation lengths in two strips of withs n and m to two different concentrations p' and p, both close to the fixed point $p_c$, with

$$(dp'/dp)_{p=p_c} = (n/m)^{1/\nu} \qquad (2).$$

A numerical evaluation of the LHS of (2) gives estimates for $\nu$ depending on m and n; they have to be extrapolated suitably to $m,n \to \infty$. Thus the technique is similar to series expansions. Ref.18 increased drastically the accuracy and gave $1/\nu = 0.750 \pm 0.001$ for two-dimensional percolation, excluding reliably an earlier suggestion [11] $1/\nu = \log(\frac{3}{2})/\log(\sqrt{3}) = 0.738$. The method is more accurate but less versatile than Monte Carlo renormalization and was restricted to two dimensions.

e) Potts models

Percolation is known[2,3] to be a special limit of the Potts model. Plausible relations have been sucessfully postulated in the last two years between the critical exponents of Potts models and other phase transitions, in two dimensions. Den Nijs[18] suggested a relation for the correlation length exponent between the Potts model and the exactly solved 8-vertex model; in the percolation limit this prediction gives $\nu = 4/3$ in two dimensions. For the other exponents in the Potts model, another relation was suggested independently by Nienhuis et al[19] and Pearson[20] (from two exponents we can determine the others by scaling laws[2,3]); for the two-dimensional percolation limit they give, for example, $\gamma = 43/18 = 2.38888$.

The above-mentioned results of Blöte et al[17] confirmed these exponents for all two-dimensional Potts models except for the very special case of the four-state Potts model, with an accuracy of about $10^{-3}$. Very recently, Herrmann[21] also found good numerical confirmation even in this remaining problem case. Thus these postulated formulas are likely to be exact generally, and therefore also in the percolation limit. In other words, since the publication of the earlier reviews[1-5] one has found the presumably exact critical exponents of two-dimensional percolation (listed below in table III) but they have not yet been proven rigorously. Note that it took more than a decade of years to transform the plausible non-rigorous $p_c = 1/2$ for square bond percolation into a rigorous result[22]; it may also take some time to prove mathematically rigorously the den Nijs - Nienhuis et al - Pearson formulas.

## III. RESULTS FOR CLUSTER NUMBERS

### a) Introduction

First we repeat the well-known definitions[2,3]. Let $n_s(p)$ be the average number (per lattice site) of clusters containing s sites each. Then the critical exponents $\alpha$, $\beta$, $\gamma$, $\delta$, $\nu$ are defined through[2,3]

$$(\sum n_s)_{sing} \propto \varepsilon^{2-\alpha};\ (\sum n_s s)_{sing} \propto \varepsilon^{\beta};\ \sum n_s s^2 \propto \varepsilon^{-\gamma};\ \xi \propto \varepsilon^{-\nu};\ \sum n_s(p_c)s(1-e^{-hs}) \propto h^{1/\delta} \quad (3)$$

where the sum goes over $s = 1,2,3,\ldots$, $\varepsilon \equiv |p-p_c| \to 0$, $h \to 0$, and $\xi$ is the correlation length (typical cluser radius); the subscript sing refers to the leading nonanalytic contribution if the value itself remains finite at $p_c$.

The scaling assumption

$$n_s(p) \propto s^{-\tau} f(\varepsilon s^{\sigma}) \quad (\varepsilon \to 0,\ s \to \infty) \qquad (4)$$

relates these exponents though the scaling laws

$$\tau = 2+1/\delta,\ \sigma = 1/\beta\delta,\ 2-\alpha = \gamma+2\beta = \beta(\delta+1)\ (= d\nu \ \text{ in d dimensions}) \qquad (5).$$

For the "susceptibility" $\sum s^2 n_s$ the "amplitude ratio" $C'/C = R$ with

$$R = \sum s^2 n_s(p=p_c+\varepsilon)/\sum s^2 n_s(p=p_c-\varepsilon) \qquad (\varepsilon \to 0) \qquad (6)$$

is of particular interest, since it is supposed to be as "universal"[23] as the critical exponents, i.e. exponents and R are (supposed to be) independent of the lattice type for all lattices of the same dimensionality d.

Another set of "non-critical" exponents is often denoted by $\zeta$, $\theta$ and $\rho$:

$$\left.\begin{array}{ll} n_s(p) \propto s^{-\theta}\exp(-\text{const}_p \cdot s^{\zeta}) & (7), \\ R_s(p) \propto s^{\rho} & (8). \end{array}\right\} s \to \infty$$

Here $R_s$ is the average radius (or gyration) of s-clusters (note $\xi^2 \propto \sum R_s^2 s^2 n_s / \sum s^2 n_s$ from Essam[3]). In contrast to the true critical exponents in eqs(3,4), these "non-critical exponents" are defined for all p, not only near $p_c$, and differ on different sides of the phase transition.

Refs.2,3 already discussed why

$$\zeta(p<p_c) = 1;\quad \zeta(p>p_c) = 1 - 1/d \qquad (9)$$

presumably is correct. Right at $p_c$, the exponent $\zeta$ is undefined whereas $\theta(p_c) = \tau$ and $\rho(p_c) = (1+1/\delta)/d = \sigma\nu$ according to scaling assumptions[2,3]. Presumably[24,25] one has for simple percolation for $\zeta$, $\theta$ and $\rho$ one value for all p between zero and $p_c$, another value for all p between $p_c$ and unity, and except for $\zeta$ a third value right at $p_c$; in more complicated situations the behavior may be more complicated[26] but we ignore these complications in this review.

In summary, not only can we select two of the critical exponents $(\alpha,\beta,\gamma,\delta,\nu,\sigma,\tau)$ as independent in view of the scaling laws (5), but in addition we have six "non-critical" exponents $\zeta(p<p_c)$, $\zeta(p>p_c)$, $\theta(p<p_c)$, $\theta(p>p_c)$, $\rho(p<p_c)$, $\rho(p>p_c)$. Refs.2,3 discussed already $\zeta(p<p_c) = 1$, $\zeta(p>p_c) = 1-1/d$, and $\rho(p>p_c) = 1/d$. In the next section we will see which progress has been made very recently for the other exponents.

b) Exponents

For the critical exponents of eq(5), in two dimensions Nightingale renormalization turned out[17] to be more accurate than Monte Carlo (renormalization) methods, and confirmed the formulas of den Nijs[18], Nienhuis et al[19], and Pearson[20], as discussed in section II e. These presumably exact exponents are listed as rational numbers in our table III. For higher dimensions, Nakanishi and Stanley[27] looked in detail at $n_s(p)$ and "equation of state"[2,3] up to d = 7. We refer to their papers for high dimensions and list here only d = 2 and d = 3.

Table III. Predictions for universal quantities, i.e. for exponents and one amplitude ratio. (See Aharony[23] for more amplitude ratios.) Rational numbers indicate (presumably) exact result, those with a decimal point are numerical estimates.

| Exponent | d=2 | d=3 |
|---|---|---|
| $\alpha$ | -2/3 | -0.5 |
| $\beta$ | 5/36 | 0.4 |
| $\gamma$ | 43/18 | 1.7 |
| $\delta$ | 91/5 | 5. |
| $\nu$ | 4/3 | 0.8 |
| $\sigma$ | 36/91 | 0.5 |
| $\tau$ | 187/91 | 2.2 |
| $\rho(p=p_c)$ | 48/91 | 0.4 |
| $\rho(p<p_c)$ | 0.6 | 1/2 |
| $\rho(p>p_c)$ | 1/2 | 1/3 |
| $\zeta(p<p_c)$ | 1 | 1 |
| $\zeta(p>p_c)$ | 1/2 | 2/3 |
| $\theta(p<p_c)$ | 1 | 3/2 |
| $\theta(p>p_c)$ | 5/4 | -1/9 |
| C'/C | .005 | 0.1 |

For the "non-critical" exponents, Aizenman et al[28] gave more proofs on $\zeta$, consistent with eq(9). For $\theta$ and $\rho$ below $p_c$, Parisi and Sourlas[29] found a relation to the Lee-Yang edge singularity in the Ising model at d-2 dimensions. From its exact solution they find exactly:

$$\theta(d=2,\ p<p_c) = 1;\ \theta(d=3,\ p<p_c) = 3/2,\ \rho(d=3,\ p<p_c) = 1/2 \qquad (10)$$

in good agreement with earlier numerical estimates[29]. Above $p_c$ Lubensky and McKane[30] related the percolation problem to that of Ising droplets and give the same $\theta$ as in the Ising model:

$$\theta(p>p_c) = (1 + 4d - d^2)/2d \quad (d \neq 3;\ d \neq 5) \qquad (11a)$$

$$\theta(d=3,\ p>p_c) = -1/9\ ;\ \ \theta(d=5,\ p>p_c) = -449/450 \qquad (11b).$$

These results are particularly valuable since numerical attempts failed to get good $\theta$ estimates above $p_c$. (Ref.30 also predicts terms $\propto s^{1-2/d}, s^{1-3/d}, \ldots$ in $\log(n_s)$ for $s \to \infty$, which would mean that for $d > 2$ eq(7) is too simple.) These results above $p_c$ can also be related to "essential singularities" and "analytic continuations" in the complex plane of the equation of state. Table III summarized also these non-critical exponents.

(For possible logarithmic factors in two-dimensional percolation see ref.31. The number of infinite clusters in an infinite system is either zero, or one, or infinite according to ref.32; the last possibility does not seem to occur at low dimensions[27]. Ref.33 discussed fluctuations in cluster numbers and related quantities.)

c) Scaling function

Is eq(4) correct? If yes, how does the scaling function f there look like ? For low dimensionality d = 2,3 it was already known[2,3,34] to be a good approximation. Nakanishi and Stanley[27] also confirmed it by Monte Carlo investigation for d up to 7. (Nightingale renormalization has not yet given results for scaling functions.) Basically, the function $f(z) = f((p-p_c)s^\sigma)$ has a bell-like shape, though it is not exactly Gaussian. Its maximum is at negative arguments (i.e. for p below $p_c$); but the higher the dimension is the closer is this maximum to zero, i.e. to $p = p_c$. On the next page, fig.1 summarizes the results of ref.27 for all d = 2 to 7.

Assumption (4) means that the ratio $n_s(p)/n_s(p_c)$ equals f(z). We see indeed from fig.1 that all points follow the same curve in this semilogarithmic plot of this ratio versus the argument z of the scaling function. This data collapse is required by the scaling assumption (4). The same scaling behavior is known since a long time [35] for the Bethe lattice (Cayley tree); and indeed for d = 6 (the upper critical dimension for percolation) the Monte Carlo results have a strong similarity to the classical Bethe lattice results, which is a simple Gaussian[35]. For example, the ratio C'/C, which can be calculated from the scaling function and which is quite small for d = 2 and 3, is close to unity for d = 6 as it is for the Bethe lattice[27]. Above six dimensions, i.e. outside the scope of the present review, the scaling assumption (4) is questionable[27,36], and more research should be done here. Below six dimensions the result (9) for $\zeta$, combined with the scaling assumption (4), suggests that the scaling function $f(z) = f(\varepsilon s^\sigma)$ decays asymptotically as

$$\log f \propto -|z|^{(d-1)/\sigma d} \qquad (|z| \to \infty) \qquad (12a)$$

above $p_c$, whereas

$$\log f \propto -|z|^{1/\sigma} \qquad (|z| \to \infty) \qquad (12b)$$

below $p_c$. At least no evidence has been found, to my knowledge, against this simple behavior.

Fig.1. Test of scaling assumption (4) for the cluster numbers in two to seven dimensions. The solid lines are least-square fits to a cubic polynomial for d = 2 to 5, and to a parabola for d = 6 and 7. The Bethe lattice result corresponds to a parabola in this type of plot. From Nakanishi and Stanley[27].

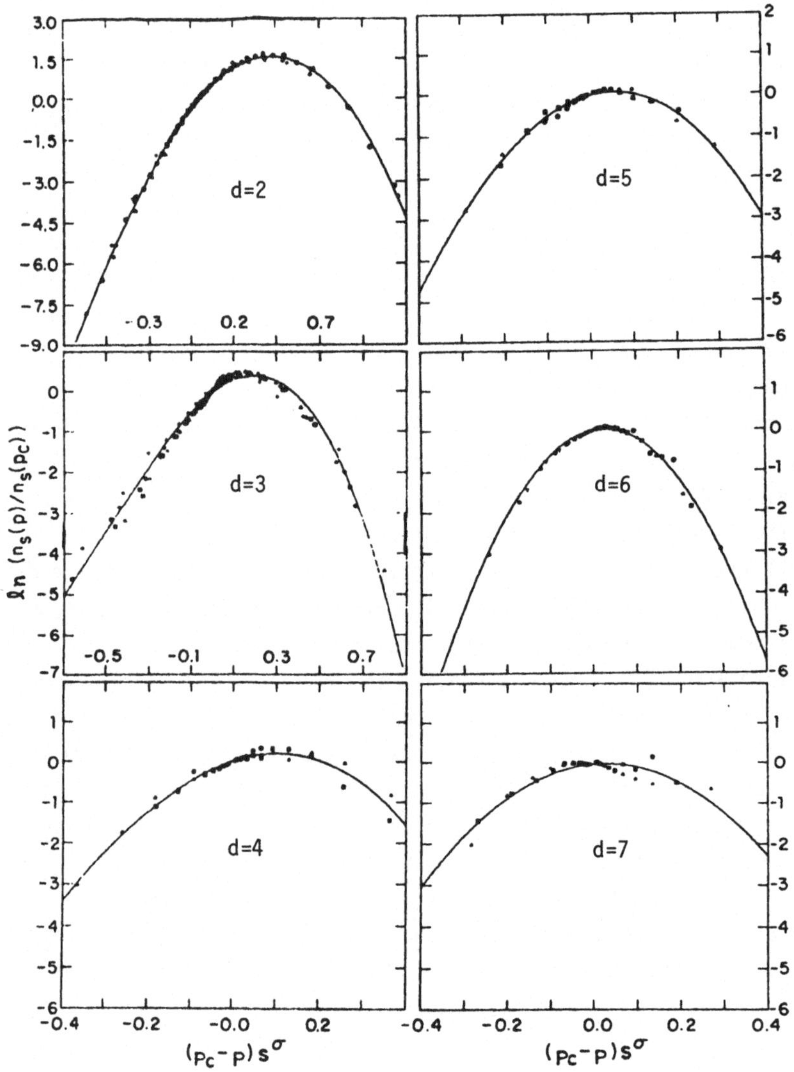

## IV RESULTS FOR CLUSTER STRUCTURE

Investigations of cluster structures can be separated roughly into two parts: Microscopic properties and coarse-grained averages. For the first type one looks at properties which take into account the details of the lattice structure; for example one finds that for very small p the average perimeter-to-size ratio of large clusters is about 2.7 for site percolation on the simple cubic lattice[37,38]. In the second "coarse grained" approach one averages over many sites to obtain information about the cluster as a whole; for example one finds in the same limit and lattice that the average cluster radius increases as $\sqrt{s}$. For further microscopic details we refer the reader to earlier reviews[1-6] and ref.8. We restrict us here to three averaged quantities: radius, density profile, and interface thickness (surface roughening).

a) Radius and density profile

In the investigation of linear polymers consisting of s monomers, one of the fundamental quantities of interest is the radius of gyration, $R_s$, and its exponent $\rho$ defined through

$$R_s = (\textstyle\sum_i r_i^2/s)^{1/2} \propto s^\rho \quad (s \to \infty) \tag{13}$$

where $r_i$ is the distance of cluster site i from the cluster center-of-mass. (For percolation, we do not recommend to denote the radius exponent by $\nu$). For percolation the linear "cluster"of polymer theory has become "branches", but the definition (13) remains the same. Note that no average over all sizes s is involved in eq(13), but of course we average over all clusters having the same size s. Our table III listed already the best values for the radius exponent $\rho$.

For small clusters, exact values for $R_s$ are tabulated in ref.39. For larger clusters, Monte Carlo simulations[8,37] at $p_c$ did not challenge the earlier conclusion[40,2,3] that $\rho(p_c) = \sigma\nu = (1+1/\delta)/d$ , as mentioned already after eq(9). Holl[37] found, with the method of simulating shape fluctuations at fixed s, that $\rho = 0.65 \pm 0.02$, $= 0.53 \pm 0.01$ and $= 0.495 \pm 0.005$ below, at and above $p_c$ in the square lattice, whereas $\rho = 0.53 \pm 0.02$, $= 0.39 \pm 0.02$, and $= 0.330 \pm 0.005$ in the simple cubic lattice. These results agree with theoretical predictions apart from errors of the order of the quoted error bars: Parisi and Sourlas gave $\rho(p<p_c) = 1/2$ in three dimensions[29], and the droplet model gives[2]

$$\rho(p>p_c) = 1/d \tag{14}$$

in d dimensions. Only the two-dimensional exponent below $p_c$ lacks a theoretical explanation.

As suggested by eq(14), above $p_c$ (as opposed to $p \le p_c$) the clusters are droplet-like with a relatively narrow interface (surface layer) separating the cluster interior from the surrounding vacuum. Figure 2 on the next page, based also on Monte Carlo simulations at fixed s, shows this density profile similar to that of a raindrop. (The density profile is the probability that a lattice site at distance r from the cluster center-of-mass belong to that cluster.) We see clearly a density plateau

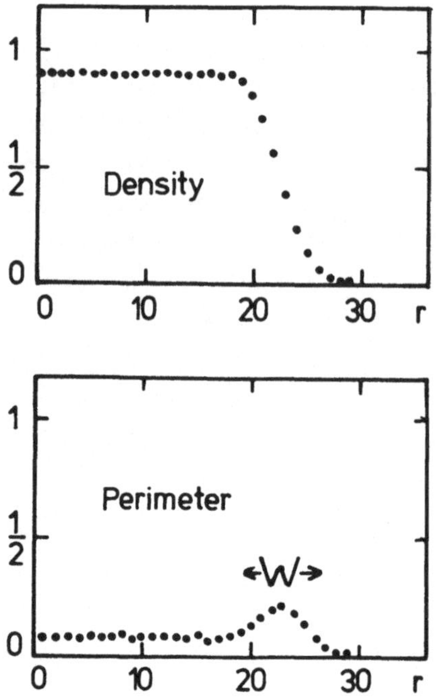


Fig.2. Monte Carlo results for density profile and perimeter profile of a very large cluster above $p_c$ (s = 1400, p = 0.9, square lattice). Similar results were observed in the simple cubic lattice. From ref.37. The surface width is indicated by W in the lower part of the figure.

in the "liquid" interior of the droplet, corresponding to the density of the infinite network with which the finite cluster would coexist at that value of p. Far away from the cluster center the density is zero; this vacuum is separated from the droplet interior by what we now call a surface. The surface width is called W.

The perimeter is the number of empty sites which are nearest neighbors to occupied cluster sites; it plays the role of an energy if one makes an analogy with thermal phenomena. Thus the perimeter profile is the probability that a site at distance r from the cluster center is empty but is neighbor to a cluster site. The lower part of fig.2 shows this perimeter profile: A constant perimeter density in the "liquid" interior of the droplet is separated from the outside vacuum by a surface with an excess amount of perimeter. This excess perimeter should be connected to the analog of surface tension for percolation.[2]

Now we want to study this surface closer: Does its width remain finite for $s \to \infty$ ?

b) Surface roughening

To study in greater detail the structure of this droplet surface it is practical to look at flat surface areas of size $L \times L$ (for three dimensions; in two dimensions the surface line has length L), instead of working with the curved surfaces of finite clusters. This flat piece of surface thus can be regarded as a part of the surface of a truely huge cluster. Instead of asking now for the limit $s \to \infty$ we look at $L \to \infty$ if we want to know if the surface thickness remains finite. In a Monte Carlo calculation, one keeps all sides of the sample fixed by rigid boundaries, except for one side where fluctuations of a free surface can develop.

For real liquids, a surface wave (ripplon, capillary wave) of wavevector q has an "elastic" free energy $F_q \propto q^2 \cdot (\text{Amplitude})^2$, for q much smaller than the reciprocal molecular distance 1/a. (The proportionality factor contains the surface tension.) In thermal equilibrium, without quantum effects, we have $F_q \sim k_B T$, or $(\text{amplitude})^2 \propto k_B T/q^2$. Summation over all possible wavevectors $\vec{q}$ thus gives the thermally averaged width W of the surface, i.e. the average displacement of the surface from the minimum energy position:

$$W^2 \propto k_B T \sum_{\vec{q}} q^{-2} \tag{15}$$

The sum can be replaced by a (d-1)-dimensional integral over $\vec{q}$ (the wavevector is always parallel to the surface), with $1/L \leq q \leq 1/a$ apart from unimportant constants. For very large L, this integral remains finite for dimensionality d above 3; it diverges logarithmically, $W^2 \propto \log(L)$, for three dimensions, and increases even stronger as L in two dimensions. This divergence is called surfaceroughening by crystal growers and others; does an analogous effects exist in percolation ? (See ref.41 for recent literature.)

Franke[42] investigated this problem by Monte Carlo methods (fixed s) in two and three dimensions, for one- and two-dimensional surfaces, respectively. The two-dimensional width W indeed followed the expected law:

$$W \propto \sqrt{L}$$

and fig.3 shows the logarithmic dependence in three dimensions. The computer time for

Fig.3. Surface width W versus log(L) in three dimensions[42].

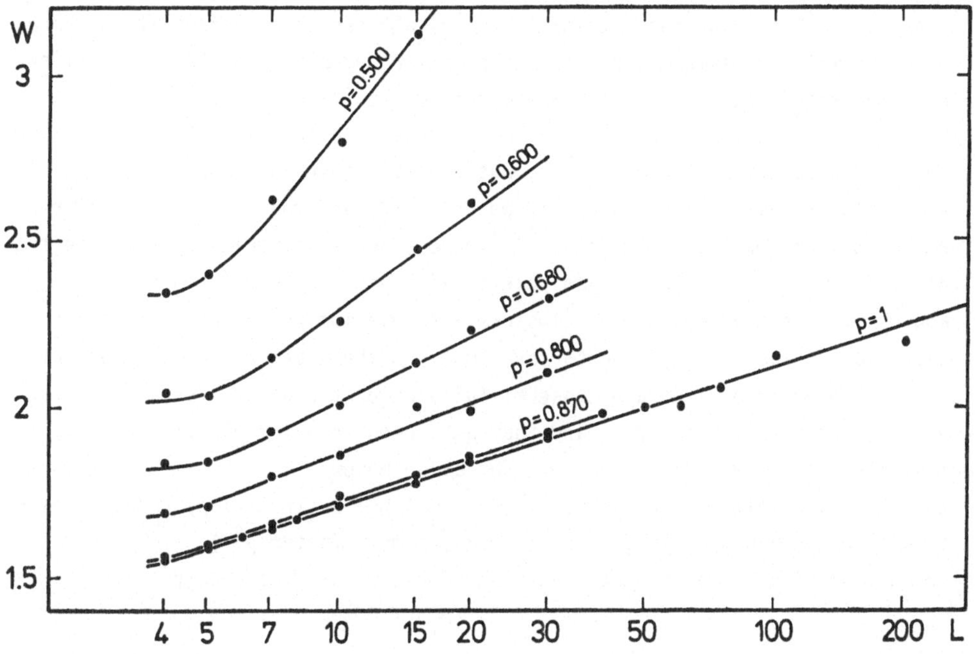

fig.3 was about hundred hours on a CDC Cyber 76 M. The limiting case $p = 1$ means that only those shape fluctuations are allowed which do not increase at all the perimeter; similarly in thermal problems at $T = 0$ only those shapes with minimal energy need to be averaged over. Fig.3 shows that even at $p = 1$ the surface remains rough, i.e. $W(L\to\infty) \to \infty$ within the limitations of this Monte Carlo experiment; for two dimensions the stronger increase was also derived[42] analytically at $p = 1$. Thus no "roughening transition" was found at an intermediate p value.

Overall, the analogy with liquid droplets was resonably confirmed by this roughening study of percolation clusters. Note that for droplets with s sites, the radius $R_s \propto s^{1/d}$ plays the role of L; thus e.g. in two dimensions the width W should increase asymptotically as $R_s^{1/2} \propto s^{1/4}$ and therefore become much smaller than the droplet radius in this limit of large s. Otherwise the droplet picture would not make much sense.

## V. RESULTS FOR ANIMALS

If $n_s(p)$ is the average number of clusters which are formed at concentration p, and if $g_{st}$ is the total number of geometrically different shapes for a cluster with s sites and perimeter t, then the fundamental randomness assumption gives

$$n_s = \sum_t g_{st} p^s (1-p)^t \qquad (16).$$

The determination of $g_{st}$ is called the animal problem (or the question of dilute branched polymers); apparently some researchers had nightmares of spider-like cluster configurations threatening their work. Besides the number also the structure of animals can be studied; e.g. for the animal radius all possible configurations have to be summed up with the same weight. Tables of animal numbers (ref.43 with earlier literature) and animal radii[39] are available for not too large s.

In ancient times[2,3] relation (16) lead to the hope of expressing all animal properties through percolation problems at $p_c$. But the Philadelphia group[44] started to free the animals of their domestication by percolation; and they all ran to their own fixed point. Renormalization methods[11,24] have shown that besides $p = p_c$, in the whole interval $0 \le p \le p_c$ no other fixed points exist except $p = 0$ (in the animal region[2]) and $p = 1$ (the one for compact clusters with minimum perimeter). Therefore[25] our "non-critical" exponents $\zeta$, $\theta$, and $\rho$ for cluster numbers and radii are those of animals for all p below $p_c$, and are those of fully compact clusters for all p above $p_c$. And the exact results of Parisi and Sourlas[29] used in our table III actually were animal results and were employed here for all p below $p_c$.

(Eq(16) shows why animals correspond to the limit $p \to 0$: Then $(1-p)^t$ equals unity and $n_s$ equals the number $g_s = \sum_t g_{st}$ apart from a simple factor $p^s$.)

Family and Coniglio[25] look at the generating function or free energy

$$F(q,h) = \sum_s n_s e^{-hs} = \sum_{st} g_{st} p^s e^{-hs} q^t = \sum_{st} g_{st} K^s q^t \qquad (17)$$

with $K = pe^{-h}$ and $q = 1-p$. A simple renormalization argument for the two variables K and q gives three fixed points for the triangular lattice: $q = K = 1/2$ (percolation threshold), $q = 1$, $K = 0.3$ (animals), and $q = 0$, $K = 1$ (compact clusters). Again animals have their own fixed point, i.e. their own set of critical exponents etc, without scaling relations to the percolation exponents. And again all percolation clusters below $p_c$ run to the animal shelter if they are renormalized.

Somewhat earlier, in a similar spirit, the Philadelphia group[45] explained that animals and percolation thresholds are related to each other similar as critical points are related to tricritical phenomena. They set $h = 0$ in the above free energy but instead allow $q \neq 1-p$, which seems equivalent. The percolation threshold then appears as a special case for $p + q = 1$ of the more general theory with p and q as independent variables. Similarly, the tricritical point in He 3 - He 4 mixtures is a special case for one particular mixing ratio, whereas the superfluid phase transition occurs also for more general mixing ratios in helium. The tricritical point corresponds in this analogy to the percolation threshold, the superfluid transition to the animal $g_{st}$. "Classical" theories (Landau expansions) become valid for d above 3 at tricritical thermal phase transitions, and above 4 dimensions for general thermal phase transitions. Classical percolation theory becomes valid for d larger than 6; thus not surprisingly one needs 8 dimensions to observe classical results (like $\theta = 5/2$) for animals[44-46].

For radii of animals, percolation clusters, and other configurations, readers acquainted with polymers may enjoy reading the Flory-type approximation of the Philadelphia group[46]. Using mean-field-like concepts similar to Flory's treatment of the excluded volume for linear polymers, they derive numerous radius exponents which agree exactly or within 0.03 with the values quoted here.

(No explanation was found for asymptotic exponents in animal numbers classified by perimeter t instead of size s.[47])

## VI CONCLUSION

This review may have given the impression that nearly everything is solved in simple percolation. Indeed, refs. 17, 29, 30 and 46 pushed us much closer to an exact solution. But a lot of problems were simply ignored in this review, like time-dependent effects[1], conductivity in random resistor networks[8], or the behavior near lattice boundaries and lattice defects[49]. Nevertheless I believe that future research will center not so much on the simple percolation theory but on modifications and applications: Correlated percolation[50], directed percolation[51], continuum percolation[52], water[50], gels[53], suspensions[54] and ants[55]. The figure on our title page does not give the impression that nothing is left to do.

I thank J.Kertesz for a critical reading of the manuscript.

## REFERENCES


1. S.Kirkpatrick, in: Ill Condensed Matter (Les Houches 1978), p.321, ed. by R.Balian, R.Maynard and G.Toulouse, North Holland, Amsterdam 1979; and AIP Conf.Proc. 58, 79 (1979)
2. D.Stauffer, Phys.Repts. 54, 1 (1979)
3. J.W.Essam, Repts.Progr.Phys. 43, 833 (1980)
4. C.Domb, E.Stoll and T.Schneider, Contemp.Phys. 20, 577 (1980)
5. J.M.Hammerley and D.J.A.Welsh, Contemp.Phys. 20, 593 (1980)
6. P.L.Leath, Phys.Rev. B 15, 5046 (1976)
7. Z.Alexandrowitz, Phys.Lett. 80 A, 284 (1980)
8. R.Pike and H.E.Stanley, J.Phys. A 14, L 169 (1981)
9. P.G. de Gennes, Compt.Rend.Acad.Sci., Ser. II, 292, 9 (1981)
10. M.E.Levinshtein, B.I.Shklovskii, M.S.Shur and A.L.Efros, Sov.Phys. JETP 42, 197 (1976)
11. P.J.Reynolds, H.E.Stanley and W.Klein, J.Phys. A 11, L 199 (1978) and Phys.Rev. B 21, 1223 (1980)
12. P.D.Eschbach, D.Stauffer and H.J.Herrmann, Phys.Rev. B 23, 422 (1981)
13. C.J.Lobb and K.R.Karasek, J.Phys. C 13, L 245 (1980)
14. J.Hoshen and R.Kopelman, Phys.Rev. B 14, 3438 (1976) and unpublished
15. D.W.Heerman and D.Stauffer, Z.Physik B 40, 133 (1980)
16. B.Derrida and J.Vannimenus, J.Physique 41, L 473 (1980)
17. H.W.J.Blöte, M.P.Nightingale and B.Derrida, J.Phys. A 14, L 45 (1981)
18. M.P.M. den Nijs, J.Phys. A 12, 1857 (1979)
19. B.Nienhuis, E.K.Riedel and M.Schick, J.Phys. A 13, L 189 (1980)
20. R.P.Pearson, Phys.Rev. B 22, 2579 (1980)
21. H.J.Herrmann, Z.Physik B, in press (1981)
22. M.F.Sykes and J.W.Essam, J.Math.Phys. 5, 1117 (1964); H.Kesten, Comm.Math.Phys. 74, 41 (1980)
23. A.Aharony, Phys.Rev. B 22, 400 (1980)
24. W.Klein and D.Stauffer, Phys.Lett. 78 A, 217 (1980)
25. F.Family and A.Coniglio, J.Phys. A 13, L 403 (1980)
26. F.Delyon, B.Souillard and D.Stauffer, subm. to J.Phys. A Lett. (1981); see also J.Roussenq, J.Aersosol Sci 12, No.5 (1981) and in preparation.
27. H.Nakanishi and H.E.Stanley, Phys.Rev. B 22, 2466 (1980) and J.Phys. A 14,No.3 (1981)
28. M.Aizenman, F.Delyon and B.Souillard, J.Statist.Phys. 23, 267 (1980)
29. G.Parisi and N.Sourlas, Phys.Rev. Lett. 46, 871 (1981)
30. T.C.Lubensky and A.J.McKane, J.Phys. A 14, L 157 (1981)
31. D.Andelman and A.N.Berker, J.Phys. A 14, L 91; D.Stauffer, Phys.Lett. A, in press; H.W.J.Blöte, private communication, February 1981.
32. C.M.Newman and L.S.Shulman, J.Phys. A 14, in press (1981)

33. A.Coniglio, H.E.Stanley and D.Stauffer, J.Phys. A 12, L 323 (1979) and Lett.Nuovo Cim. 28, 33 (1980)
34. J.Hoshen, D.Stauffer, R.J.Harrison, G.H.Bishop and G.D.Quinn, J.Phys. A 12, 1285 (1979)
35. J.W.Essam and K.M.Gwilym, J.Phys. C 4, L 228 (1971)
36. A.B.Harris and T.C.Lubensky, Phys.Rev. B 23, June (1981)
37. K.Holl, Staatsexamensarbeit, Cologne University, August 1981; K.Holl and H.Gould, in preparation
38. J.A.M.S.Duarte and H.J.Ruskin, Physica, in press (1981)
39. H.P.Peters, D.Stauffer, H.P.Hölters and K.Loewenich, Z.Physik B 34, 399 (1979)
40. R.J.Harrison, G.H.Bishop and G.D.Quinn, J.Statist.Phys. 19, 53 (1978)
41. D.J.Wallace and R.K.P.Zia, Phys.Rev.Lett. 43, 808 (1979)
42. H.Franke, Z.Physik B 40, 61 and in preparation; and Thesis, Cologne University (1981)
43. M.F.Sykes, D.S.Gaunt and M.Glen, J.Phys. A 14, 287 (1981)
44. T.C.Lubensky and J.Isaacson, Phys.Rev.Lett. 41, 829 and 42, 410 (1978)
45. A.B.Harris and T.C.Lubensky, Phys.Rev. B 23, 3591 (1981)
46. J.Isaacson and T.C.Lubensky, J.Physique 41, L 469 (1980)
47. J.A.M.S.Duarte, Z.Naturf. 33a, 1404 (1978) and Z.Physik B 33, 97 (1978); B.Rocksloh and D.Stauffer, Z.Naturf. 34a, 1140 (1979)
48. C.J.Lobb and D.J.Frank, J.Phys. C 12, L 827 (1979) and AIP Conf.Proc. 58, 308 (1980)
49. K.De'Bell and J.W.Essam, J.Phys. C 13, 4811 (1980); J.P.Clerc, J.Carton, J.Roussenq and D.Stauffer, subm. to J.Physique L.(1981)
50. H.E.Stanley, J.Teixeira, A.Geiger and R.L.Blumberg, Physica 106 A, 260 (1981) and this Conference
51. W.Kinzel and J.M.Yeomans, J.Phys. A 14, L 163 (1981) with further literature.
52. S.W.Haan and R.Zwanzig, J.Phys. A 10, 1547 (1977); T.Vicsek and J.Kertesz, J. Phys. A 14, L 31 (1977); E.T.Gawlinsky and H.E.Stanley, subm. to J.Phys. A L.
53. D.Stauffer, Physica 106 A, 177 (1981)
54 J.L.Bouillot, C.Camoin, M.Belzons, R.Blanc, and E.Guyon, IUTAM-IUPAC Symposium on Interactions of Particles in Colloidal Dispersions, Canberra (Australia) March 1981
55. J.Roussenq, Thèse, Université de Provence, 1980.

NOTE ADDED AT CONFERENCE

Derrida and de Seze found $\rho(p < p_c) = 0.6408 \pm 0.0003$ in two dimensions.

# EXPERIMENTAL RELEVANCE OF PERCOLATION

Guy Deutscher [a)]
Department of Physics and Astronomy
Tel Aviv University
Ramat Aviv, Tel Aviv, Israel

## Abstract

There is good experimental evidence that in random metal-insulator mixtures, percolation processes determine the critical metal volume fraction at the conductance threshold, the behaviour of the conduction above the threshold and that of the dielectric constant below the threshold. In the superconducting state, the critical current density is determined by the percolation correlation length. Other properties are sensitive to the clusters topology. Application of percolation to other systems is also discussed.

Until recently, most percolation "experiments" were performed on computers. However, it is now becoming clear that a broad range of concepts developed by percolation theories, ranging from the existence of a well defined percolation threshold to more difficult issues such as that of the structure of large clusters, can be very useful in understanding the properties of a variety of physical systems. To mention only a few: metal-insulator mixtures, superconducting-insulator mixtures, gels, water droplets in suspension, and maybe spiral arms galaxies. We shall review some of the concepts involved, and then discuss their applications.

A) The percolation threshold

When sites—or bonds—are occupied randomly on a lattice, clusters of different sizes are formed[1]. Computer experiments on large lattices indicate that two different situations can be realized, depending on the value of the occupation probability p. When p is smaller than a critical value $p_c$, only finite clusters are formed. When $p > p_c$, a cluster extending throughout the lattice coexists with smaller clusters. In the limit of an infinite lattice, only finite clusters exist below $p_c$ while it is thought that one—and only one—infinite cluster coexists with finite clusters above $p_c$. The value of $p_c$ depends on the type of percolation considered (site or bond), on the symmetry of the lattice, and on its dimensionality.

The relevance of the percolation threshold $p_c$ for actual physical systems is complicated by the fact that one often deals with a random continuum rather than with a lattice. However the following remark, due to Scheer and Zallen[2], is very useful for the latter case: if spheres with a radius equal to half the lattice spacing are located at the occupied sites of a lattice, the fraction $v_c$ of the volume that they occupy at the percolation threshold is nearly independent from the symmetry of the lattice. In two dimensions (2D) $v_c \simeq 0.45$ and in three dimensions (3D) $v_c \simeq 0.15$. If one views a random continuum as a superposition of different lattice structures, the above numbers give the critical volume fractions at which percolation occurs in 2D and 3D.

A1) The case of metal-insulator mixtures

The obvious application is the case of (unmiscible) metal-insulator mixtures: they should be insulating when the metal volume fraction $v < v_c$, and metallic when $v > v_c$. Unfortunately, most metal-insulator mixtures (such as Au-$SiO_2$, Ni-$SiO_2$, Aℓ-$SiO_2$ and similar alloys using $Aℓ_2O_3$ as the insulator[3]) display a 3D metal-insulator transition at v ~ 50% rather than the predicted 15%. These early results raised some doubts as to the applicability of percolation concepts to the conduction mechanisms of metal-insulator mixtures. But more recent experiments have shown that the above discrepancy comes from the fact that the above alloys are not random, but rather granular, i.e., the (amorphous) insulator coats the (crystalline) metallic particles in a fairly systematic fashion. There is evidence that this granular structure persists up to high metallic volume fractions. It is likely that in that case the value of the threshold is controlled by tunneling mechanisms rather than by the percolation processes.

There exist, however, other mixtures where the metal and the insulator appear to be distributed at random. Such is the case for In-Ge and Pb-Ge[(4)]. Fig. 1 shows the contrast between the granular and the random case, and the resemblance between the latter and random ordering on a network. And, indeed, the critical metal volume fraction is found to be about equal to 15% for In-Ge and Pb-Ge.

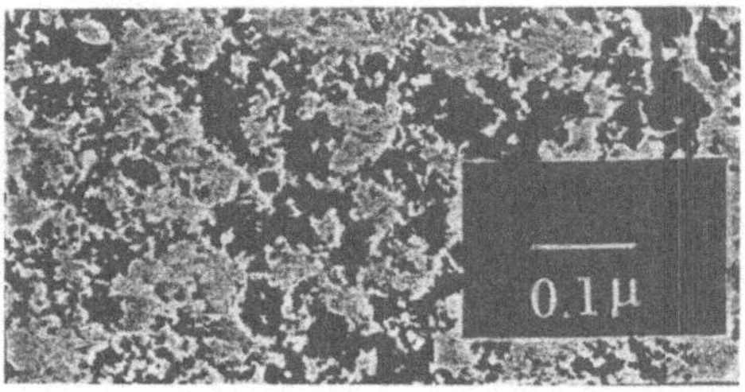


Fig. 1a: InGe

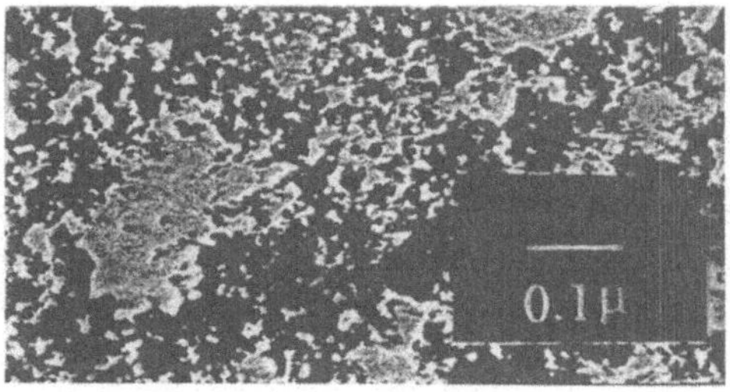


Fig. 1b: PbGe

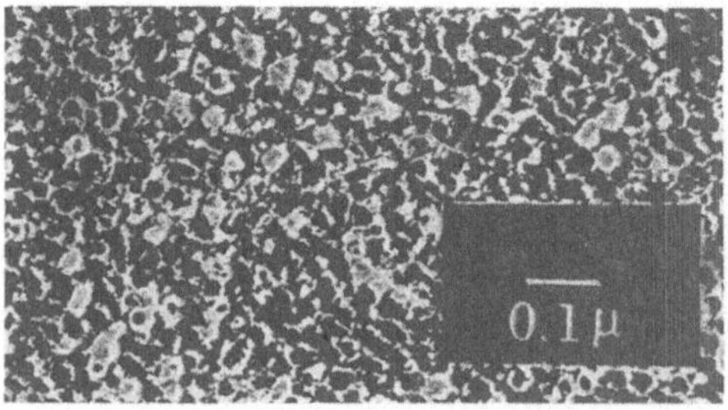


Fig. 1c: AℓGe

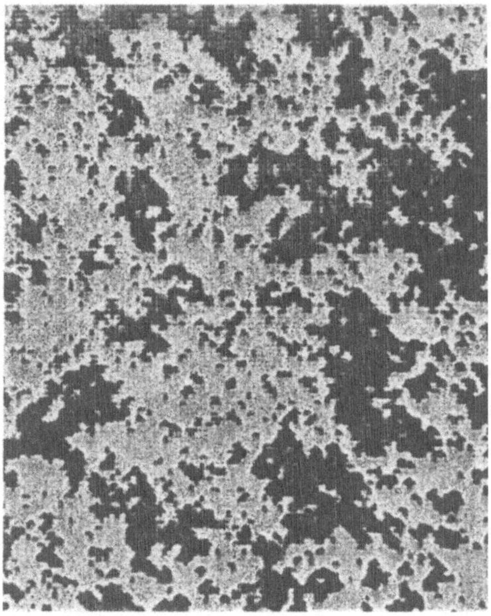

Fig. 1d: (Collective phenomena, front covers by H. Ogita, F. Yonesawa, A. Veda, T. Matsubara and H. Matsuba). Clusters formed in the 2D Ising model transition.

A comparison between the resistivity data of a granular (Aℓ-Ge) and a random (In-Ge) mixture reveals another interesting difference (Fig. 2). While the resistivity of Aℓ-Ge is almost temperature independent in the metallic regime, that of In-Ge shows a resistivity ratio significantly larger than unity down to v values quite close to $v_c$. This difference indicates clearly that in the granular case the conduction is largely controlled by tunneling (which is temperature independent, at least in the metallic regime), while it proceeds along macroscopic metallic channels in the random case. This is an interesting qualitative indication that percolation processes are relevant to the case of the random mixtures.

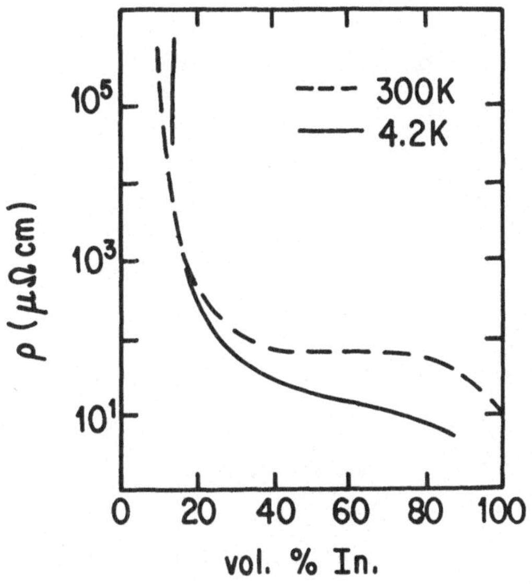


Fig. 2a: Resistivity of InGe

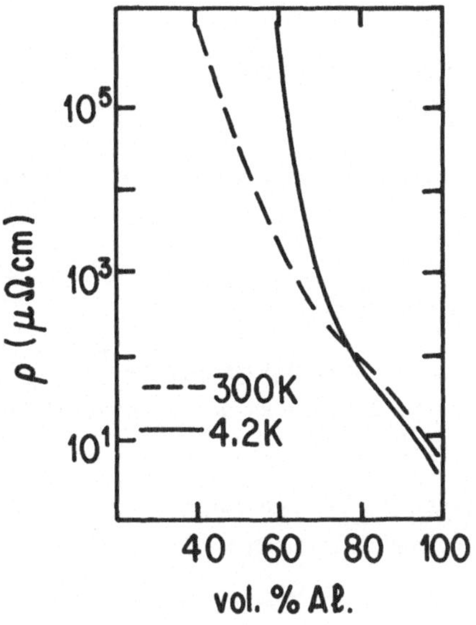


Fig. 2b: Resistivity of AℓGe

A2) <u>Other examples</u>

Although the case of the metal-insulator mixtures may be the best understood quantitatively, the concept of the percolation threshold is certainly applicable to other systems. Amongst them are the well known disease propagation in an orchard, and the more recently proposed model for star birth propagation in galaxies[5]. In the first case, a disease affecting a tree has a finite probability p of spreading to the nearest neighbours. If p is smaller than a critical value $p_c$ the disease will remain local (finite clusters), while in the other case it will spread throughout the orchard (infinite cluster). In order to maximize income, one should obviously select for the orchard a nearest neighbour distance such that p will be somewhat smaller than $p_c$. The exact value of p to be selected will depend—amongst other parameters—on the typical cluster size and clusters numbers for p near $p_c$.

In the star formation model proposed by Seiden and Gerolla[5], the burst of a supernova produces a shock wave in the galactic gas, which can be strong enough to trigger the condensation of new stars. Some of them will supernovae at a latter stage, etc. Depending on the probability p for new star formation when a supernova explodes, new star formation can either remain localized and actually die out ($p < p_c$) or propagate throughout the galaxy ($p \geq p_c$). Again, the value of $p_c$ depends on the dimensionality of the galaxy, which can be 3 or 2. The dependence of $p_c$ on dimensionality has been used to explain the difference between elliptical and spiral galaxies as arising from the competition between the high star formation rate in the initial 3D regime and the collapse of the gas cloud to a disk.

B) Cluster size, cluster numbers and related properties below the percolation threshold

The scaling theory of percolation[1] predicts that there exists a typical cluster size $s_\xi \propto |p - p_c|^{-1/\sigma}$, and that the number of clusters of size s is given by

$$n_s(p) \propto s^{-\tau} f\left(\frac{s}{s_\xi}\right) \qquad (1)$$

where $f(o) = 1$, so that $n_s(p = p_c) \propto s^{-\tau}$.
The function $f(x)$ introduces an exponential cut off at $x \geq 1$, i.e., there are very few clusters of size $s > s_\xi$.

The existence of large but finite clusters below $p_c$ leads to interesting physical consequences, which we shall discuss in this section.

Recent electron microscopy studies have shown that cluster size distributions as predicted by (1) can indeed be observed in metal-insulator mixtures. Fig. 3 shows a micrograph of a two dimensional $A\ell$-$A\ell_2O_3$ film (i.e., it contains only one layer of metallic grains). The metallic and insulating areas are roughly equal, so that one expects $p \sim p_c$. Indeed, in qualitative agreement with eq. (1), a very broad cluster size distribution is observed: there are both very small and very large clusters. A detailed study of the cluster size distribution shows that $n_s \propto s^{-\tau}$ over about two decades of $n_s$, with $\tau = 2.1 \pm .15$ in good agreement with Monte Carlo calculations[1] Another interesting observation is that the perimeter to area ratio of the large clusters is size independent, another prediction of percolation theory[1] linked to the "ramified" structure of the large clusters. Indeed, direct observation shows that they are not at all spherical.

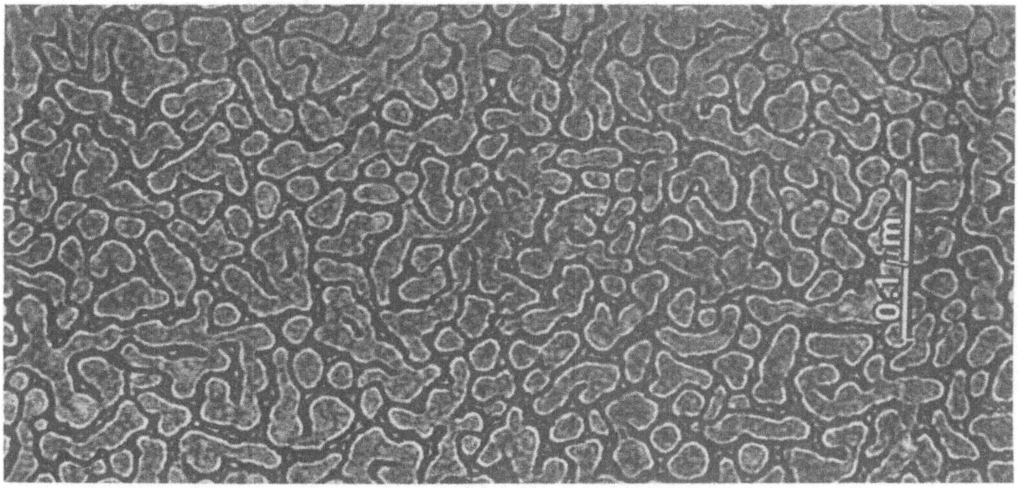


Fig. 3: Structure of a 2D $A\ell$-$A\ell_2O_3$ film (after Ref. 6)

B1) Divergence of the dielectric constant at $p_c$

The existence of large clusters at $p \lesssim p_c$ has a strong influence on the dielectrip properties of metal-insulator mixtures below the percolation threshold. These large clusters can be viewed as an infinite cluster interrupted at intervals of length $\xi$. When a d.c. electric field is applied at the ends of the specimen, the resulting local fields are highly inhomogeneous, being essentially concentrated in the thin dielectric regions that separate the almost touching large metallic clusters. This effect, which leads to a divergence of the macroscopic dielectric constant of the specimen near $p_c$ [7], $\varepsilon \; \alpha \; (p_c - p)^{-s}$, has been observed experimentally in a composite consisting of Ag particles in $KC\ell$ matrix[8].

B2) Divergence of the conductivity of a superconductor-normal mixture at $p_c$

A closely related phenomenon is the conductivity behaviour of a superconductor-normal metal mixture. Here slightly below $p_c$ large superconducting clusters short out most of the sample, with only short normal bridges contributing to the resistance of the specimen. Also here the theory[9] predicts $\sigma \; \alpha \; (p_c - p)^{-s}$. Measurements of the superconducting transition of inhomogeneous superconductors have provided the first experimental determination of the exponent s[10]. For the 3D case, the theoretical value s = 0.7 agrees with the observed behaviour of the dielectric constant and of the superconducting transition (Fig. 4)

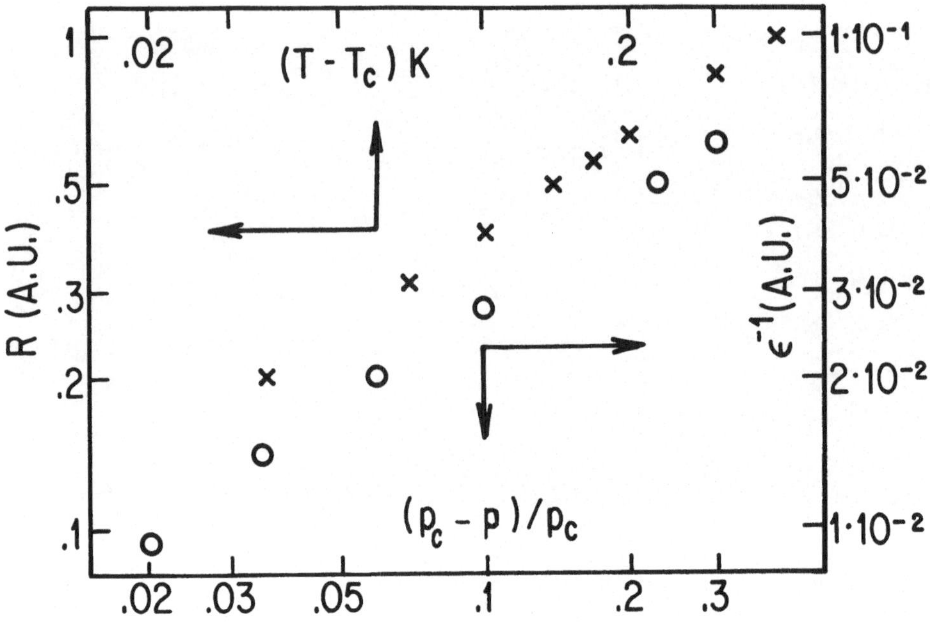


Fig. 4: The exponent s determined by the divergence of σ (Aℓ-Ge) and of ε(KCℓ)

B3) Electrostatic charging below $p_c$

Another physical quantity influenced by the presence of large clusters is the activation energy for the conductivity of a metal insulator mixture below $p_c$. This activation process is the electrostatic charging of metallic clusters. One can expect the electrostatic charging energy to be inversely proportional to an effective cluster's radius $r_{eff}$ as to the effective dielectric constant, $E_c \propto \varepsilon^{-1} r_{eff}^{-1}$. Experimentally, one observes (Fig. 5) $E_c \propto (p_c - p)^{1.45\pm.1}$.

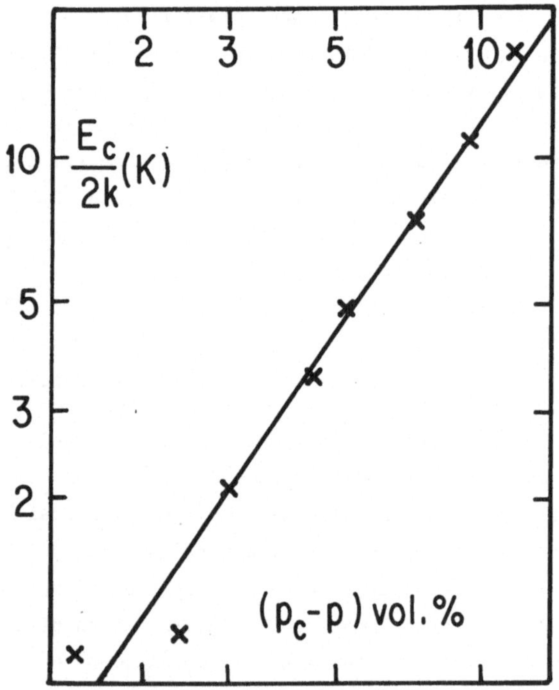


Fig. 5: Electrostatic energy in AℓGe (Data Ref. 11)

The critical exponent for $\varepsilon$ being known to be equal to .7 from the above determinations, one concludes that $r_{eff}$ diverges at $p_c$ with a critical exponent equal to .75 $\pm$ .1. This is close to the critical exponent for the diameter of the typical cluster (known as the percolation correlation length) $\xi \alpha |p - p_c|^{-\nu}$, with $\nu = .85$ in 3D [12]. The important observation is that, in any case, the divergence of the dielectric constant at $p_c$ is not sufficient to explain the variation of $E_c$ with concentration: the existence of large clusters must be taken into account as well.

C) Susceptibility of finite superconducting clusters

It was recently noticed that certain superconducting properties are sensitive to the clusters topology [13], and amongst them the susceptibility of finite clusters [14]. This comes from the fact that when the individual superconducting grains have a diameter $D < \lambda_L(T)$, where $\lambda_L(T)$ is the London penetration depth, only loops of grains contribute to the diamagnetism of the clusters (dead ends do not). This situation is realized in the random InGe mixtures [4]. For clusters consisting of n grains, it is possible to define a fraction $f_n$ of grains located in loops, and an effective average total loops area $S_n$. The susceptibility of the sample is then given by

$$\chi = - 2\pi\ C\ \Delta^2(T)\ \sum_n \phi_n S_n f_n \tag{2}$$

where $\phi_n$ is the probability for a grain to belong to a cluster of size n, $\Delta(T)$ is as defined in Ref. 15, and

$$C = \frac{8}{\pi}\ N(o)\ \hbar\ D_o (k_g T_c)^{-1} \tag{3}$$

where $D_o$ is the diffusion coefficient along a chain of grains and N(o) is the electronic density of states at the Fermi level per grain. $\phi_n$ is known [16] to vary as:

$$\phi_n = n^{-1-\beta/\beta+\gamma} \tag{4}$$

up to a cut-off $N^* \simeq (p_c - p)^{-\beta+\gamma}$

De Gennes proposes

$$f_n \ \alpha\ n^{-x} \tag{5}$$

and

$$S_n \ \alpha\ n^{y} \tag{6}$$

With these assumptions

$$\frac{\chi}{2\pi C\Delta^2} = \sum_{n=1}^{n=N^*} n^{-(1+u)} \tag{7}$$

where $u = \frac{\beta}{\beta+\gamma} + x - y$.

If $u > o$, the convergence of the sum [7] is fast and the result only weakly dependent on N*. If $u < o$, $-\chi \;\alpha (p_c - p)^{u(\beta + \gamma)}$ and the susceptibility will display a critical behaviour near $p_c$. Numerical estimates of x and y are currently being made,[17] as well as measurements of $\chi$[18] on random InGe mixtures.

D) Properties of the infinite cluster

The most obvious property to study above $p_c$ is the conductivity of the infinite cluster. This can be done by random resistor network computer experiments[12] or by measuring the conductivity of metal-insulator mixtures[19]. In both cases, one finds

$$\sigma = \sigma_o (p - p_c)^t \tag{8}$$

with $t \simeq 1.7$ in 3D and $t \simeq 1.0$ in 2D[12]. These exponents are strikingly different from that for the order parameter of the percolation problem

$$p_\infty = (p - p_c)^\beta \tag{9}$$

$p_\infty$ measures the probability for an occupied bond (or site) to be part of the infinite cluster. Accepted values for $\beta$ are $\beta(3D) \simeq 0.4$ and $\beta(2D) \simeq 0.14$ [1].

The behaviour of the conductivity is not easily related to any simple topological property of the infinite cluster, and in that sense conductivity measurements (although quite easy to perform) are not by themselves sufficient to gain a good understanding of the percolation problem above $p_c$. The same remark applies to the elasticity of gels.[22]

D1) The percolation correlation length above $p_c$

One of the fundamental concepts of percolation theory is—as in other phase transition problems—the existence of a length scale $\xi$ that diverges at $p_c$

$$\xi = a|p - p_c|^{-\nu} \tag{10}$$

where a is the elementary length scale, i.e. the lattice parameter for percolation on a lattice or the diameter of the elementary crystallites in metal-insulator mixtures.

The physical meaning of the correlation length $\xi$ below $p_c$ is simply that it measures the spatial extension of the typical clusters (the clusters of size $s_\xi$). As noted by Stauffer[16] it is different from the radius of gyration R of these clusters

$$R^2 = a\,(p_c - p)^{-2\nu + \beta} \tag{11}$$

The divergence of $\xi$ is closely related to that of the dielectric constant, while that of $R^2$ gives the effective cross-section of polymer coils[20].

A geometrical interpretation of $\xi$ above $p_c$ was proposed independently by Skal and Shklovskii[21] and de Gennes[22] to try and relate the behaviour of the conductivity above $p_c$ to the structure of the infinite cluster. After deleting all dead ends (which do not carry a current) from the infinite cluster, one is left with "macrobonds" connected at "nodes," a node being the linking point of at least three macrobonds. In this picture, macrobonds form an irregular superlattice, the "typical" lattice parameter or internode distance being precisely $\xi$. The conductivity of the infinite cluster is then proportional to the number of macrobonds per unit cross-section

$$\sigma \propto \xi^{-(d-1)}\left(\frac{\xi}{l}\right) \tag{12}$$

where d is the dimensionality of the sample, and $(\frac{\xi}{l})$ a corrective factor taking into account that the physical $l$ of a macrobond can be larger than the geometrical distance between nodes: $l \propto \xi(p - p_c)^{-\delta}$, where $\delta$ is a twistedness index. A comparison between (10) and (12) lends

$$t = (d - 1)\nu + \delta \tag{13}$$

with $\delta \gtrsim 0$.

D2) Experimental checks of the Skal and Shklovskii model

Numerical values for t and $\nu$ obtained from computer work[9] are compatible with (13) in 3D ($t_{3D} \simeq 1.6 - 1.7$, $\nu_{3D} \simeq 0.8 - 0.9$) if $\delta \simeq 0$, but there seems to be a problem in 2D because $t_{2D} \simeq 1.0$[12] and $\nu_{2D} \simeq 1.3$[1], which is incompatible with (13) since $\delta$ cannot be negative (although the disagreement is marginal since error margins on computer determined exponents can be of the order of 10%).

An experimental check of eq. (13) was performed[19] by measuring, on the same samples (PbGe and Aℓ Ge), the normal state conductance and the critical current density $j_c$ in the superconducting state. The idea here is that the critical current density (unlike the normal state conductance) should only depend on the number of macrobonds per unit cross section and not on their effective length, so that $j_c \propto (p - p_c)^v$ with

$$v = (d - 1)\nu \tag{14}$$

Comparison between experimental values of t and v thus provide a direct test of (13). The main error in the experimental determination of t and v comes from the uncertainty on the composition of the samples. However, since t and v are measured on the same series of samples, their relative values are known quite accurately. The results: $t \simeq v = 1.7 \pm .15$ in 3D, $t = 0.9 \pm .1$ and $v = 1.3 \pm .1$ in 2D (Fig. 6), definitely indicate that the Skal and Shklovskii model of the infinite cluster is not consistent with experimental data.

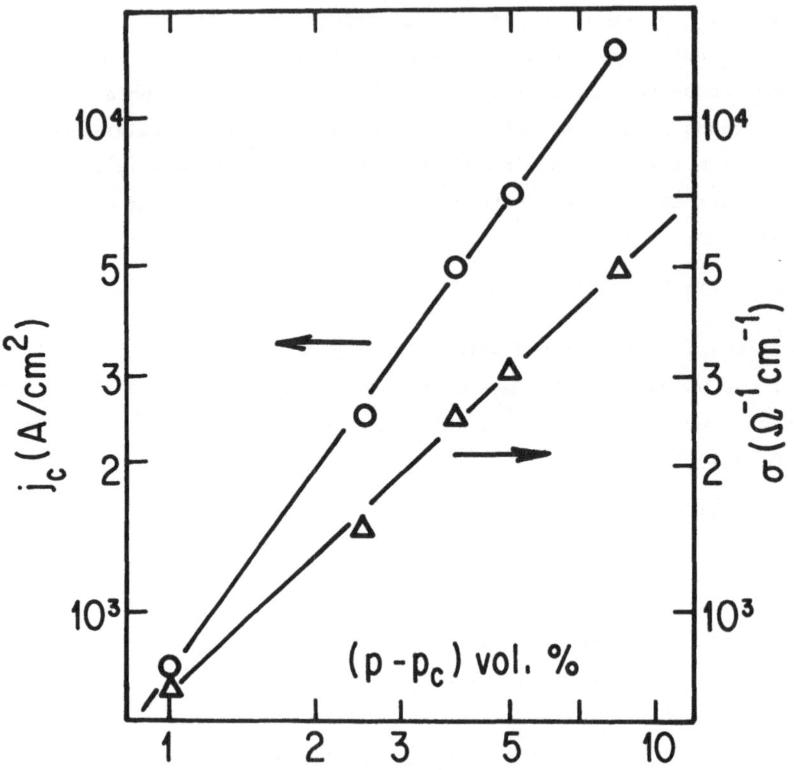


Fig. 6: Critical current density $j_c$ and normal state conductivity $\sigma$ of PbGe data from Ref. 10).

D3) A modified interpretation of the Skal and Shklovskii model

It may however be observed that the experimental values of $v^{(19)}$ and the calculated values of $\nu^{(1)}$ verify eq. (14) quite well, both in 3D and in 2D. One may therefore wonder whether the model, although clearly not applicable to the conductance, may not be actually valid for the critical current.

Our present understanding is that this is indeed the case. Applicability of the model to the conductance rests on the assumption that macrobonds of length $\xi$ link nodes at which shorter macrobonds can be effectively lumped together: visual inspection of computer generated clusters[23] immediately invalidates this assumption in 2D; no long streches can in general be observed linking well defined modes. On the other hand, short streches of width ~ unity can be observed linking denser regions of the infinite cluster

We propose that the correlation length above $p_c$ should be considered as the typical distance between these bottlenecks, rather than as the typical distance between (fairly ill defined) nodes. Since a current will quench first the superconducting order parameter at the points of highest current density, the number of bottlenecks per unit cross-section, $\xi^{-(d-1)}$, will determine the macroscopic critical current density in accordance with eq. (14). Therefore a measurement of $j_c$ versus concentration gives a direct determination of the critical index for the percolation length.

In this interpretation, the infinite cluster is considered as a network of clusters connected by bottlenecks rather than as a network of macrobonds (with attached dangling bonds) connected at nodes. The behaviour of $\sigma$ then depends not only on that of $\xi$ (eventually corrected by a twistedness effect) but also on the structure of the clusters, and therefore is not amenable to a simple analysis.

The peculiar behaviour of $\sigma$ in 2D may find its origin in the fact that in that dimensionality the clusters radius of gyration as given by (11) has a critical index very close to that of $\xi$, because of the small value of $\beta$.

D4) Short range versus long range correlations in the infinite cluster

While long range correlations in the infinite cluster, described by $\xi$, dominate the behaviour of the superconducting critical current density, the above analysis suggests that short range correlations have a non-negligible effect on the conductivity.

It has been proposed recently[24] that the behaviour of the upper critical field of a percolating superconductor is dominated by these short range correlations near the percolation threshold. The reason is that, in the dirty limit, the upper critical field can be written as[15]:

$$H_{c2} = \phi_o (D\tau)^{-1} \tag{15}$$

where D is the diffusion coefficient in the percolating netowrk and $\tau = \frac{\hbar}{k_B T_c}$. Near $p_c$, D becomes small and $(D\tau)^{1/2}$ smaller than the percolation correlation length. In this limit, D is actually time dependent because at short time diffusion proceeds faster[25] than in the long time limit where $D \propto \sigma$. (This is due to the relatively high density of the bottleneck connected clusters, as discussed in the preceding section).

Experimental results on the critical field of InGe films indicate that in 2D $H_{c2}^{-1} \propto D_{\text{short time}} \propto (p - p_c)^{0.6}$, in contrast with $D_{\text{long time}} \propto (p - p_c)^{1.1}$ (Fig. 7).

Notice that near $p_c$ a magnetic field quenches superconductivity preferentially in the high density regions of the infinite cluster, while a current does just the opposite. Therefore superconducting measurements are of particular interest because they can help us distinguish and measure separately long range and short range correlations—in contrast with the conductivity which depends on both.

We now use the critical field and critical current data to establish a relation

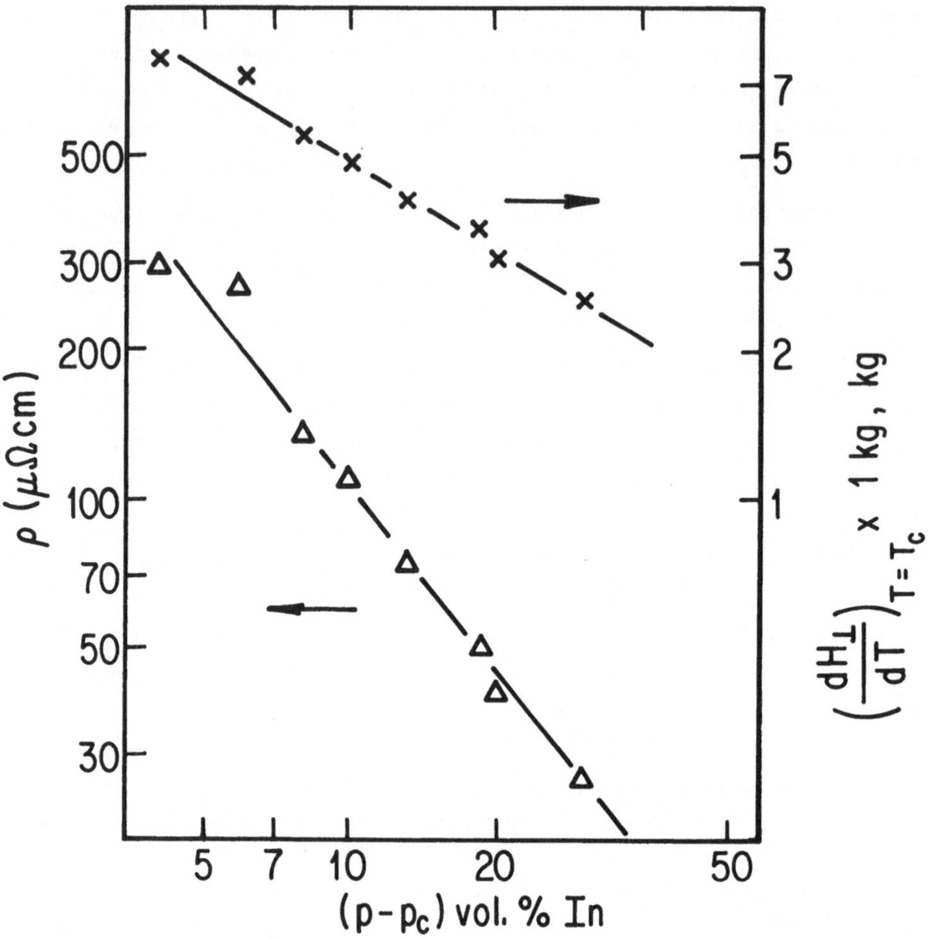


Fig. 7: Normal state resistivity ρ and upper critical field slope near $T_c$ for InGe

for the conductivity in 2D.

As stated above, our model for the infinite cluster is that of an average lattice of period $\xi$ where the nodes are occupied by clusters having a radius of gyration R. The infinite cluster has a density p $P_\infty$. According to the definition of R (11) the clusters have therefore a density $\rho = p$, since p $P_\infty = (R/\xi)^2\rho$. Since screening currents are induced only in loops, we shall assume that the critical index for $H_{c2}$ is equal to the index for the loops density in these clusters. According to (15) this is also the index for the inverse of the short time diffusion constant. With each cell of the network consisting of a cluster of radius R (diffusion coefficient $D_{\text{short time}} \propto (p - p_c)^{0.6}$) and a macrobond of length $(\xi - R)$ in series the resistance per square of the network is obtained as

$$R_\square \propto (p - p_c)^{-0.6} - \frac{\beta}{2} (p - p_c)^{-\nu} \log (p - p_c) \qquad (16)$$

with the 2D values $\nu = 1.3$ and $\beta = .14$. In this expression we have used the approximation $(p - p_c)^{\beta/2} \simeq 1 + \beta/_2 \log (p - p_c)$, valid except very close to $p_c$ since $\beta/_2 \ll 1$. It is only for $(p - p_c) < 1.10^{-2}$ that the second term of the right-hand side of (16) dominates. In the range $2.10^{-2} < (p - p_c) < 2.10^{-1}$ typical of most measurements, eq. (12) gives a behaviour very close to linear in agreement with experimental data.

## Conclusions

Most of the examples presented in this review refer to metal-insulator mixtures. As mentioned in the introduction, the same concepts can be applied to other systems sometimes in a straightforward fashion: for the gelation process, the analog of p is the fraction of reacted monomers and the elasticity follows a power law with exponent t[21] while the viscosity diverges with exponent s below $p_c$[27]; in granular superconductors, it is the superfluid density (whose inverse is proportional to the square of the penetration depth $\lambda_L$) which varies with exponent t[21], as verified experimentally in $A\ell$-$A\ell_2O_3$ films[28].

There are also modified percolation problems which are of interest. In the star formation model of Ref. 5, one can define a correlation in space (typical extension of a finite cluster) and in time (typical life-time of a finite cluster). The exponent for the latter has been found[29] to be larger than unity, i.e. it is different from the 3D exponent for the correlation length ($\nu_{3D} = 0.85$) although the system is in some respect 3D (2 space and 1 time) and has indeed a threshold $p_c \simeq 0.17$ close to the 3D value.

In microemulsions of water droplets in suspensions[26] there is evidence that dynamical phenomena are important. The value of the water volume fraction at the threshold (~ 0.14) and the exponent for the conductance above the threshold (~ 1.55) are very close to values obtained for 3D random metal-insulator mixtures, but the value of the exponent below the threshold (~ 1.3) is much larger than the value of s obtained in static systems (~ 0.7, see fig.4 ).

Most of the properties studied up to now are not directly sensitive to the structure of the clusters, but there is some hope that more detailed information on their topology can be obtained by studying their superconducting properties.

## Acknowledgements

We have greatly benefited from conversations with O. Entin-Wohlman, S. Alexander, P. G. de Gennes and R. Orbach. This research was partially supported by the U.S. Israel BiNational Science Foundation and by N.S.F. Grant DMR 78-27129.

References


a) On sabbatical leave at the Physics Dept., U.C.L.A.
1) For a review, see D. Stauffer, Physics Reports 54, (1979).
2) H. Scher and R. Zallen, J. Chem. Phys. 53, 3759 (1980).
3) B. Abeles in Applied Solid State Science, edited by R. Wolfe (Academic Press, New York) 1976, Vol. 6, p. 1 and Adv. Phys. 24,407 (1975).
4) G. Deutscher, M. Rappaport and Z. Ovadyahu, Solid State Commun. 28, 593 (1978).
5) P. E. Seiden, L. S. Schulman and H. Gerolla, Ap. J. 232, 702 (1979).
6) A. Alessandrini, G. Deutscher and R. Leibovitz, to be published.
7) D. Bergman, Phys. Rev. Lett. 44, 1285 (1980).
8) D. M. Grannan, J. C. Garland and D. B. Tanner, Phys. Rev. Lett. 46, 375 (1981).
9) J. P. Straley, AIP Conference Proceedings n°40, New York 1978 (p. 108).
10) G. Deutscher and M. Rappaport, J. Phys. C6,581 (1978).
11) Y. Shapira, Ph.D. Thesis, Tel Aviv University (1981).
12) J. P. Straley, Phys. Rev. B15, 5733 (1977).
13) G. Deutscher, I. Gravé and S. Alexander, to be published.
14) P. G. de Gennes, C.R. Acad. Sc. Paris, to appear.
15) P. G. de Gennes, "Superconductivity of Metals and Alloys," W.A. Benjamin (N.Y., 1966).
16) D. Stauffer, J. Chem. Soc., (Faraday Trans. II) 72, 1354 (1976).
17) H. E. Stanley, private communication.
18) Tel Aviv group, to be published.
19) G. Deutscher and M. Rappaport, J. Phys. Lett. 40, L-219 (1979).
20) M. Daoud and P. G. de Gennes, J. Phys. 38, 85 (1977), and P. G. de Gennes, J. Phys. Colloque C3, 17 (1980).
21) A. Skal and B. Shklovskii, Friz. Tekh. Poluprodn. 8, 1586 (1974); [Sov. Phys. Semicond. 8, 1029 (1975)].
22) P. G. de Gennes, J. Phys. Lett. 37, L-1 (1976).
23) S. Kirkpatrick in Electrical Transport and Optical Properties of Inhomogeneous Media - 1977, A.I.P. Conference Proceedings No. 58 (New York, 1979), p. 79.
24) G. Deutscher, I. Gravé and S. Alexander, to be published.
25) C. Mitescu and J. Rousseng, C.R. Acad. Sc. Paris 238A, 999 (1976), and A.I.P. Conference Proceedings 40, 377 (1978).
26) M. Lagües, R. Ober and C. Taupin, J. Phys. Lett. 39, L-487 (1978).
27) P. G. de Gennes, J. Phys. Lett. 40, L-197 (1979).
28) D. Avraham, G. Deutscher, R. Rosenbaum and S. Wolf, J. Phys. C6, 586 (1978).
29) L. Schulman and P. Seiden, to be published.

# THEORY OF DILUTE ANISOTROPIC MAGNETS

R.B. Stinchcombe
Department of Theoretical Physics
1 Keble Road Oxford OX1 3NP

Abstract. Diluted anisotropic magnets are treated by renormalisation group scaling methods. The dependence of correlation lengths on anisotropy, concentration and temperature, and of transition temperature on anisotropy and concentration, are discussed.

In diluted magnets the transition temperature $T_c$ falls to zero as the magnetic concentration p is reduced to the percolation threshold $p_c$. Such critical curves may be calculated by real space renormalisation group methods[1,2]. Resulting curves[2] for spin ½ Ising and Heisenberg models on the simple cubic lattice are shown in Figure 1, and they compare well with experiment[3].

The divergence of the correlation length $\xi$ as the critical curve is approached typically crosses over between the percolative behaviour and the thermal behaviours characteristic of the pure magnet or the magnet at $p_c$. This crossover behaviour and also that induced by anisotropy can be discussed by the real space methods[4,5], as outlined below.

The model considered is the uniaxially anisotropic Heisenberg ferromagnet with "Hamiltonian"

$$- \beta H = \Sigma K \, [\alpha(\sigma_i^x \sigma_j^x + \sigma_i^y \sigma_j^y) + \sigma_i^z \sigma_j^z] \tag{1}$$

where $\alpha \equiv 1-\delta$ ranges from 0 (Ising limit) to 1 (isotropic Heisenberg limit). The longitudinal and transverse ($/\!/ , \perp$) inverse correlation lengths ($K_1 = 1/\xi_1$) of the pure one-dimensional version[6] of this model are shown in Figure 2. Anisotropy-induced crossover occurs when the reduced temperature $\tau \equiv 1/K$ takes the value $\tau^* = (1-\alpha^2)^{\frac{1}{2}}$.

In the corresponding dilute two and three-dimensional cases at the percolation threshold, the inverse correlation lengths look very like Figure 2, again exhibiting anisotropy crossover[7,8]; near $p_c$, thermal-percolation and anisotropy crossover may occur together.

Such effects can be treated by considering the transformation of parameters (concentration, temperature, anisotropy) under a length dilatation by b. We use a simple Migdal method[9] which combines b bonds in series and $b^{d-1}$ bonds in parallel (d = dimensionality). The parallel combination adds components of the exchange interaction. Combination in series is easy if one or more of the combined interactions vanishes, giving zero resultant. If the distribution of exchange interactions,

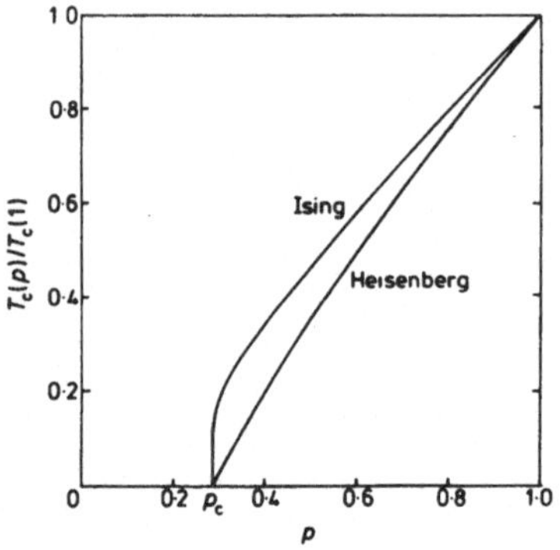


Figure 1. Reduced transition temperature versus concentration for diluted magnets.

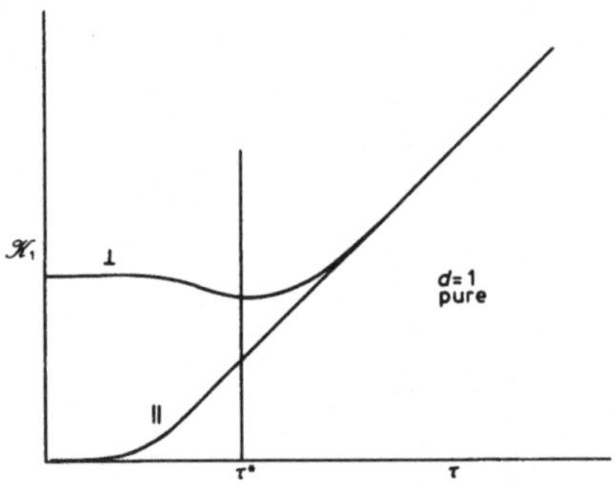


Figure 2. Inverse correlation lengths versus reduced temperature for pure anisotropic chain.

initially binary ($K_i$ = K,0 with respective probabilities p, 1-p in the bond-diluted case) is artificially kept binary[10] only series combination of equal non-zero interactions is required. Commutation difficulties arise here, but can be overcome by approximate methods[11] for spin $\frac{1}{2}$ (by which the curves of figure 1 were obtained[2]) or by going to the infinite spin (classical) model. There is no essential difference between the functional forms obtained[2,4,5], and explicit results quoted hereafter are for the classical model[5,6].

The series step takes θ to bθ and leaves u unchanged where, for low temperatures

$$\theta = \cosh^{-1} 1/\alpha \tag{2}$$

$$u = K(1-\alpha^2)^{\frac{1}{2}} + F(\theta). \tag{3}$$

F(θ) is proportional to θ and $\ell n\theta$ at small and large θ, and is negligible at low temperature except where temperature scalings are marginal.

The scaling implies that for the pure anisotropic chain at low temperatures, the inverse longitudinal correlation length is

$$K_1 = 1/\xi_1 = \theta\, \Phi_1(u)\ . \tag{4}$$

Asymptotic forms for $\Phi_1$ can be obtained from the limits $\theta \to 0,\infty$ at $T \neq 0$ (where $\xi_1$ remains finite), yielding the power law ($K_1 \propto T$) and exponential behaviour ($K_1 \propto \exp(-2\tau^*/\tau)$) seen in Figure 2 above and below the crossover temperature τ* given by (3).

The marginality of the variable u is responsible for the anisotropy crossover. This marginality also occurs automatically in the proper scaling of the distribution of exchange interactions in the d ≥ 2 dilute case, but is only maintained in the binary approximation if the scaled and initial binary distributions are related by matching moments of an appropriate variable[5], which turns out to be $K_1$.

The scalings resulting from the Migdal approximation and the binary approximation matching moments of $K_1$ are (for the example of the dilatation b=2, in two dimensions)[5]

$$\theta' = 2\theta \tag{5}$$

$$p' = 2p^2 - p^4 \tag{6}$$

$$p'\Phi_1(u') = p^4\Phi_1(2u-F(2\theta)) + 2p^2(1-p^2)\Phi_1(u). \tag{7}$$

In the pure limit these imply that for weak anisotropies ($\delta = (1-\alpha)$ small) and low temperatures the resulting scaling variables are $\delta^{\frac{1}{2}}$ and a power of $e^{-K}$ so that the inverse correlation length is

$$1/\xi = e^{-\nu K} f(\delta^{\frac{1}{2}}/e^{-\nu K}). \tag{8}$$

This includes the usual exponential dependence for the pure isotropic case[9] and the crossover form implies that the transition temperature behaves like[4]

$$T_c \propto 1/\ln(1/\delta). \tag{9}$$

Other pure results for this and for the three-dimensional case can be obtained similarly.

Hereafter we concentrate on the critical regime near the percolation threshold ($p \sim p_c$, low temperatures). The above discussion then results in the following being scaling variables (i.e. scaling by the factor b):

$$\theta,\ \delta p^{\nu_p},\ K_1{}^{\nu_p}/(\tanh\theta)^{1/\Phi_H-1} \tag{10}$$

where $\delta p \equiv |p-p_c|$. $\nu_p$ is the usual percolation exponent (arising from (6)) and $\Phi_H$ will turn out to be the Heisenberg percolation-thermal crossover exponent. Particular scalings give particular values of $\nu_p$, $\Phi_H$, but always $\delta p$ and $K_1$ scale with the same power, which makes the Ising crossover exponent unity (the exact result[12]). $\Phi_H$ is still subject to discussion[5,13].

It follows from (10) that the inverse correlation lengths of the dilute d-dimensional system near the percolation threshold have the form

$$1/\xi_{//,\perp} = \theta, \Phi_{//,\perp}(K_1{}^{\nu_p}/\theta(\tanh\theta)^{1/\Phi_H-1},\ \delta p^{\nu_p}/\theta). \tag{11}$$

For $\theta \to 0$ this yields for the weakly anisotropic Heisenberg model above its crossover temperature

$$1/\xi_{//,\perp} = \delta p^{\nu_p} f_{//,\perp}(\tau^{\Phi_H}/\delta p). \tag{12}$$

A linear dependence of $f_{//}$ on its argument is consistent with its exact asymptotic forms; and the linear function is exact in the special case d=1 (where $\nu_p = \Phi_H = p_c = 1$). Such considerations suggest the scaling plots shown in figures 3 and 4, in which experimental results for Heisenberg systems above the anisotropy crossover temperature for various $\delta p$ fall on universal curves. Figure 3 is for the d=1 magnet TMMC[14] and Figure 4 for the layer magnet $Rb_2Mn_pMg_{1-p}F_4$[15].

The limit $\theta \to \infty$ in (11) gives the behaviour of $\Phi$ when both its variables are small, from which the correlation lengths of the dilute Ising or anisotropic Heisen-

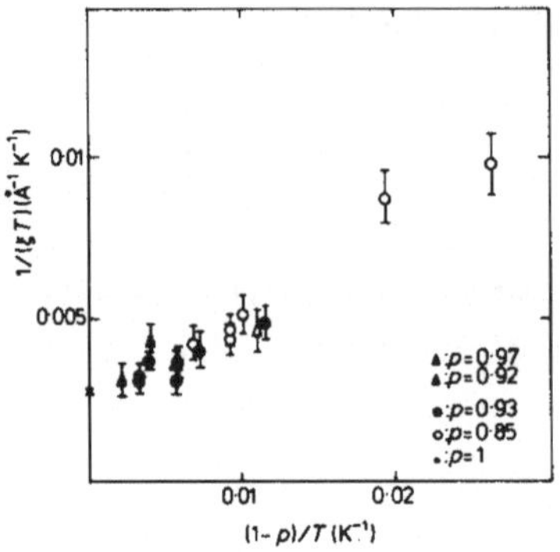


Figure 3. Universal plot for diluted Heisenberg chain.

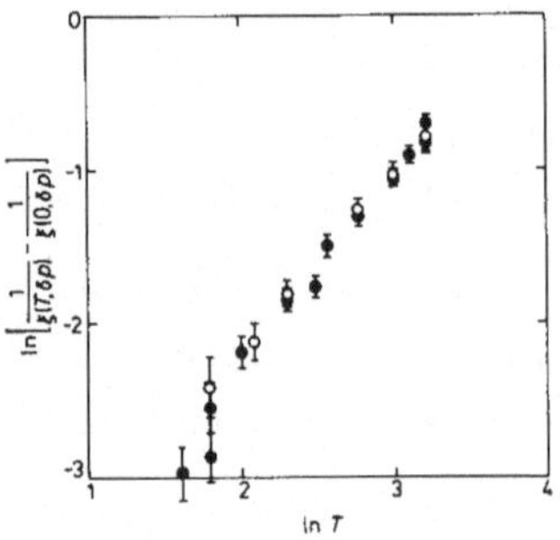


Figure 4. Universal plot for diluted Heisenberg layer magnet.

berg model below its crossover temperature can be shown to be

$$1/\xi_{//,\perp} = \delta p^{\nu_p} g_{//,\perp} (K_1/\delta p \tanh \theta^{(1/\Phi_H-1)/\nu_p}) . \tag{13}$$

The appropriate limit of (11) also yields for $p=p_c$ the inverse longitudinal correlation length vanishing like $K_1$ raised to the powers $\nu_p$ and $\nu_p\Phi_H$, respectively below and above the anisotropy crossover, as used in the analysis of experimental results[8].

(12) and (13) imply the well known[15,1,2] power law and exponential dependences of reduced transition temperature $\tau_c$ on $p-p_c$ already seen for Heisenberg and Ising cases in Figure 1, and for the weakly anisotropic Heisenberg model below its crossover give

$$\exp(-2\tau^*/\tau_c) \propto p-p_c \; . \tag{14}$$

Many other specific results can be obtained from the scaling forms derived here; and the method can be generalised to treat other types of crossover mechanisms, or other critical properties amenable to observation.

References

1. A P Young and R B Stinchcombe J. Phys. C 9 4419 (1976); C Jayaprakash, E K Riedel and M Wortis Phys. Rev. B 18 2244 (1978); J M Yeomans and R B Stinchcombe J. Phys. C 11 L525 (1978), J. Phys. C 12 347 (1979); R B Stinchcombe J. Phys. C 12, L41, 2625 (1979); M Barma, K Kumar and R B Pandey J. Phys. C 12 L283 (1979).
2. R B Stinchcombe J. Phys. C 12 L389 4533 (1979).
3. D J Breed, K Gilijamse, J W E Sterkenburg and A R Miedema J. Appl. Phys. 41 1267 (1970); E Lagendijk and W J Huiskamp Physica 62 444 (1972).
4. R B Stinchcombe J. Phys. C 14 397 (1981).
5. R B Stinchcombe J. Phys. C 13 3713, 3723 (1980).
6. R B Stinchcombe J. Phys. C 13 L133 (1980).

7. R J Birgeneau, R A Cowley, G Shirane and H J Guggenheim Phys. Rev. Lett. 37 940 (1976).
8. R J Birgeneau, R A Cowley, G Shirane, J A Tarvin and H H Guggenheim Phys.Rev. 21 317 (1980) - and other references therein.
9. A A Migdal Soviet Phys. - JETP 42 743 (1976).
10. R B Stinchcombe and B P Watson J. Phys. C 9 3221 (1976).
11. M Suzuki and H Takano Phys. Lett. 69A 426 (1979).
12. D J Wallace and A P Young Phys. Rev. B 17 2384 (1978).
13. A Coniglio Phys. Rev. Lett. 46, 250 (1981).
14. J P Boucher, C Dupas, W J Fitzgerald, K Knorr and J P Renard J. Physique 39 L86 (1978); Y Endoh, G Shirane, B J Birgeneau and Y Ajiro Phys. Rev. B 19 1476 (1979).
15. D Stauffer Z. Phys. B 22 161 (1975); T C Lubensky Phys. Rev. B 15 311 (1977).

# FINITE SIZE SCALING AND PHENOMENOLOGICAL RENORMALIZATION

B. Derrida, L. de Seze
SPT and SPSRM, CEN Saclay, 91190 Gif-sur-Yvette, France

and

J. Vannimenus
GPS, Ecole Normale Supérieure, 24 rue Lhomond 75005 Paris, France

Abstract : The basic equations of the phenomenological renormalization method are recalled. A simple derivation using finite-size scaling is presented. The convergence of the method is studied analytically for the Ising model. Using this method we give predictions for the 2d bond percolation. Finally we discuss how the method can be applied to random systems.

## I Introduction

The phenomenological renormalization (PR) method, first introduced by Nightingale[1], is a very powerful tool for studying the critical properties of a large class of models in statistical mechanics. The method consists in calculating the thermodynamic properties of one dimensional systems (like infinite strips of finite width) and in extracting from this information the critical properties of infinite systems in higher dimension. The main interests of the method are the following :

- it gives satisfactory results with a very reasonable amount of calculation
- the results can be improved systematically by increasing the width of the strips
- the convergence of the results is much more rapid than that of the Monte Carlo renormalization (a power law convergence instead of a logarithmic one)
- in contrast to most real-space renormalizations where there is a proliferation of interactions, here, only one parameter is renormalized.

We first recall the basic equations of the PR method. Then we explain how these equations can be derived from the finite size scaling hypothesis. For the 2d Ising model, we calculate analytically how the estimations of the critical temperature and the critical exponent $\nu$ converge when the width of the strip increases. Numerical results on 2d bond percolation are presented. Finally the application of the method to random systems is discussed.

## II The phenomelogical renormalization method

Suppose that one wants to calculate the critical temperature and the critical exponents of a two dimensional system (for example the Ising model). For simplicity, the only parameter in the model is the temperature T. Using the transfer matrix, one can calculate any thermodynamic quantity $Q_n(T)$ (like the correlation length $\xi_n(T)$, the magnetic susceptibility $\chi_n(T)$...) for an infinite strip of finite width n. Following Nightingale[1], let us write the fundamental equation of the PR method :

$$\frac{\xi_n(T)}{\xi_m(T')} = \frac{n}{m} \tag{1}$$

This equation establishes a correspondence between two strips : the strip of width n at temperature T is related by a scaling transformation to a strip of width m at temperature T' : this transformation is a contraction of ratio n/m of the width of the strip. Thus the correlation length which is the characteristic length along the strip has to be contracted in the same ratio.

*The main hypothesis of the PR method is to assume that the relation (1) between* T *and* T' *depends only on the ratio* n/m. Therefore if $\xi_\infty(T)$ is the correlation length of the infinite system, one has

$$\frac{\xi_n(T)}{\xi_m(T')} = \frac{n}{m} = \frac{\xi_\infty(T)}{\xi_\infty(T')} \tag{2}$$

Then from (2), one can find the critical temperature $T_c$ and the critical exponent $\nu$ ($\xi_\infty(T) \sim |T - T_c|^{-\nu}$) as usual with real space renormalizations

$$\frac{\xi_n(T_c)}{\xi_m(T_c)} = \frac{n}{m} \tag{3}$$

$$1 + \frac{1}{\nu} = \frac{\log\left[\frac{d\xi_n}{dT}(T_c)/\frac{d\xi_m}{dT}(T_c)\right]}{\log[n/m]} \tag{4}$$

Suppose that we want to calculate another critical exponent $\omega$ which describes the critical behaviour of a quantity $Q_\infty(T)$ of the 2d system. [$Q_\infty(T) \sim |T - T_c|^{-\omega}$]. Assuming again that $Q_n(T)/Q_m(T')$ depends only on the ratio n/m, one has :

$$\frac{Q_n(T)}{Q_m(T')} = \frac{Q_\infty(T)}{Q_\infty(T')} \tag{5}$$

Then the critical exponent $\omega$ can be obtained by :

$$\frac{\omega}{\nu} = \frac{\log[Q_n(T_c)/Q_m(T_c)]}{\log[n/m]} \tag{6}$$

where $T_c$ and $\nu$ are given by (3) and (4).

## III Derivation of the PR equations from the finite-size scaling hypothesis

We shall show now that the basic equations (2) and (5) of the PR method are consequences of finite-size scaling[2,3]. The content of the finite-size scaling hypothesis is to assume the existence of scaling functions $F_Q$ such that

$$Q_n(T) \sim Q_\infty(T) F_Q[n/\xi_\infty(T)] \tag{7}$$

This relation is expected to be valid when the size n of the system is large and when T is in the neighbourhood of $T_c$. Notice that when T approaches $T_c$, $Q_\infty(T)$ and $\xi_\infty(T)$ are singular whereas $Q_n(T)$ remains regular. Therefore the function $F_Q(z)$ must behave like $z^{\omega/\nu}$ for $z \to 0$ in order to eliminate the singularities of $Q_\infty$ and $\xi_\infty$.

Consider now two temperatures T and T' such that

$$\frac{\xi_\infty(T)}{\xi_\infty(T')} = \lambda \tag{8}$$

For any choice of $\lambda$, equation (8) provides a relation between T and T'.
Using the expression (7), we can write the ratio $Q_n(T)/Q_m(T')$ as :

$$\frac{Q_n(T)}{Q_m(T')} = \frac{Q_\infty(T)}{Q_\infty(T')} \frac{F_Q[n/\xi_\infty(T)]}{F_Q[m/\xi_\infty(T')]} \tag{9}$$

If we choose the ratio $\lambda = n/m$, the arguments of the two functions $F_Q$ in (9) are identical. Then, the function $F_Q$ is eliminated in (9) and one recovers equation (5). Equation (2) follows as a particular case of equation (5), when Q is equal to $\xi$.

## IV Application to the Ising Model

It is never possible to make numerical calculations on very large strips. So the corrections to finite-size scaling are not negligible. This is why the estimations $T_c(n,m)$ and $\nu(n,m)$ which are calculated by the equations (3) and (4) differ from the exact values $T_c$ and $\nu$ and converge to these exact values only when n or m becomes infinite. In order to estimate the accuracy of the method, it is interesting to know the oonvergence law of $T_c(n,m)$ and $\nu(n,m)$. This convergence law can be calculated analytically for the 2d Ising model. The Hamiltonian of the 2d Ising model is

$$H = -K \sum_{<ij>} \sigma_i \sigma_j \quad \text{with} \quad \sigma_i = \pm 1 \tag{10}$$

From the exact solution, one knows that :

$$K_c = \frac{1}{2} \log(1+\sqrt{2}) \quad \text{and} \quad \nu = 1 \tag{11}$$

Using the exact expression (that one can find in reference 1) of the correlation lengths $\xi_n$ for strips of width n with periodic boundary conditions, one can show that the estimations of $K_c$ and $\nu$ obtained by solving (3) and (4) for large n and m are :

$$K_c(n,m) = \frac{1}{2} \log(1+\sqrt{2}) - \frac{\pi^3}{192} \frac{\lambda(\lambda+1)}{n^3} + \ldots \tag{12}$$

$$\nu(n,m) = 1 - \frac{\pi^2 \log 2}{24} \frac{\lambda^2-1}{n^2 \log\lambda} + \ldots \tag{13}$$

where $m = n/\lambda$. One sees that if the ratio n/m remains finite when n and m increase, the convergence is rather rapid (a power law). This convergence is optimal for $\lambda$ as close to 1 as possible. Therefore the best choice for m is $m = n-1$. One can notice that if only $n \to \infty$ and m remains finite, then $\lambda \sim n$ and the convergence for $\nu$ becomes logarithmic as in Monte Carlo renormalizations[16,17].

## V Application to percolation

Following Nightingale[1], the Phenomenological Renormalization Method has been used to study a large class of models : generalized Ising models[4,5], Ising antiferromagnets in a magnetic field[6], lattice gas models[7,8], quantum spin systems[9], Lee and Yang singularities[10], percolation[11,12], directed percolation[13], self avoiding walks[14], lattice animals[12]...

We present in table I results recently obtained[12] for bond percolation on a square lattice. The correlation lengths were calculated for strips of width n with periodic boundary conditions by the transfer matrix method[11]. For each choice of n and m, the estimations of $p_c$ and $\nu$ were obtained by solving the equation (3) and (4)

| n | m | $p_c$ | $\nu$ |
|---|---|---|---|
| 2 | 1 | .50260 | 1.2410 |
| 3 | 2 | .48559 | 1.2015 |
| 4 | 3 | .49133 | 1.2374 |
| 5 | 4 | .49563 | 1.2710 |
| 6 | 5 | .49774 | 1.2922 |
| 7 | 6 | .49873 | 1.3047 |
| 8 | 7 | .49921 | 1.3121 |
| 9 | 8 | .49948 | 1.3169 |
| Extrapolation | | .5000 ±.0002 | 1.332 ± .003 |
| Expected values | | .5(exact) | 4/3[15] |

Table I

The extrapolations[12] were obtained by assuming a power law convergence of the results[7]. We see here the advantages of the PR method : first, even when n and m are small, the estimations of $p_c$ or $\nu$ are rather satisfactory ; Second the extrapolation ($n \to \infty$) is much easier than in Monte Carlo renormalizations[16,17] where the numbers to extrapolate are always obtained with errorbars. Moreover the convergence [$4/3 - \nu(n) \sim n^{-x}$ with $x = 2.3 \pm .4$] for large n of the exponent $\nu(n)$ is more rapid than in the Monte Carlo renormalizations [$4/3 - \nu(n) \sim \log^{-1} n$]. Last but not least, the calculations do not cost much computer time (the PR for widths n = 9 and m = 8 took less than 5 minutes of IBM 3033).

## VI Application to models with random interactions

For pure systems, the thermodynamic properties of infinite strips can be calculated from the eigenvalues of transfer matrices. For random systems (e.g. random magnets) the transfer matrices become random and the eigenvalues of the transfer matrices are replaced by the Liapounov exponents which describe the asymptotic behaviour of a product of a large number of transfer matrices. Although a lot is known about the existence of these Liapounov exponents, there does not exist rules to calculate them. Therefore one must use the brute-force method of multiplying a large number N of random matrices (the error on the Liapounov exponent is at least of order $N^{-1/2}$). These statistical errors make more difficult the calculations of the derivatives in equation (4). However the ideas of finite-size scaling remain in principle valid for random systems. Recently Pichard and Sarma[18] used this idea to study the problem of localization and could find rather accurate estimations of the critical behaviour of correlation lengths. In the same spirit, a polymer in a random medium has been studied recently[19].

## References

1. Nightingale M.P. Physica 83A (1976) 561
2. Fischer M.E. 1971. Int. School of Physics "Enrico Fermi". Varenna 1970 course n°51 ed M.S. Green (N.Y. Academic)
3. Hamer C.J. and Barber N.B. Journ. Phys. A14 (1981) 241, 259
4. Nightingale M.P. Proc. Kon. Ned. Ak. Wet. B82 (1979) 235
5. Sneddon L. Journ. Phys. C11 (1978) 2823
6. Sneddon L. Journ. Phys. C12 (1979) 3051
7. Ràcz Z. Phys. Rev. B21 (1980) 4012
8. Kinzel W. and Schick M. Phys. Rev. B23 (1981) 3435
9. Sneddon L. and Stinchcombe R.B. Journ. Phys. C12 (1979) 3761
10. Uzelac K. and Jullien R. Journ. Phys. A14 (1981) L151
11. Derrida B. and Vannimenus J. Journ. de Physique Lettres Paris 41 (1980) L473
12. Derrida B. and de Seze L. (in preparation)
13. Kinzel W. and Yeomans J.M. Journ. Phys. A14 (1981) L163
14. Derrida B. Journ. Phys. A14 (1981) L5
15. den Nijs M.P.M. Journ. Phys. A12 (1979) 1857
16. Reynolds P.J., Stanley H.E. and Klein W. Phys. Rev. B21 (1980) 1223
17. Esbach P.D., Stauffer D. and Hermann H.J. Phys. Rev. B23 (1981) 422
18. Pichard J.L. and Sarma G. J. Phys. C14 (1981) L127 and preprint 1981
19. Derrida B. (in preparation)

# GEOMETRICAL STRUCTURE AND THERMAL PHASE TRANSITION OF THE DILUTE s-STATE POTTS AND n-VECTOR MODEL AT THE PERCOLATION THRESHOLD

Antonio Coniglio

G.N.S.M. and Istituto di Fisica Teorica, Mostra d'Oltremare pad. 19, 80125 Napoli, Italy*, and Center for Polymer Studies**, Boston University, Boston MA. 02215, U.S.A.

ABSTRACT

A new relation is given in percolation theory from which follows that the backbone of the incipient infinite cluster is made of singly-connected links and "blobs", the number of links diverge with an universal exponent 1. It is also shown that this exponent characterizes the crossover exponent of the dilute s-state Potts model while for the n-vector model is given by the low density resistivity exponent.

## 1. Introduction

In percolation theory, the structure of the very large cluster, the incipient infinite cluster (IIC), is rather complex. One important task is to individualize those geometrical properties which are relevant to related problems such as disordered systems. Here exact relations are presented which relate the number of singly-connected links and the electrical resistance of the backbone of the IIC to the thermal properties respectively of the dilute s-state Potts model and n-vector model near T=0.

A model for the IIC, based on exact relations, is also proposed.

## 2. Cluster Structure

In random bond percolation with bond probability $p \leqslant p_c$ the following relation can be proved[1] for any lattice and dimensionality d

$$p \frac{dp_{ij}}{dp} = l_{ij} \tag{1}$$

where $p_{ij}$ is the pair connectedness function, i.e. the probability that site i and j belong to the same cluster and $l_{ij}$ is the average number of cutting bonds, also called singly connected links, between i and j. A cutting bond is such that if one is cut i and j are no longer connected.

*Permanent Address

**Supported in part by ARO and AFOSR

Define the following quantity

$$L = \frac{\Sigma_j l_{ij}}{\Sigma_j p_{ij}} \tag{2}$$

This is roughly the average number of cutting bonds between sites separated by a distance of connectedness length $\xi_p$ . From (1) and (2) follows

$$L = \frac{p}{S}\,\frac{dS}{dp}$$

where $S = \Sigma_j\, p_{ij}$ is the mean cluster size which diverges as $S \sim \varepsilon^{-\gamma}$, $\varepsilon = |p-p_c|/p_c$
Consequently

$$L \sim \varepsilon^{-1} \tag{3}$$

This result is valid in any dimension. It was already predicted in a different way in a previous paper[2] and confirmed by Monte Carlo methods for the square lattice by Pike and Stanley[3]. It shows that the backbone of the IIC is a linear chain made of cutting bonds interrupted by "blobs" (multiply connected bonds). (Fig. 1a). This distinction of singly-connected and multiply-connected bonds was originally introduced by Stanley[5].

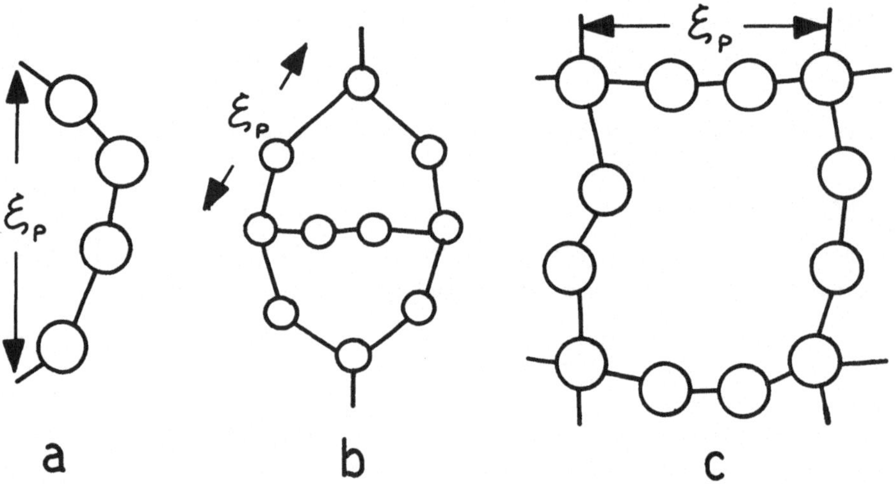

Fig. 1a) Schematic structure of the backbone of the incipient infinite cluster below $p_c$ ; b) structure of a "blob": two points separated by a distance of the order of $\xi_p$ are connected by a chain of cutting bonds and "blobs" (self similarity). c) Picture of the "nodes, links and blobs" model, a generalization of the well known "nodes and links" model[7,9] for the infinite cluster above $p_c$.

From (3) using a simple argument it is possible to show that for low dimensionality the "blobs" are important. In fact if there were no "blobs" the backbone of the IIC would have been made only of cutting bonds with the obvious relation $L \geqslant \xi_p$ since the end-to-end distance must be smaller than any other path. Since $\xi \sim \varepsilon^{-\nu_p}$ from (1) would follow $\nu_p < 1$ which contradicts the well established result $\nu_p \simeq 1.33$ in 2 dimensions. Therefore the blobs must necessarily present in 2 dimensions so that L can be smaller than $\xi_p$.Of course as the dimensionality goes up they become less relevant until they disappear in 6 dimensions, where the backbone becomes a random walk[7].

The "blobs" themselves have a structure. In fact a relation similar to (1) for the biconnected part of the IIC can be proved from which it can be shown that the "blobs" are made of cutting bonds and "blobs" (Fig. 1b). These new "blobs" of course also have a similar structure. The process ends until one reaches "blobs" whose linear dimension is of order of magnitude much smaller than $\xi_p$ .The self similar structure which comes out has been the basic ingredient for a recently proposed fractal model of the infinite cluster[8].

Above $p_c$ other relations similar to eq. 1 can be obtained which will suggest a model for the backbone of the infinite cluster, made of nodes separated by a distance of the order of $\xi_p$ , connected by chains made of links and "blobs" with the same self similar structure as the IIC below $p_c$. (Fig. 1c). This model is a generalization of the nodes and links model proposed by Skal and Shklovskii[9] and indipendently by de Gennes[7].

3. Crossover exponent of the dilute s-state Potts model and n-vector model at the percolation threshold

Consider the bond dilute s-state Potts model and n-vector model near the multicritical point $p=p_c$ , T=0. If one approaches this point along the direction T=0, $p \to p_c$ , the correlation length for both models diverge with the percolation exponent $\nu_p$ while in the perpendicular direction $p=p_c$ $T \to 0$ the correlation length $\xi$ diverge as $\xi \sim \tau^{-\nu_T}$ where $\tau = e^{-sJ/KT}$ for the s-state Potts model and $\tau = \frac{KT}{J}$ for the n-vector model, $\nu_T$ is the thermal exponent which is different in the two models, J is the nearest neighbor coupling constant.

Using scaling[10], one can define the crossover exponent $\phi = \frac{\nu_p}{\nu_T}$ via

$$\chi = \varepsilon^{-\gamma_p} \quad F(\tau/\varepsilon^{\phi}) \tag{4}$$

where $\chi$ is the susceptibility near the multicritical point.
For $\tau/\varepsilon^{\phi} \ll 1$ (4) becomes

$$\chi = A\,\varepsilon^{-\gamma_p}\,[\,1 + B\,\tau/\varepsilon^{\phi} + \ldots.] \tag{5}$$

with A and B constants.

In order to find the crossover exponent $\phi$ we have calculated[1] to the first order in $\tau$ for $p \leqslant p_c$ the pair correlation function $g_{ij}$ for both models. We find

$$g_{ij} = p_{ij} + C_s \tau l_{ij} + \dots \text{ (s-state Potts model)} \tag{6}$$

$$g_{ij} = p_{ij} + C_n \tau R_{ij} + \dots \text{ (n-vector model)} \tag{7}$$

where $C_s$ and $C_n$ are constants, $l_{ij}$ is the average number of cutting bonds between i and j introduced in Sect. 2 and $R_{ij}$ is the average resistance between i and j if a unit electrical resistance is associated to each bond. This two point resistance has been introduced by Fisch and Harris[11] who also defined

$$R = \frac{\Sigma_j R_{ij}}{\Sigma_j p_{ij}} \tag{8}$$

which is roughly the resistance between two points separated by a distance of the order of $\xi_p$. Since the "blobs" do contribute to the electrical resistance $R_{ij} \geqslant l_{ij}$. Therefore from (2) and (8) $R \geqslant L$ and $R \sim \varepsilon^{-Z_R}$ with $Z_R \geqslant 1$ which agrees with the series expansion numerical values[11].

Summing eqs. (6) and (7) over j and taking into account (2), (3) and (8) we have

$$\chi = A\,\varepsilon^{-\gamma_p} [\,1 + C_s \tau \varepsilon^{-1} + \cdot\cdot\,] \qquad \text{(s-state Potts model)}$$

$$\chi = A\,\varepsilon^{-\gamma_p} [\,1 + C_n \tau \varepsilon^{-Z_R} + \dots\,] \qquad \text{(n-vector model)}$$

comparing with (5) we find $\phi = 1$ for the s-state Potts model and $\phi = Z_R$ for the n-vector model. This result was predicted in a different way in a previous paper[2]. $\phi = 1$ agrees with other theoretical[12] and experimental[13] results for s = 2 (Ising). For a connection between dilute n-vector model and random resistor see also Ref. 14. Recent experiments on dilute Heisenberg antiferromagnetic systems give $\phi = 1.48 \pm .15$ and $0.88 \pm .04$ respectively for d = 2 and 3 to be compared with the series expansion value[11] of $Z_R = 1.43 \pm .02$ and $1.12 \pm .02$. The agreement in two dimensions is very good while in 3 dimensions the presence of dipolar interaction in the experimental system may be the cause of the discrepancy.

## CONCLUSIONS

In conclusion two geometrical features of the IIC, the cutting bonds and the electrical resistance, have been found to be relevant to the thermal behaviour of the dilute s-state Potts model and n-vector model near $p_c$. One is characterized by an universal exponent 1, the other by an exponent $Z_R \geqslant 1$ which can be evaluated numerically.

## REFERENCES


1. A. Coniglio, to be published
2. A. Coniglio, Phys. Rev. Lett. (1981)
3. R. Pike and H.E. Stanley, J. Phys. A 14, L169 (1981)
4. In the "links and nodes" model of the backbone of the infinite cluster it is argued by Skal and Shklovskii[9] and postulated by de Gennes[7] that the length L of the macrobonds connecting two nodes diverges as $L \sim \varepsilon^{-1}$, but no blobs are present in this model.
5. H.E. Stanley, J. Phys. A 10, L211 (1977)
6. See for example P.D. Eschbach, D. Stauffer and H.J. Herrmann, Phys. Rev. B 23, 422 (1981)
7. P.G. de Gennes, Jour. de Phys. Lett. 37, L1 (1976)
8. Y. Gefen, A. Aharony, B.M. Mandelbrot and S. Kirkpatrick, preprint 1981 and this conference
9. A. Skal and B.I. Shklovskii, Fiz. Tekh. Popuprovdn. 8, 1586 (1974) [Sov. Phys. Semicond. 8, 1029 (1975)]
10. D. Stauffer, Z. Phys. B 22, 161 (1975)

    H.E. Stanley, R.J. Birgeneau, P.J. Reynolds and J.F. Nicoll, J. Phys. C 9, L553 (1976)

    T. Lubenski, Phys. Rev. B 15, 311 (1977)
11. R.Fisch and A.B. Harris, Phys. Rev. B 18, 416 (1978)
12. M.J. Stephen and G.S. Grest, Phys. Rev. Lett. 38, 567 (1977), D.J. Wallace and A.P. Young, Phys. Rev. B 17, 2384 (1978)
13. R.A. Cowley, R.J. Birgeneau, G. Shirane, H.G. Guggenheim and H. Ikeda, Phys.Rev. B 21, 4038 (1980)
14. S. Kirkpatrick, Solid State Comm., 12, 1279 (1973);R.B.Stinchkombe,J.Phys.C12, 2625 (1979)
15. R.A. Cowley, G. Shirane, R.J. Birgeneau, E.C. Swenson, H.J. Guggenheim, Phys. Rev. B 22, 4412 (1980).

# PERCOLATION, CRITICAL PHENOMENA AND FRACTALS

A. Aharony and Y. Gefen, Dept. of Physics and Astronomy, Tel Aviv University, Ramat Aviv (Israel)

B. Mandelbrot and S. Kirkpatrick, IBM T.J. Watson Research Center, Yorktown Heights, N.Y. 10598 USA

## Abstract

Physical and geometrical properties are studied on self similar fractal lattices. Properties of spin systems are shown to depend on various topological factors, in addition to the fractal dimensionality. A (non random) fractal model is proposed for the backbone of the infinite cluster near percolation in d dimensions, and its properties agree with those of the backbone for $d \lesssim 4$.

## 1. Introduction

The study of the dependence of critical phenomena on the dimensionality of space, d, has been very useful in our understanding of these phenomena.[1] In particular, Ising spin models with short range interactions are believed to have a lower critical dimensionality equal to one, at which the correlation length diverges as $\xi \sim (e^{-2K})^{-\nu}$, with $K = J/kT$ (J = exchange, k = Boltzmann constant, T = temperature) and $\nu = 1$. At higher dimensionalities $\xi \sim |T-T_c|^{-\nu}$, and the universal exponent $\nu$ decreases monotomically with d down to $\nu = \frac{1}{2}$ for $d > 4$. These statements have been based on various calculations at integer d or on $\varepsilon$-expansions, and it has not been clear if their extrapolation to non integer d is unique. It has been recently realized,[2-6] that the infinite cluster, which occurs at the percolation threshold, is self similar and may thus be described as having a non-integer fractal dimensionality.[3] The fractal dimensionality D of a self similar structure is defined as follows: If each unit in the structure is replaced by N similar units, with a length scale smaller by a factor b, then $b^D = N$. If the probability to belong to the infinite cluster is $P(p) \sim (p-p_c)^\beta$ (p = concentration of non-zero bonds, $p_c$ = percolation threshold), then $D = d-\beta/\nu$. Similarly, if the probability to belong to the infinite backbone (ignoring "dangling" bonds) is $B(p) \sim (p-p_c)^{\beta'}$ then its fractal dimensionality is $D = d-\beta'/\nu$.

The aim of the present study is to understand the dependence of critical phenomena on fractal properties, and to use this understanding for calculating physical quantities near the percolation threshold.

## 2. Critical Phenomena on Fractal Lattices

We have recently[7] put spins on the sites of various self similar lattices, and applied the renormalization group to find their critical properties. The lattice is assumed to be self similar down to some microscopic length scale, at which the spins have nearest neighbor interactions. As an example, Fig. 1 shows two stages of the Sierpinski gasket, whose fractal dimensionality is $D = \ln 3/\ln 2 \simeq 1.585$. We showed exactly that the Ising model on this gasket has no ordering at finite temperatures, and that its correlation length diverges when $T \to 0$ as $\xi \sim \exp[4 \exp(4K)]$. More examples are listed in Ref. 7. The important conclusions were that in addition to the fractal dimensionality D, critical properties depend on many other geometrical properties, e.g. the minimum order of ramification R (measuring the number of significant interactions which one must cut in order to isolate an arbitrarily small bounded part of the system; R = 3 in Fig. 1) and its homogeneity, the connectivity (measuring the minimum fractal dimensionality of the "cut" required to isolate a bounded infinite part of the system when $R = \infty$), the lacunarity (measuring the extent of the failure of a fractal to be translationally invariant), etc. There exists no lower critical fractal dimensionality, and $T_c = 0$ whenever R is finite. These results imply a generalization of the notion of universality. Similar results were more recently established for Heisenberg-like spin models.[8]

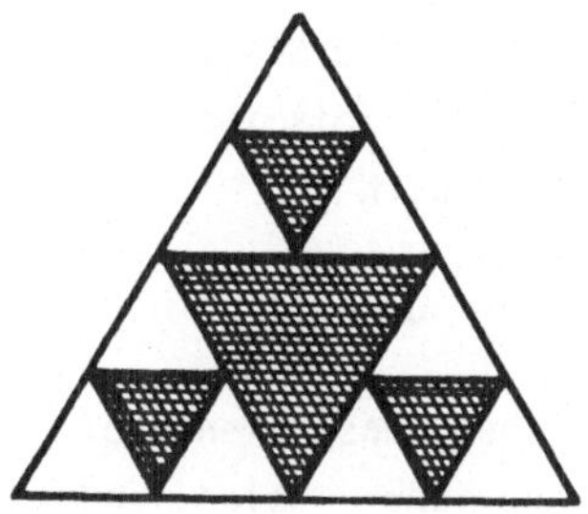

Fig. 1. Sierpinski gasket: the shaded areas are successively eliminated.

## 3. A Fractal Model for the Backbone

The fractal dimensionality of the Sierpiński gasket (Fig. 1), $D \simeq 1.585$, is very close to the numerical value found for the fractal dimensionality of the backbone of the infinite two dimensional cluster near percolation.[6] Their orders of ramification are also similar. If the random nature of the real backbone is not important, then calculations on the gasket (which can be done exactly) may yield physical properties which are relevant for the backbone.[9]

We have generalized the Sierpiński gasket to d dimensions, by starting with a d-dimensional hypertetrahedron and by successively eliminating the (central) volume bounded by the lines connecting the mid-points of the edges. The resulting structures have

$$D = \ln(d+1)/\ln 2,$$

i.e. D = 1., 1.585, 2., 2.322 and 2.807 for d = 1, 2, 3, 4 and 6. The values for $d \leq 4$ agree very well with available values of $\beta'$.[5,6,9] The agreement breaks down at d = 6,[9] where the simpler Skal-Shklovskii[10] "links and nodes" picture probably applies. This simpler picture does not give a consistent description for $d \lesssim 4$.

We have also put resistors on the (smallest scale) links of these lattices, and found exactly[9] that their resistance scales as $\rho(ba) = \frac{d+3}{d+1}\rho(a)$ when the length scale increases by a factor b = 2. This implies that the conductivity measured on a scale L behaves as $\sigma(L) \sim L^{-\tilde{t}}$, with

$$\tilde{t} = d - 2 + \tilde{\zeta} \ , \qquad \tilde{\zeta} = \ln\,[(d+3)/(d+1)]/\ln 2.$$

If $L < \xi \sim (p-p_c)^{-\nu}$, then measurements of $\sigma$ will depend on L. If $L > \xi$ then the self similar picture no longer holds, and we have

$$\sigma \sim \xi^{-\tilde{t}} \sim (p-p_c)^t \ , \qquad t = \nu\tilde{t}.$$

We find $\tilde{t}$ = 0, 0.737, 1.585, 2.485 and 4.363 for d = 1, 2, 3, 4 and 6, and the values for $d \lesssim 3$ are in reasonable agreement with known values of t.[9] At higher d one seems to need more than self similar loops.

The above calculation also yields the crossover exponent for Heisenberg spin systems, $\phi_H \equiv \nu\tilde{\zeta}$. We find that this relation between $\phi_H$ and t should hold for any model.

## 4. Conclusion

Fractal model systems prove to be very helpful in understanding critical phenomena and percolation. Once the relevant geometrical characteristic of a system (e.g. the backbone) are identified, one can construct a simple fractal lattice, and calculate any wanted physical property. The fractal models presented above give much better estimates for $\beta'$ and for t than any available alternative model.

We have enjoyed discussions with S. Alexander, D. J. Bergman and Y. Shapir. This work was supported by the U.S.-Israel Binational Science Foundation.

## References

1. A. Aharony, in Phase Transitions and Critical Phenomena, edited by C. Domb and M. S. Green, (Academic, N.Y. 1976), Vol. 6, p. 357.
2. H. E. Stanley et al, J. Phys. C9, L553 (1976).
3. B. B. Mandelbrot, "Fractals: Form, Chance and Dimension" (Freeman, San Francisco 1977).
4. B. B. Mandelbrot, Ann. Israel Phys. Soc. 2, 226 (1978).
5. D. Stauffer, Phys. Repts. 54, 1 (1979).
6. S. Kirkpatrick, A. I. P. Conf. Proc. 40, 99 (1978).
7. Y. Gefen et al, Phys. Rev. Lett. 45, 855 (1980).
8. Details will be published.
9. Y. Gefen et al, preprint.
10. A. Skal and B. I. Shklovskii, Sov. Phys.-Semicond. 8, 1029 (1975).

# NEW DIRECTIONS IN PERCOLATION, INCLUDING SOME POSSIBLE APPLICATIONS OF CONNECTIVITY CONCEPTS TO THE REAL WORLD

H. Eugene Stanley*

Center for Polymer Studies and Department of Physics
Boston University, Boston, Massachusetts 02215 USA

Abstract. This talk is designed to complement that of D. Stauffer; together both seek to review recent work in percolation theory that has taken place after completion of the two recent reviews by Stauffer (1979) and Essam (1980). Stauffer's talk focusses on new results concerning percolations clusters, while this talk concerns some less well understood topics, including possible applications of percolation to the real world. The organization of this talk is presented in the following outline. After presenting a word of philosophy, we shall describe several topics and exemplify each with a particular system:

| TOPIC | EXAMPLE |
|---|---|
| 1. Pure percolation | incipient infinite cluster topology |
| 2. Generalizations of pure perc. | |
| A. No solvent (random-bond perc.) | Flory gel |
| B. Solvent (correlated site-bond perc.) | Tanaka gel |
| 3. The model itself | Ising droplets |
| 4. The solvent itself | $H_2O$ and $D_2O$ |
| A. "Puzzle of liquid water" | |
| B. Clues | |
| C. Hypothesis | |
| D. Tests | |
| (i) Computer water | |
| (ii) Real water | |
| E. Summary and outlook | |

* * * * *

A few yearsago, Victor Weisskopf remarked that the 1980's may become known as the "decade of disorder," and for this reason it is noteworthy that Carlo di Castro has organized the first international conference of the 1980's explicitly on the subject of disorder. It is therefore with some timidity that I accepted Carlo's request to project some of the new directions that this decade may take.

Before beginning, I should acknowledge my collaborators in this research, R. Pike, A. Coniglio, W. Klein, J. Teixeira, A. Geiger, L. Bosio, and R. L. Blumberg. The theoretical models to be presented here were motivated strongly by discussions of experimental phenomena with C. A. Angell, R. Bansil, R. Birgeneau, J. Leblond,

*John Simon Guggenheim Memorial Fellow, 1980-81.

P. Papon and T. Tanaka. I must also acknowledge fruitful interactions with M. Daoud, F. Family, E. Gawlinski, A. Gonzales, H. Gould, S. Muto, H. Nakanishi, S. Redner, P. J. Reynolds, P. Ruiz, G. Shlifer, and D. Stauffer. Finally, I should state my indebtedness to D. Stauffer for useful suggestions on how the two of us could partition the topic of percolation at this meeting, and comments on this manuscript in its preliminary form.

The main point of this talk is to exemplify the basic "philosophy" that a theorist can be of use by exhaustively studying relatively simple models--suitably generalized--since the insights so gained can lead to eventual clarification and understanding of the subtle phenomena that occur in the real world. This philosophy has been strikingly illustrated in the field of phase transitions and critical phenomena, as indicated in Fig. 1. There I schematize the abstraction process in which one begins with a real system, identifies the essential physics of the system, and then formulates the simplest model that incorporates that essential physics. Two familiar examples are

(i) a fluid near its critical point, for which the essential physics is an interparticle interaction potential characterized by a hard-core repulsion and short-range attraction, and

(ii) a dilute polymer solution, for which the essential physics is the hard-core repulsion (or "excluded volume") alone.

Useful progress has resulted on system (i) and system (ii) using, respectively, two cases of the n-vector model--n=1 and n=0.

The third example shown in Fig. 1 is that of polymer gelation. The essential physical feature of a gel is connectivity, and to this end we expect the s=1 case of the s-state Potts hierarchy to be relevant. On the other hand, temperature-dependent effects such as those due to the presence of solvent are excluded from this simple "pure percolation" case--and we shall see below that some straightforward extensions are necessary.

At the risk of oversimplification, we could say that in 1920 there was only the Lenz-Ising station on the "Metro map" of Fig. 1. Then the East-West branch and the North-South branch were constructed, thereby greatly enriching the range of physical phenomena that could be described by simple models. As concerns the "percolation" station on the North-South Metro line, we can anticipate that the range of phenomena will be extended as we study various generalizations of "pure" percolation.

## 1. PURE PERCOLATION (Example: incipient infinite cluster topology)

One of the first seminars I ever heard on the Ising model was in the early 1960's from my statistical mechanics professor Roy Glauber. He described a generalization of the Ising model (which itself has no dynamics) to what has come to be called the Glauber model [1]. At the beginning of his seminar, Glauber actually apologized for having an "Ising disease," for at that time workers on the Ising model were thought

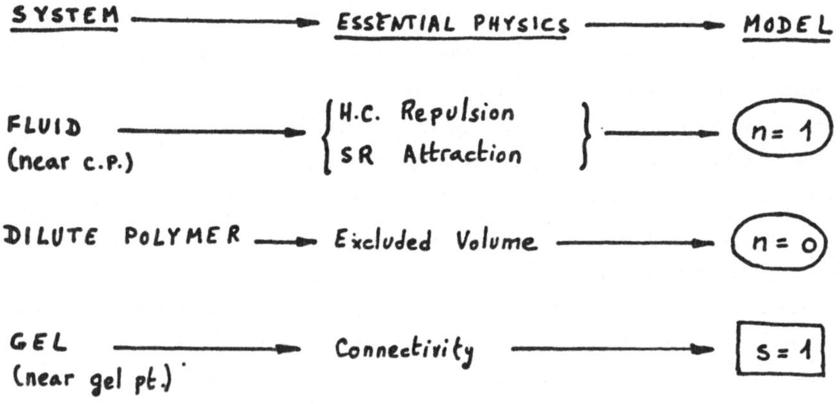


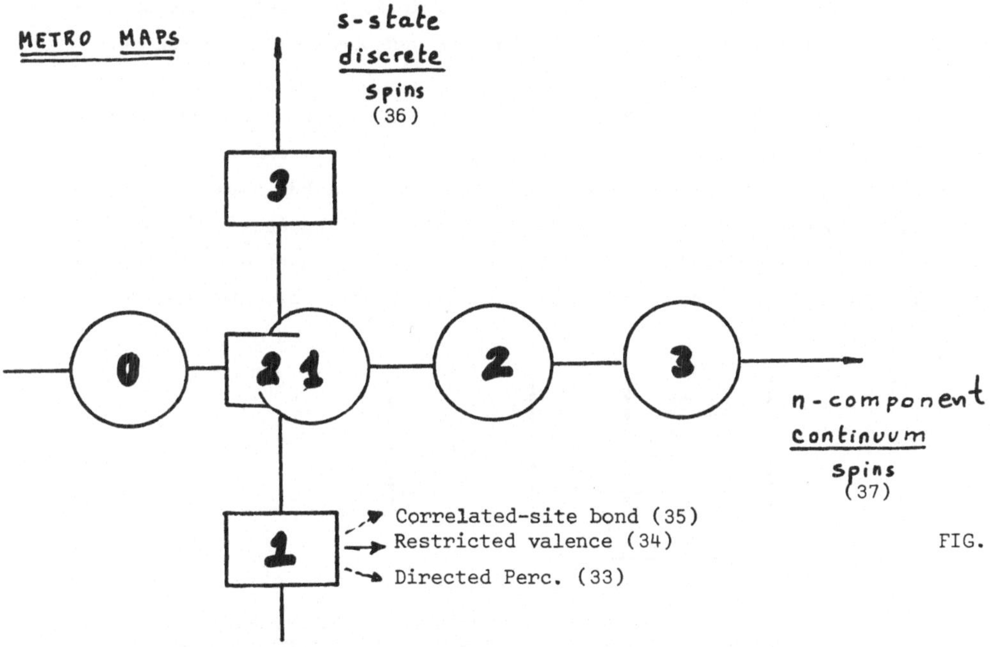


FIG. 1

to be "hooked" on the apparent simplicity of this simple model. Nowadays we appreciate the fact that it was not the researchers who had the Ising disease, but rather the critics who invented the term! Indeed, it would be difficult to imagine how the recent revolution in phase transitions and critical phenomena understanding could have occurred were it not for the firm foundation of understanding of the "simple" Ising model.

By the end of this conference, the non-practition of percolation may be wondering if perhaps we have a mutant strain of the Ising disease. You may be right! However it is my hope that our studies of this simple model will eventually provide the underpinnings of true progress on systems for which the essential physics would appear to be connectivity. A very recent example is highly significant: the study of the distribution of zeros of the d-dimensional Ising model partition function on the imaginary H axis [2], which has recently been seen to be directly related to the statistics of a branched polymer in dimension d+2 [3].

As an example, we discuss in this section one particularly fascinating aspect of the "percolation disease" that shows signs of yielding to solution. This is the question of how one describes the incipient infinite cluster that appears as the percolation threshold is approached from below. Consider, e.g., a site-dilute random magnet. When the fraction p of magnetic sites is very small, the system consists of small disconnected clusters of magnetically correlated sites. As $p \to p_c^-$, the mean cluster size increases until at $p=p_c$ a single cluster spans the entire lattice. The percolation threshold $p_c$ is a critical point. By studying the propagation of magnetic correlations through the incipient infinite cluster that appears in the dilute magnet at its percolation threshold, one can obtain information about order propagation near this critical point.

Since the incipient infinite cluster dominates the behavior of the system it is important to be able to describe its structure. If a cluster is considered as a network of wires carrying electrical current between two parallel bus bars, it can be decomposed into a conducting 'backbone' [4,5] and many 'dangling ends' that do not contribute to the electrical conductivity (and hence order propagation) between the ends [Fig. 2]. Describing the topology of the backbone is a formidable unsolved problem.

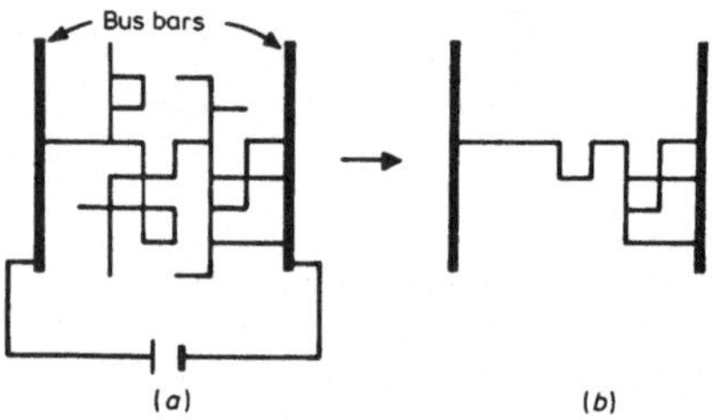


FIG. 2

**Bus bars are attached to the extremities of a cluster and a potential difference applied (*a*). Current flows in the backbone bonds (*b*), but not through 'dangling ends' which are attached to the backbone at only one vertex.** After Ref.5.

The backbone bonds may be divided into two classes, conveniently visualized as 'red' and 'blue' [6]. Red (blue) bonds are singly connected (multiply-connected); removing a single red (blue) bond breaks (does not break) the connection between two parallel 'east-west' bus bars touching the extreme northernmost and extreme southernmost bonds. An example of this backbone decomposition is shown in Fig. 3, where 'red' bonds are shown as heavy lines and 'blue' bonds as light lines.

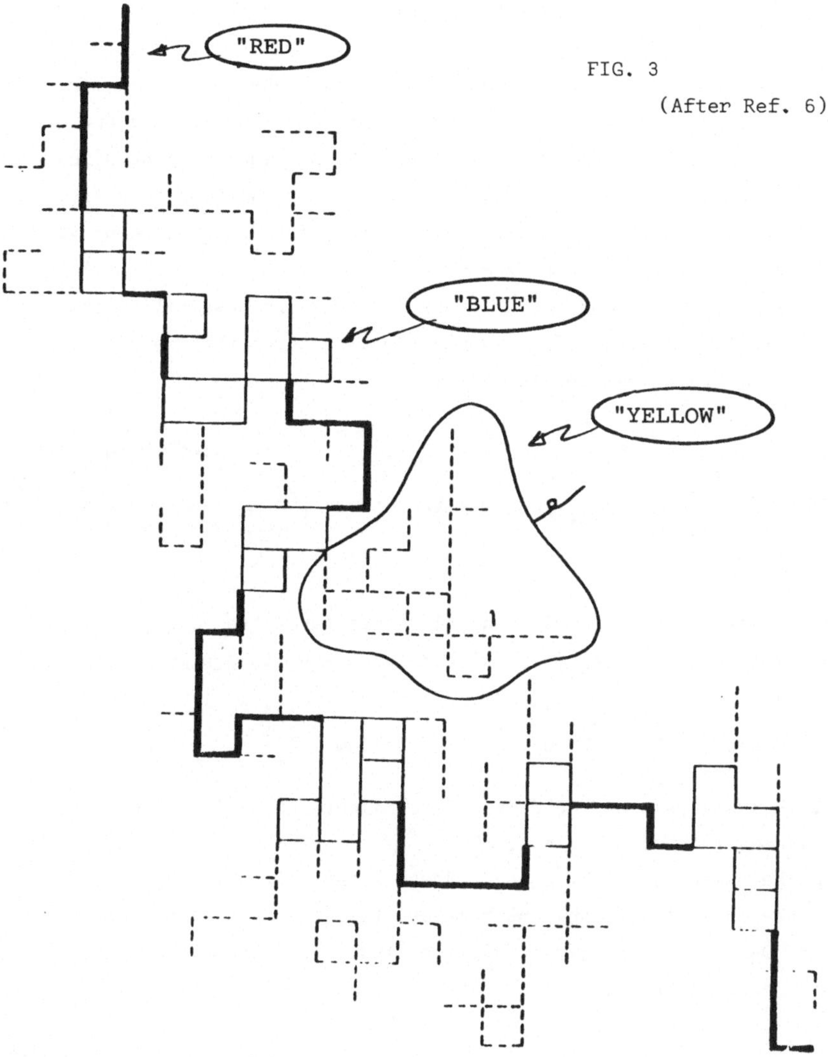


FIG. 3
(After Ref. 6)

The distinction between red and blue bonds seemed important in connection with the dilute Ising magnet--e.g., propagation of order along a string of red bonds would be analogous to propagation in a simple one-dimensional system, while propagation or order in a "blue blob" is more like propagation in the d-dimensional magnet [7]. For this reason, I posed the "red-blue" problem in a talk given at the annual Canadian Undergraduate Physics Conference in Toronto in October 1977, and an undergraduate member of the audience, Rob Pike, came up afterward to announce that he thought he could solve the problem.

Pike succeeded in formulating a computer algorithm [8] that partitions bonds into three separate colors: red, blue, and yellow (the dangling ends). His algorithm is free of boundary effects, and gives the <u>weighted</u> cluster distribution function

$$n_1(b,p) \propto b\, n_o(b,p), \tag{1}$$

where $n_o(b,p)$ is the number of b-bond clusters per site. The factor b makes the physically-important large clusters much more numerous; e.g., to generate one 60,000 bond cluster near $p_c$ by traditional "box-filling methods" requires generating 60,000 times as many one-bond clusters, 30,000 times as many two-bond clusters, and so forth. Thus for a given number of clusters generated, the Pike method provides much more accurate statistics for large clusters.

One of the worries that clouded the Pike project was the possibility that as $p \to p_c^-$, the mean number of red bonds $S_{Red}$ would approach zero; this would occur, e.g., by the joining up of yellow dangling ends:

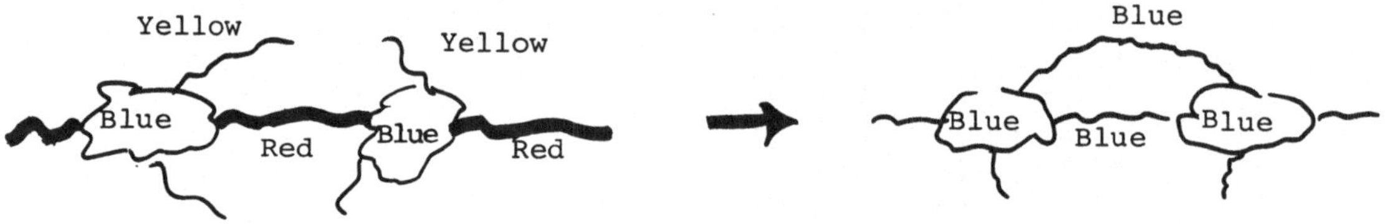


Fortunately, Pike's data showed clearly that all three functions $S_{Red}$, $S_{Blue}$, and $S_{Yellow}$ diverge. The corresponding critical exponents are, to 2 significant figures,

$$\gamma_{Red} = 1.0, \quad \gamma_{Blue} = 1.7, \quad \gamma_{Yellow} = 2.4. \tag{2}$$

In connection with his analysis of the dilute ferromagnet, Coniglio [9] has very recently proved rigorously that $\gamma_{Red} = 1$.

From (2) and the result that the connectedness length $\xi_p \sim |p - p_c|^{-\nu}$ with $\nu = 4/3$ for both the full cluster <u>and</u> the backbone [5], we can evaluate the fractal dimensions $D^\dagger$ and $D_B{}^\dagger$ of the incipient infinite cluster and of the incipient infinite backbone. From the results [6]

$$D^\dagger = y_h = (\gamma + d\nu)/2\nu \qquad D_B{}^\dagger = y_{h_B} = (\gamma_B + d\nu)/2\nu \quad , \tag{3a}$$

where h is the occupation probability of a family of ghost bonds that connect every <u>cluster</u> bond to the ghost site, and $h_B$ is the analogous quantity that connects every <u>backbone</u> bond to the ghost site, we find

$$D^\dagger \simeq 91/48 \simeq 1.9 \qquad D_B{}^\dagger \simeq 157/96 \simeq 1.6 \quad , \tag{3b}$$

where we have used the extended den Nijs conjecture $\gamma = 43/18 = 2.4$ and $\gamma_{Blue} = (\gamma_{Red} + \gamma_{Yellow})/2$.

The utility of the red-blue decomposition of the backbone to the classic problem of order propagation in the dilute ferromagnet has recently been demonstrated [10]. However there remain interesting questions relating to cluster topology. An important open question is the relation to cluster properties of the exponent $t$ describing the approach to zero as $p \to p_c^+$ of the electrical conductivity, $\sigma_{el} \sim (p - p_c)^t$. The beautiful experiments of Deutcher, Lagües and collaborators [11] should serve as a strong stimulus for a program designed to put the exponent $t$ on a firmer conceptual foundation. Another question concerns the new exponent $\gamma_{min}$ defined through $L_{min} \sim (p_c - p)^{-\gamma_{min}}$, where $L_{min}$ is the shortest "cow path length" through the cluster backbone (from one bus bar to the other). Pike and Stanley [8] find $\gamma_{min} = 1.35 \pm 0.02$ when averaging over all clusters, and $\gamma_{min} = 1.49 \pm 0.02$ when averaging over only the largest clusters.

## 2. GENERALIZATIONS OF PURE PERCOLATION (Example: gels)

In this section, we motivate and then describe the generalization of the pure percolation problem necessary to describe the essential physics of polymer gelation. The simplest example that illustrates the basic phenomenon of gelation is polyfunctional condensation of monomers. Suppose all our monomers are identical, and that each has $f$ functional groups that can react with one of the $f$ groups of another monomer. The simplest case-- $f = 0$ --produces no reactions at all; an example might be argon! The next simplest case, $f = 1$, results in dimers only. If $f = 2$, we can have unbranched linear polymers. For $f \geq 3$, we form branched polymers, as illustrated in Fig. 4 for particular $f = 3$ example, trimethoyl benzene. Each benzene ring has

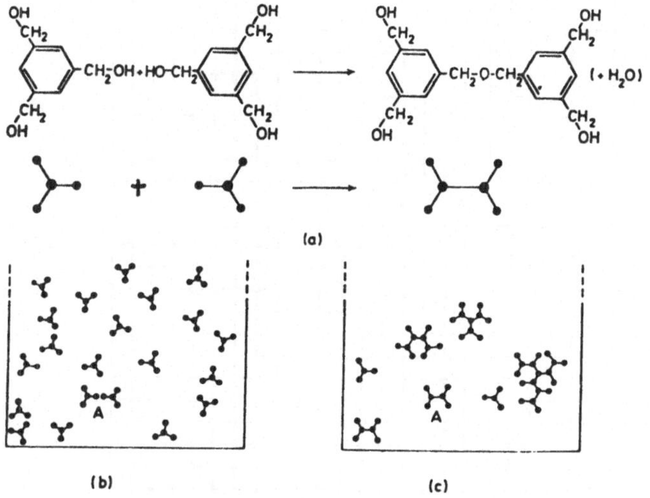


FIG.4
[after M.Gordon]

three reactive methyl groups. Methyl groups from two different monomers can react to form an ether linkage. This process is characterized by a parameter $\alpha$, termed the conversion, which is the fraction of reacted methyl groups. Clearly if $\alpha=0$ only monomers are present. If $\alpha>0$, there exist finite polymers of all sizes. However the probability of an infinite polymer is zero for all $\alpha$ less than a critical value $\alpha_c$. For $\alpha > \alpha_c$, however, there is non-zero probability of occurrence of a single polymer that is infinite in spatial extent. The probability of this infinite molecule to occur jumps from zero (for $\alpha<\alpha_c$) to unity (for $\alpha>\alpha_c$). Thus the connectivity of the system changes drastically at $\alpha=\alpha_c$, and this "phase transition" is termed the gelation threshold.

A. NO SOLVENT (Model #1: Random-bond percolation)

The first successful model to capture the essential physics of the gelation threshold was proposed 40 years ago by P. J. Flory, and developed in a series of classic papers by both Flory and W. H. Stockmayer [12]. The Flory-Stockmayer (FS) model not only predicts the gelation threshold, $\alpha=\alpha_c$, but also provides the "critical exponents" characterizing the behavior of various quantities in the immediate vicinity of $\alpha_c$.

The FS model is most easily explained after first introducing the problem of random bond percolation. Random bond percolation can be easily illustrated with the use of Fig. 5. Suppose we have an infinitely high and infinitely long wire fence.

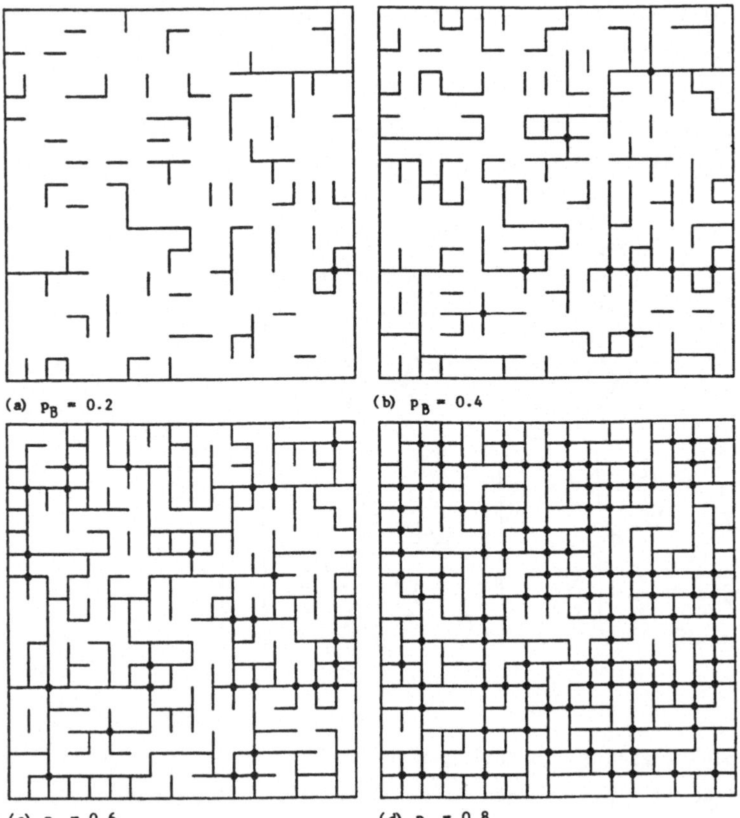

(a) $p_B = 0.2$ (b) $p_B = 0.4$ (c) $p_B = 0.6$ (d) $p_B = 0.8$

FIG.5

Imagine also that a randomly chosen fraction $p_B$ of the links of this fence are conducting while the remaining fraction $(1 - p_B)$ are insulating. Computer simulations of a finite (16 x 16) section of this fence are shown in Fig. 5 for

$p_B$ = 0.2, 0.4, 0.6, and 0.8. Clearly for $p_B$ small, as in part (a), the system consists of small clusters of conducting bonds. In (b) the conducting fraction $p_B$ has doubled, yet the system still consists of only finite clusters . . . the "scale" has increased, but not the essential macroscopic conductivity. In (c), $p_B = 0.6$, and the system is macroscopically different: in addition to the finite clusters, there is a single cluster that is infinite in spatial extent (of course, the fence must be infinite if the cluster is to be infinite!). For some value of $p_B$ in between part (b) and (c), there is a threshold $p_B^c$; below $p_B^c$, the fence cannot conduct, while above $p_B$ it can. Thus its macroscopic properties change suddenly as a microscopic parameter $p_B$ increases infinitesimally from $p_B^c - \delta$ to $p_B^c + \delta$ (Fig. 6a).

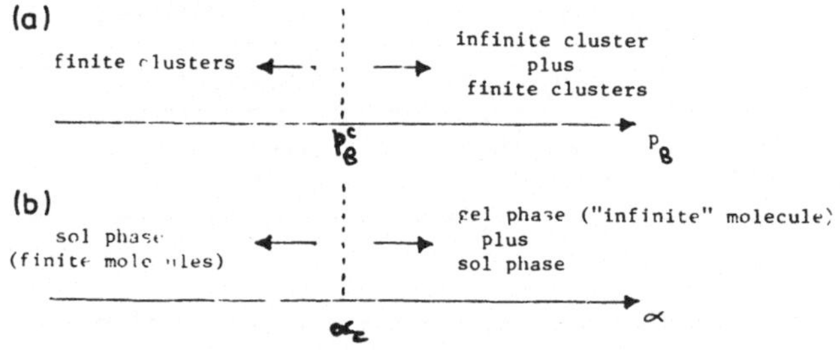


FIG.6

Similarly, below the gelation threshold $\alpha_c$, the system of Fig. 4 consists of only finite-size polymers; it cannot, e.g., propagate a shear stress. Above the threshold it can. Thus the macroscopic properties change suddenly as a microscopic parameter $\alpha$, the extent of reaction (or equivalently, the fraction of formed crosslinks) increases infinitesimally from $\alpha_c - \delta$ to $\alpha_c + \delta$ (Fig. 6b).

The FS model was formulated in a fashion that at first sight seems to be lattice independent: one merely requires that a given polymer be forbidden to loop back upon itself--in short, intramolecular interactions are excluded. Today we recognize this assumption as fully equivalent (as far as critical behavior is concerned) to the statement that the polyfunctional monomers be required to occupy the sites of a Cayley tree pseudolattice [13]: to each and every configuration of the f-functional monomers there is a one-to-one correspondence with a configuration of bonds on the Cayley tree with coordination number $z=f$.

What is the effect of allowing for loops? Clearly the threshold is expected to increase, since extra bonds will be formed that will merely create a loop rather than contributing to the formation of an infinite branched network. Moreover, we expect that the behavior of the system in the immediate vicinity of the gel point to be characterized by different critical exponents. In fact, if we are to believe the utility of lattice models, then it turns out that exponents are shifted from their Cayley tree values quite considerably (Fig. 7). Of course, one could well question the appropriateness of a lattice model to represent a continuum system, but recent calculations for "continuum percolation" suggest that at least the exponents are unchanged [14].

| (a) THERMAL | (b) PERCOLATION | (c) GEL |
|---|---|---|

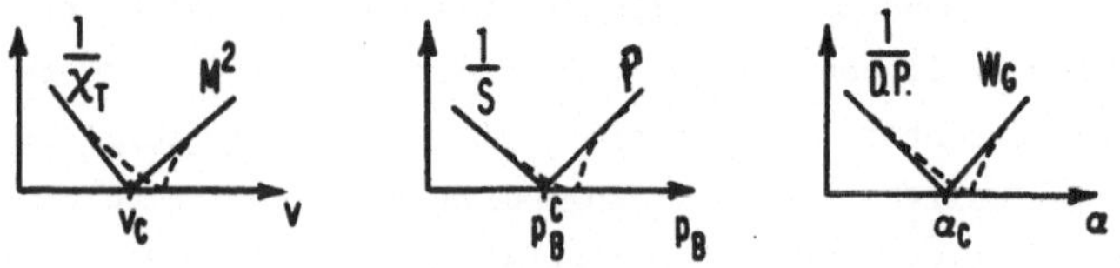


| | | |
|---|---|---|
| coupling constant | bond probability | "conversion" |
| $v_c = \tanh(J/kT_c)$ | $p_B^c$ = percolation pt. | $\alpha_c$ = gel point |
| droplet | bond cluster | polymer |
| M<br>$\chi_T$ | P(p)<br>S(p) | $W_G$ = wt. fraction<br>D.P. = deg. of poly. |
| G(H,T) | $\langle N_{clusters} \rangle$ | $\langle N_{molecules} \rangle$ |
| Husimi-Temperley<br>$d_+ = (2\beta+\gamma)/\nu = 4$ | Fisher-Essam<br>$d_+ = 6$ | Flory<br>$d_+ = 6$ |

FIG.7

B. SOLVENT (Model #2: Correlated-site, random-bond percolation)

In Model #1, solvent effects are not included. Nor are temperature effects included in any statistical mechanical fashion: all $2^N$ states of a system consisting of N bonds are equally probable. This simplifying feature has great merit in that Model #1 is extremely tractable. However we know that solvent effects are quite profound in gelating systems.

In FS theory, all sites are occupied by monomers. Two simplifying assumptions of the FS theory are (i) the absence of solvent molecules, and (ii) the absence of correlations between the molecules. Recently both solvent effects and correlations have been taken into effect via "Model #2" [15].

(a) Solvent Effects. Suppose we allow the sites to be of two sorts, M and S. M sites are occupied by monomers and S sites by solvent. The original "random-bond" percolation problem (Fig. 8a) is now a "random-site-bond" problem (Fig. 8b). The FS

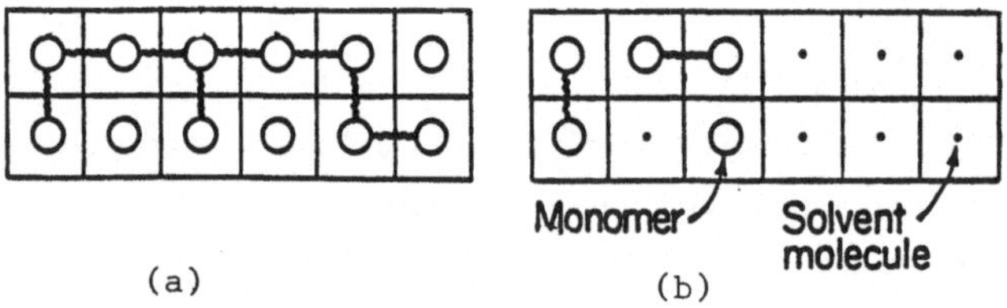


(a) (b)

FIG.8

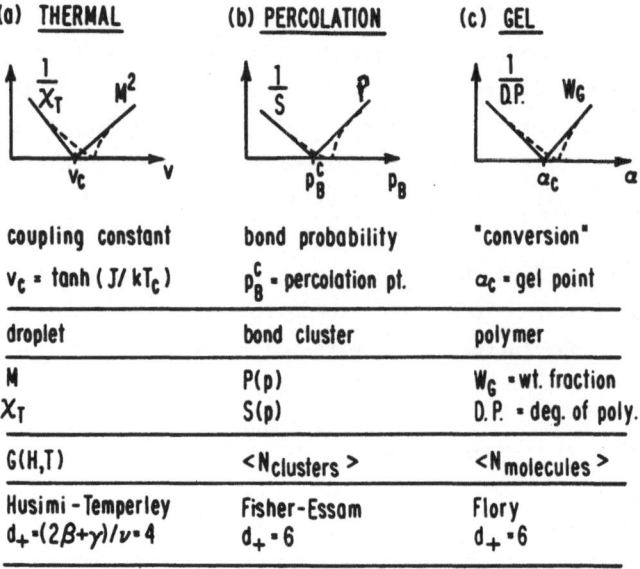

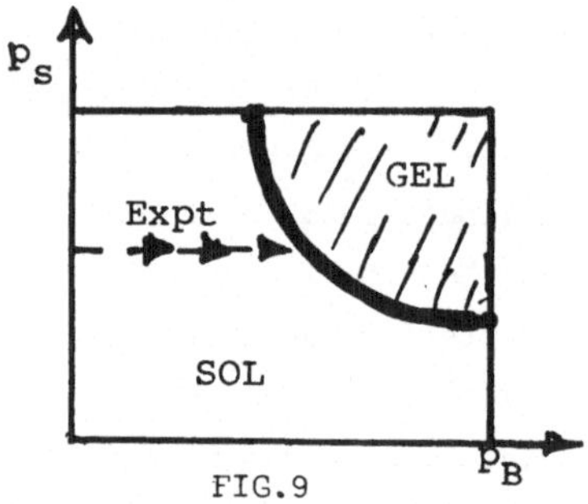


FIG.9

critical point in the simple FS phase diagram of Fig. 6 a is now an entire "line" of critical points in Fig. 9b. If $p_M$ is the density of M sites, then FS theory corresponds to the special case $p_M = 1$ (heavy solid line). A typical experiment corresponds to moving to the right along the horizontal dashed line.

(b) Correlations among molecules. We shall assume that the monomers and solvent molecules are not randomly distributed among the sites. Rather, we shall assume a correlation of the standard lattice-gas model sort. In specifying the interactions, we must consider that the monomers can interact with each other in two ways. One is the usual van der Waals interaction, and the other is a directional interaction that leads to chemical bonds. The particle-particle interaction of this system is reasonably approximated by the following nearest-neighbor interactions

$$-W_{SS} = \text{solvent-solvent}$$

$$-W_{SM} = \text{solvent-monomer}$$

$$-\epsilon_{MM} = \begin{cases} -W_{MM} & \text{van der Waals (with weight } \rho_u) \\ -E & \text{bonding energy (with weight } 1-\rho_u) \end{cases} .$$

The energy of a given configuration is then [15]

$$H_{CSK} = -W_{SS}\sum_{<ij>} \pi_i^S \pi_j^S - W_{SM}\sum_{<ij>} \pi_i^S \pi_j^M - \epsilon_{MM}\sum_{<ij>} \pi_i^M \pi_j^M \quad , \tag{4}$$

where the summations are over nearest-neighbor parts of sites <ij> and $\pi_i^S(\pi_i^M) = 1$ if site i is occupied by a solvent molecule (monomer), and zero otherwise; obviously $\pi_i^S + \pi_i^M = 1$ since each cell must be occupied by either a solvent molecule or monomer.

Finally, we take into account the fact that the bond probability is temperature dependent (Fig. 10).

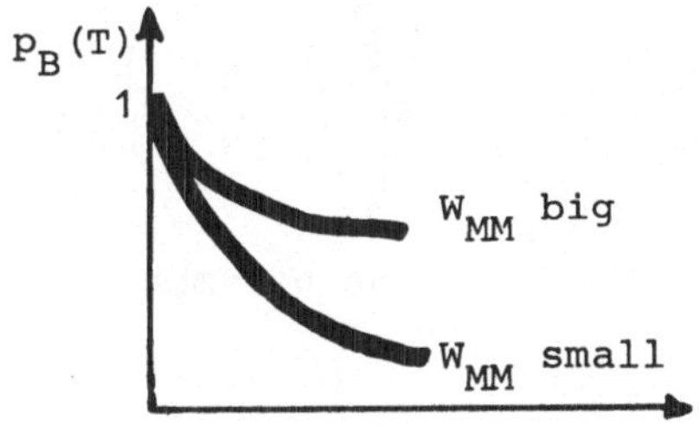


FIG.10

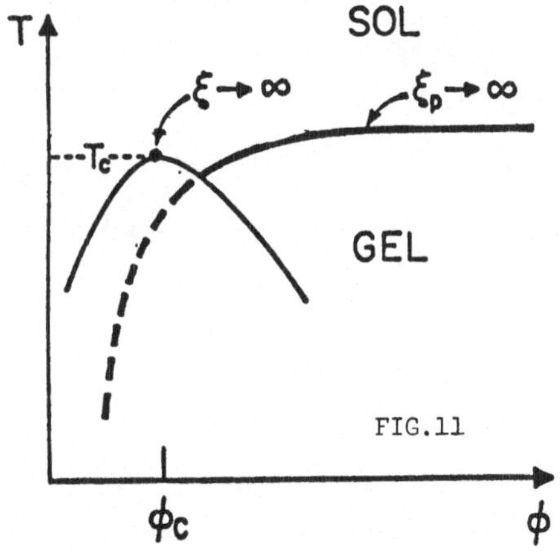


FIG.11

The resulting model has been solved for the Cayley tree--i.e., in the absence of intramolecular interactions. The resulting phase diagram is shown in Fig. 11. We see that in addition to (i) a line of connectivity critical points separating the high-temperature sol phase from the low-temperature gel phase, there exists also (ii) a phase separation curve characterized by a consolute temperature $T_c$ below which the system splits into two separate phases. We should emphasize that curve (i) corresponds to looking at the system through the eyeglasses of a "percolation freak," whose only concern is connectivity; curve (ii) corresponds to looking at the same system wearing now the spectacles of an Onsager whose concern is with ordinary thermal phase transitions and critical phenomenon.

The calculations are in qualitative agreement with recent experiments of Tanaka and co-workers [16]. Quantitative agreement is not to be expected due to the complete neglect of intramolecular interactions.

It is possible to adjust the experimental parameters in such a fashion that the sol-gel phase boundary intersects the consolute point. At this point, labeled Q in Fig. 12, we expect that both connectivity fluctuations and concentration fluctuations are simultaneously critical. Percolation is relevant to the former, while the lattice-gas (Ising) model is relevant to the latter. A particularly intriguing open question concerns the sort of phenomena that one might expect in the immediate neighborhood of such a "queer" point. There is an increasing body of evidence that the Q point has a much richer structure than the ordinary percolation point [9,10].

Very recently, Barrett has generalized the CSK model to incorporate the features displayed by the polymeric system hyaluronic acid, which undergoes an important order-disorder transition [17]. This system is schematized in Fig. 13; there are two types of side groups, amide and carboxyl (labeled A and C) as well as two types of 'solvent molecules,' potassium and phosphate ions (labeled + and -).

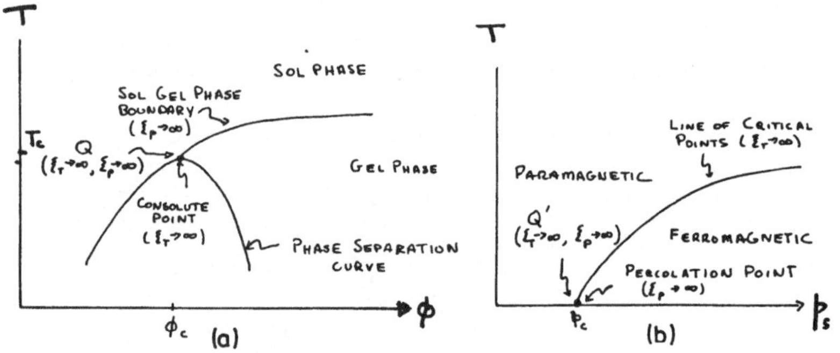


FIG.12

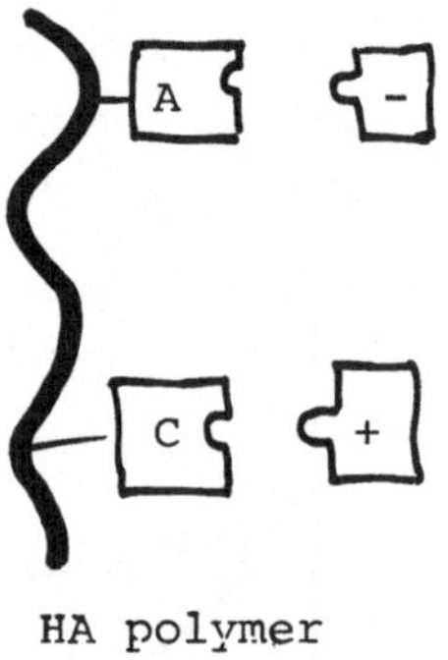


FIG.13

3. THE MODEL ITSELF (Example: Ising droplets)

Now we shall discuss a surprise that is always rather pleasant to observe when it occurs in theoretical physics: a model introduced in one context turns out to be relevant in another context. In Fig. 8b we partitioned the cells of a lattice-gas model into monomers and solvent molecules; suppose we return to the original Ising model where the sites are considered to be up spins and down spins. Moreover, suppose the bonds are considered to be 'active' with probability $p_B$ just as in Fig. 8b.

Recently, Delyon, Souillard, and Stauffer [18] found that the CSK model displays an essential singularity* in the region shown shaded in Fig. 14a; the 'DSS line'

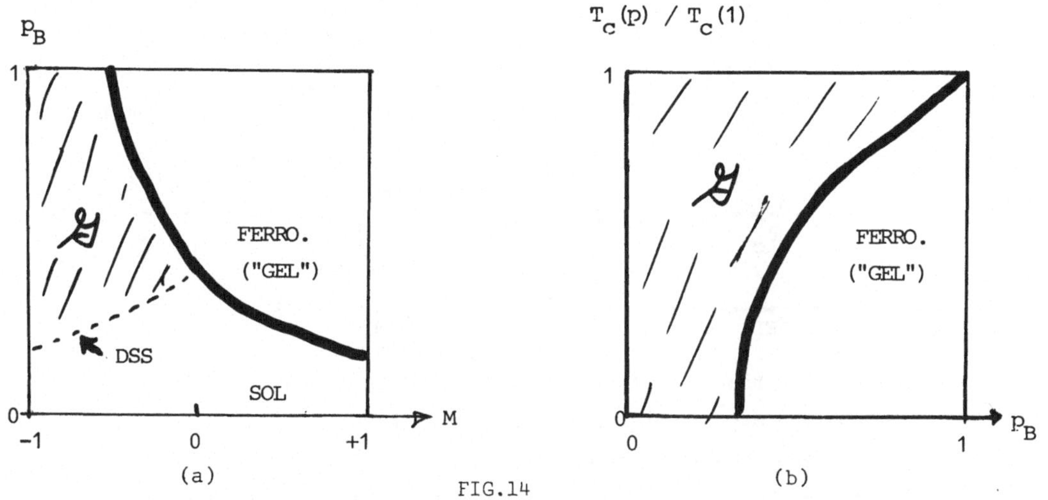


FIG.14

bordering that region from below is the mirror image of the sol-gel phase boundary reflected to negative M. The shaded region seems possibly analogous to the Griffiths phase [19] of a randomly-dilute Ising magnet (Fig. 14b), about which very recent interesting results have been obtained by Lubensky and McKane [20].

A natural question to ask is whether one can use the Hamiltonian $H_{CSK}$ of (4) to describe Ising droplets near the critical point [21]. It is certainly not the case that percolation clusters are candidates for Ising droplets. To see this, consider the magnetization curve of d=3 Ising ferromagnet, sketched in Fig. 15. At T=0, all the spins are up. As T increases from zero, small clusters of down spins appear and these percolation at a temperature $T_p$ for which the density of down spins, $\rho_\downarrow$, equals the percolation threshold for Ising-correlated percolation. Müller-Krumbhaar [22] demonstrated by Monte-Carlo simulation that $T_p < T_c$, a result that is certainly intuitively plausible since at $T_c$ $\rho_\downarrow = 0.5$. Simply stated, percolation clusters are 'too big' for Ising droplets!

The 'red-blue' picture of the incipient infinite cluster presented above in Sec. 1 is useful in understanding why it is that percolation cluster are 'too big' to be Ising droplets. Fig. 16a shows a tiny portion of the incipient infinite cluster just below the percolation threshold, where now 'red' spins (shown as 'heavy

*Specifically, $\log P_s \sim s^{1-1/d}$, where $P_s$ is the probability of the origin to belong to a s-site cluster. The DSS phase transition is characterized by a change in the asymptotic behavior of $P_s$, but there is no divergence of the mean cluster size and no appearance of an infinite network.

spins') are those which, if removed, break the cluster into two separate clusters. Such breaks will occur with probability $\exp(-2J/kT)$, since $2J$ is the energy required to mis-align an Ising spin in an ordered linear chain segment.

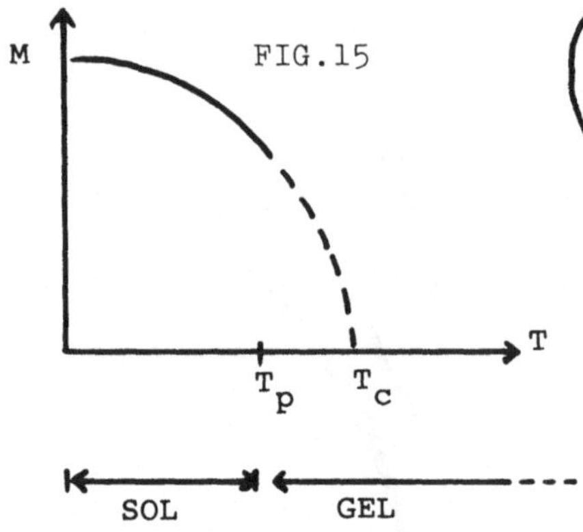


FIG.15

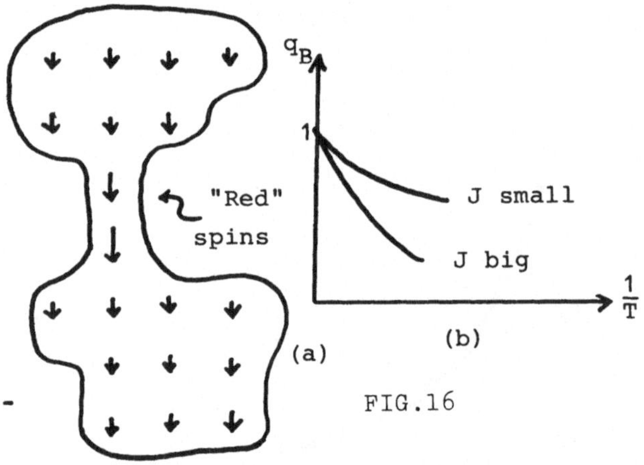


FIG.16

One possibility is that percolation clusters have nothing to do with Ising droplets! A second and more promising possibility is to find an 'algorithm' whereby the percolation clusters can be 'broken up' into smaller droplets in such a fashion that the droplets will be critical quantities at the critical point [21]. The second possibility seems to work: suppose we use the CSK model to define Ising droplets, by considering Ising droplets to be identical to percolation clusters in this model, with $p_B$ chosen such that two spins have an 'inactive' bond between them with probability

$$q_B \equiv 1 - p_B = \exp(-2J/kT). \tag{5}$$

Thus at $T=0$, none of the bonds is inactive and the Ising droplets are the same as ordinary site percolation clusters. However as $T$ increases from zero, an increasing fraction of the bonds become inactive (Fig. 16b). It is important to emphasize that the Ising interaction is always present between two neighbors; the bonds are introduced as an artifice for the purpose of defining droplets.Numerical studies in two [23] and three [24] dimensions indicate that these site-bond percolation clusters in fact do behave just like Ising droplets in that they diverge at the critical point. Thus we see that a model introduced in connection with polymer gelation would appear to be of potential relevance to a completely different problem, that of defining Ising droplets near the critical point.

4. THE SOLVENT ITSELF (Example: Water)

In Sec. 2 we began with a physical system, polyfunctional condensation in a solvent. We then formulated a Hamiltonian (4) designed to describe the essential physical features. In Sec. 3, we described how a special case of the model was useful in describing apparently an apparently quite different system, an Ising ferromagnet near its critical point. In this final section, we consider the solvent itself which, for the gels we considered, was water.

We shall present this work like a detective story, describing first the 'puzzle,' then identifying some 'clues.' Next we shall make a 'hunch' or hypothesis, and finally we shall briefly describe some ongoing efforts designed to test this hunch.

## A. THE PUZZLE

Almost every property of liquid water has some aspect that is paradoxical. Three simple static functions that are highly anomalous are displayed in Fig. 17.

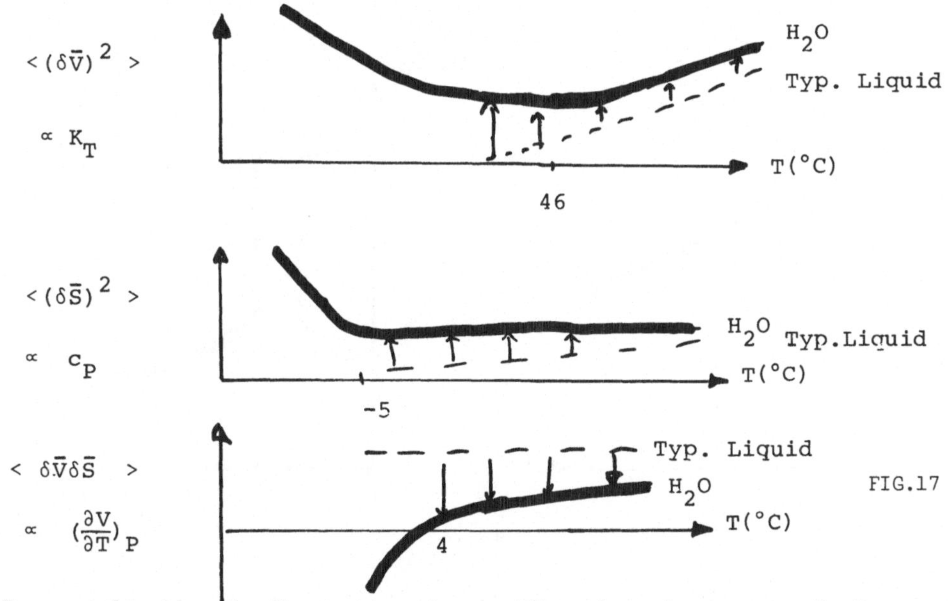


FIG.17

(i) For most liquids, the fluctuations in specific volume decrease as T is decreased; for water this is true at high temperature, but not for $T \lesssim 46°C$.

(ii) For most liquids, the fluctuations in specific entropy decrease as T is decreased. For water, this is true at high temperature, but not at sufficiently low temperature $(T \lesssim -5°C)$.

(iii) For most liquids, the cross fluctuations of specific volume and specific entropy (proportional to the coefficient of thermal expansion $(\partial V/\partial T)_P$ are positive.*
For water, this is true at high temperature, but not below 4°C (11°C for $D_2O$).

The list of strange properties could occupy this entire talk. Moreover, all the anomalies seem to be greatly accentuated on reducing T below $T_m$, the normal melting temperature. The important point is that despite tremendous accomplishments of recent years in obtaining experimental information on the detailed properties of water--even

*This is intuitively plausible, since when there is a positive specific volume fluctuation there are more arrangements and there is a corresponding positive fluctuation in the specific entropy. The product of these two positive fluctuations is positive.

down to the lowest attainable temperatures (roughly -40°C)--no physical picture has emerged that even qualitatively encompasses all the experimental facts. What is the essential physical mechanism (or mechanisms) underlying these unusual phenomena?

B. CLUES

The first place to look for a physical mechanism is the microscopic properties. The principal difference, microscopically, between water and most other liquids is that water has an intermolecular potential that is believed to strongly favor a highly directional tetrahedral network of hydrogen bonds. Melting ice at $T_m$ is believed to result in breakage of only about 20% of these bonds (Fig. 18). Moreover, heating

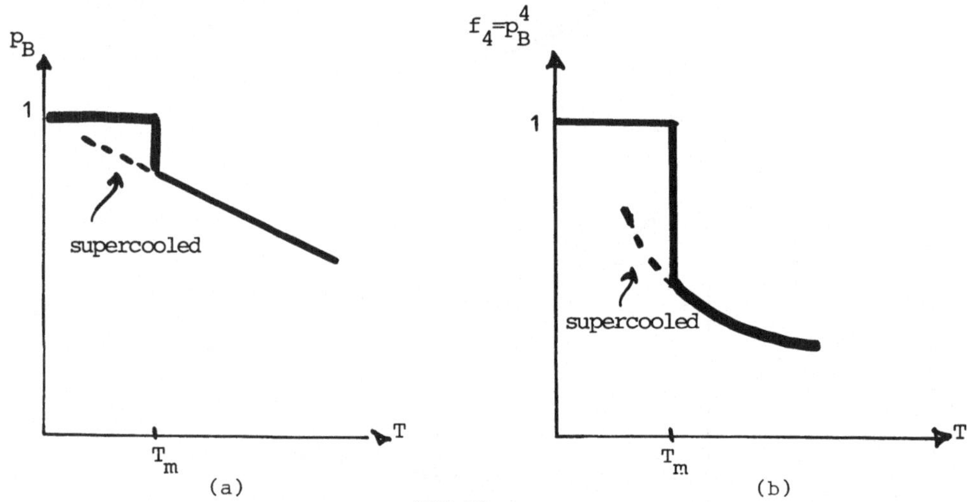


FIG.18

water above $T_m$ results in a gradual (almost linear) decrease in the fraction of intact bonds $p_B$. In particular, nothing seems to occur in the known microscopic parameters $p_B$ near the temperatures at which thermodynamic functions become unusual (e.g., 46°C for $K_T$ and 4°C for $\alpha_P$ ). Thus we can conclude that the 'puzzle of liquid water' will require some sort of mechanism (i) whereby the hydrogen bonds among the four-functional monomers play a dominant role, and (ii) which can amplify the smoothly-varying bond parameter $p_B$.

C. HYPOTHESIS ("A HUNCH")

When one considers bonding among four-functional monomers, one immediately thinks of the FS theory of polyfunctional condensation ('random-bond percolation') discussed above in Sec. 2. Could the anomalies observed in liquid water be associated with its bond percolation threshold? Two of the reasons for rejecting this simple possibility are the following:

(i) Water is well above the bond percolation threshold for any three-dimensional

network (e.g., $p_c \approx 0.4$ for the loose-packed ice $I_h$ lattice).
(ii) Associated with the percolation threshold are enhanced connectivity fluctuations, not enhanced density and entropy fluctuations as occur in water at low temperatures.

It is thus necessary to go beyond conventional 'pure percolation' if one is to find a physical mechanism germane to the unusual behavior displayed by this particular hydrogen-bonded gel, liquid water. In the time remaining, I will describe a zeroth order model which I term a 'polychromatic correlated-site percolation picture' [25]. We shall for convenience assume that the bonds are randomly intact or broken. Suppose, then, that I randomly break 20% of the bonds of this ice $I_h$ lattice, and color the oxygen atoms 5 different colors, according to whether there are 0, 1, 2, 3, 4 intact hydrogen bonds (cf. Fig. 19). Fig. 19 shows that situation, where for convenience the $d = 3$ ice $I_h$ lattice is replaced by a $d = 2$ square lattice.

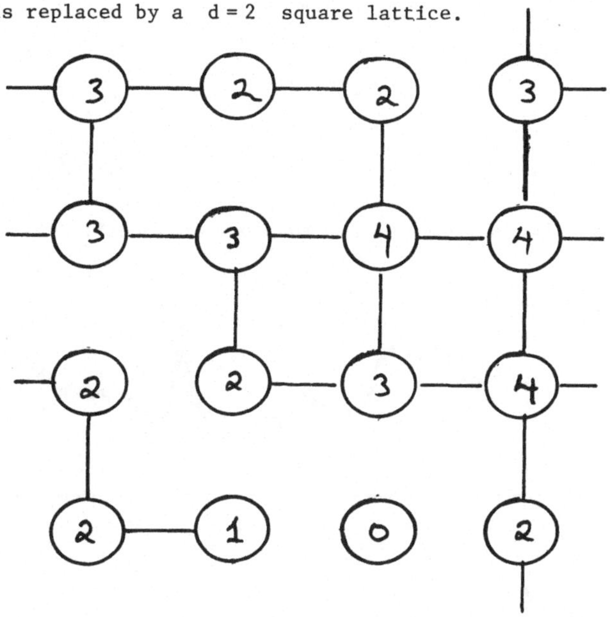


FIG.19

The fraction $f_j$ of oxygen atoms belonging to each of the 5 species is simply given by the binomial theorem,

$$f_j = \binom{4}{j} p_B^j (1 - p_B)^{4-j} \quad . \tag{6}$$

Although $f_j$ is determined by simple statistics, the connectivity of each of the five species of oxygens is far from trivial. In fact, the positions are strongly correlated, just as if there were an energy term leading to a tendency for like-colored oxygens to 'clump together'. This 'clumping' is most clearly seen if we calculate the cluster number distribution $n_j(s,p)$ for cluster of the same color $j$ $(j = 0, \ldots, 4)$.

Thus when looking at the bond connectivity problem, water appears as a large macroscopic space-filling network--as expected from continuum models of water. However when we focus on the oxygens, we find that water can be regarded as having certain clustering features, the clusters being not isolated 'icebergs' in a sea of dissociated liquid (as postulated in mixture models dating back to Roentgen) but rather patches of like-colored molecules embedded in a highly connected network or 'gel'.

At this state, one can ask 'So what?' That is, 'What is the relation between the patches in the infinite bond network and the observed anomalies of liquid water?'

The answer to this question must be 'NOTHING WHATSOEVER!' unless there is some local property of the like-colored patch that is different from the global properties of the hydrogenbond network as a whole. Suppose we conjecture that the local specific volumes associated with a species $j$ patch are related by

$$\bar{V}_4 > \bar{V}_3 > \bar{V}_2 > \bar{V}_1 > \bar{V}_0 \tag{7a}$$

and the local specific entropies by

$$\bar{S}_4 < \bar{S}_3 < \bar{S}_2 < \bar{S}_1 < \bar{S}_0 \quad . \tag{7b}$$

Then there will be enhanced fluctuations in specific volume and entropy (Fig. 20) that

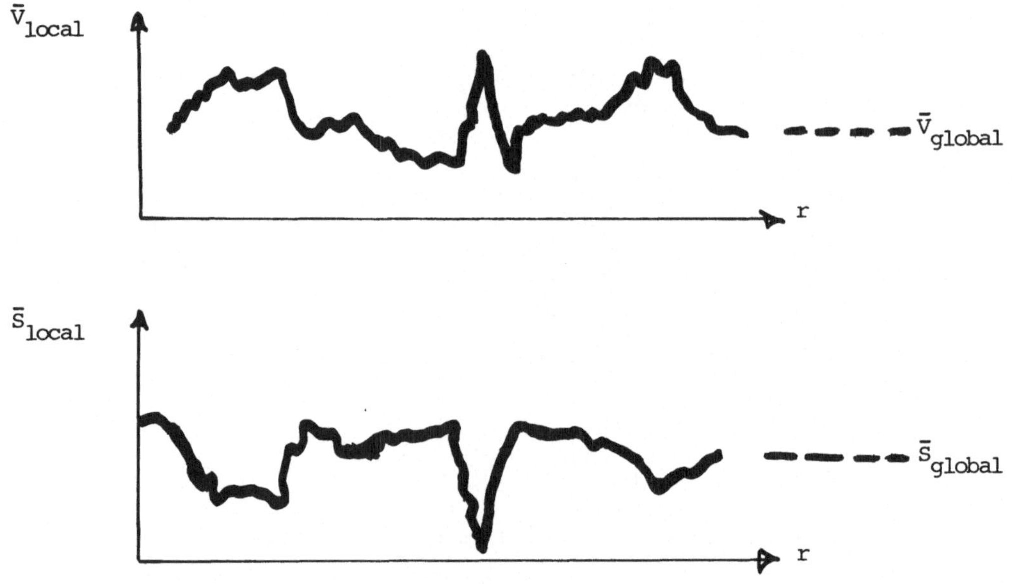


FIG. 20

are, moreover, correlated. This enhancement will increase as $p_B$ increases, leading to corrections to the functions as shown in Fig. 17 and Table 1. Moreover, a 1% increase in $p_B$ (on, e.g., lowering T) results in a 4% increase in $f_4$ [Fig.18(b)]; thus we have the desired 'amplification mechanism' mentioned above under Section B.

Table 1

**Summary of the qualitative predictions of the simple bichromatic discrete percolation model for the behavior of three static response functions**

| | $K_T$ | $C_P$ | $\alpha_P$ |
|---|---|---|---|
| Sign of anomaly | + | + | − |
| Lower $T$ | ↑ | ↑ | ↓ |
| Increase $P$ | ↓ | ↓ | ↑ |
| Dilute with $D_2O$ | ↑ | ↑ | ↓ |
| Dilute with a "patch-breaking" impurity | ↓ | ↓ | ↑ |

D. TESTS

1. Computer water

Detailed microscopic assumptions of the sort we are making can most accurately be tested using computer simulations of real water. A variety of different calculations have by now been carried out that serve to test various of the aspects of the simple percolation model, and thus far all calculations are in agreement that

(i) Bonds break randomly; e.g., Eq. (6) is verified by a wide range of data using different simulation procedures [25,26]

(ii) The bond networks agree with the detailed predictions of random-bond percolation theory [25]

(iii) The site clusters of like colored oxygens agree with the detailed predictions of correlated-site percolation theory [25]

(iv) The local density around a 4-bonded oxygen is smaller than the global density of the network [25]

It is important to emphasize that the quantitative agreement between percolation model and computer simulation summarized in (i)-(iv) utilizes no adjustable parameters.

2. Real water

The most striking experimental evidence arises from recent measurements of $S(q)$ by the technique of SAXS (small-angle x-ray scattering), and these are summarized schematically in Fig. 21 [27]. For most liquids, $S(q)$ is q-independent at small $q$, while for liquid water at sufficiently low temperature there appears to be a pronounced forward peak whose shape is consistent with the predictions of an Ornstein-Zernike theory; the resultant correlation length is consistent with the predictions of a characteristic patch diameter from the computer simulation studies. The concept of 'patches' has also been used very recently by D'Arrigo to interpret Raman spectroscopic data down to -24°C [28].

On at least a qualitative level, we can test all the predictions of Table 1. We find all 15 predictions to be consistent with observed phenomena (a detailed discussion is given in Ref. 25).

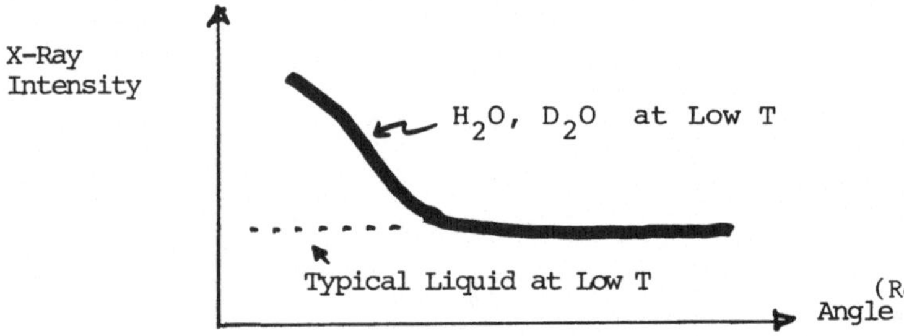


FIG.21

(Ref. 27)

E. SUMMARY AND OUTLOOK

In a one-hour talk, it has been impossible to discuss all the experimental facts and the extent to which percolation theory may or may not be relevant. It is clear, however, that theoretical elaborations are called for if the present heuristic picture is to become sufficiently developed to provide a fully quantitative description of low-temperature water. Among the most important is the creation of an appropriate Hamiltonian that describes simultaneously (i) the interactions and (ii) the connectivity between water molecules. Approaches analogous to those used for describing solvent effects on gelation are under study at the present time. Also important is a study of how serious an approximation is made by the imposition of a discrete symmetry (the assumption of two bond states, intact and broken) upon a physical function $V(r)$, that is certainly not discrete. Here one can argue that the essential physical features of the present discrete picture are not altered, just as the discrete lattice-gas adequately describes the essential physics of a fluid near its critical point. However a more refined 'polychromatic continuum percolation' (or 'rainbow percolation') is a natural and quite feasible extension [29]: each pair of atoms has an interaction energy $V(r)$ that can be put in $1:1$ correspondence with a 'wavelength'. Wavelengths in the violet end of the spectrum correspond to strong bonds, while wavelengths in the red end of the spectrum correspond to weak bonds. Next define the color of each oxygen to be the mean of the wavelengths of its four most-strongly interacted partners (Fig. 22). Thus a snapshot of the system

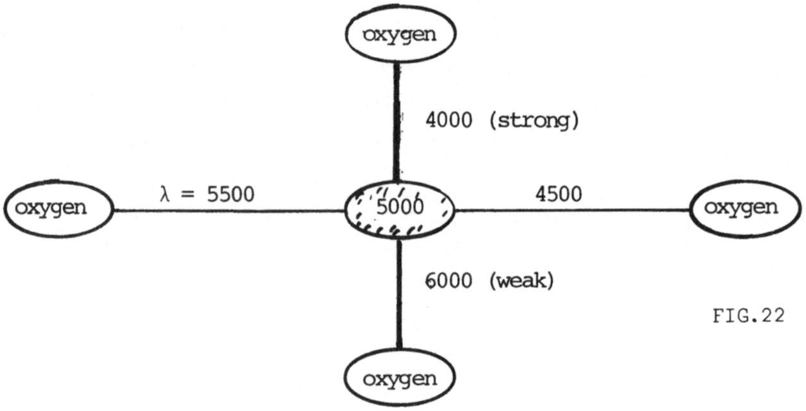


FIG.22

would not be a set of dots with 5 different discrete colors, but rather a set of dots with a continuum ('rainbow') of different colors. A Maxwell demon situated on a given oxygen would see a local density that, to lowest order, depends in a 1 : 1 fashion on the sum of the wavelengths of the incident bonds. The essential feature of the discrete polychromatic model is preserved in the continuum polychromatic model: there is a spatial correlation between the density fluctuations, since each bond is shared by two sites! The mean 'wavelength' of all the bonds decreases as T decreases and we preserve the simple physical mechanism leading to correlated fluctuations that increase as T decreases.

Thus far, we have not discussed the dynamics since time is limited. However the essential physical fact is that the bond lifetime $\tau_B$ is about a picosecond, so that on ordinary observational time scales a bowl of liquid water does not behave like a bowl of jello. Just as the essential puzzle posed by the static behavior of water is how the microscopic function $p_B$ which varies smoothly with T can lead to static functions that are amplified at low temperatures, so the essential puzzle posed by the dynamic behavior is how the smoothly-varying microscopic function $\tau_B$ can lead to dynamic functions that are greatly amplified at low temperatures. The simple mechanism proposed in Ref. [25] has recently been extended by the Pisa group [30] and applied to the interpretation of new dielectric relaxation data.

Thus much remains to be done, both theoretically and experimentally. However we are optimistic that the essential physical mechanism of the present picture (correlated fluctuations arising from random bond breaking) will not be lost by the sort of extensions and elaborations that will perforce arise. In particular, it is our hope--as physicists--that whatever the solution may be to the 'puzzle of liquid water,' it is sufficiently economical that only a single physical mechanism remains to be discovered to encompass the entire range of phenomena--ranging from the compressibility minimum at +46°C to the Angell singularity at about -46°C [31]. It would be most unsatisfactory if we would require one mechanism for, say, $T > T_m$ and a second mechanism for $T < T_m$.

What, then, does the future hold? The most optimistic possibility is that we can answer 'yes' to the question posed by the title, and our program may possibly lead to a useful foundation on which to further understanding of liquid water. At the least optimistic, percolation concepts will prove to be irrelevant to the behavior of liquid water, but nevertheless we will have learned a great deal about a simple model in which sites become correlated as a result of a random process. Even the least optimistic possibility is not without some satisfaction, however, since any well-understood model is likely to have some application in the complex physical world. This seems to be the case here, also: Brodsky of IBM Research Laboratories has recently succeeded in applying the model to hydrogenated amorphous silicon, $aSiH_x$, a prototype material for energy conversion [32]. Specifically, he proposes that pure Si corresponds to our model with $p_B = 1$. Hydrogenation corresponds to breaking covalent Si-Si bonds, and therefore corresponds to increasing temperature, pressure, or patch-breaking impurity concentration (Fig. 23). By doing calculations

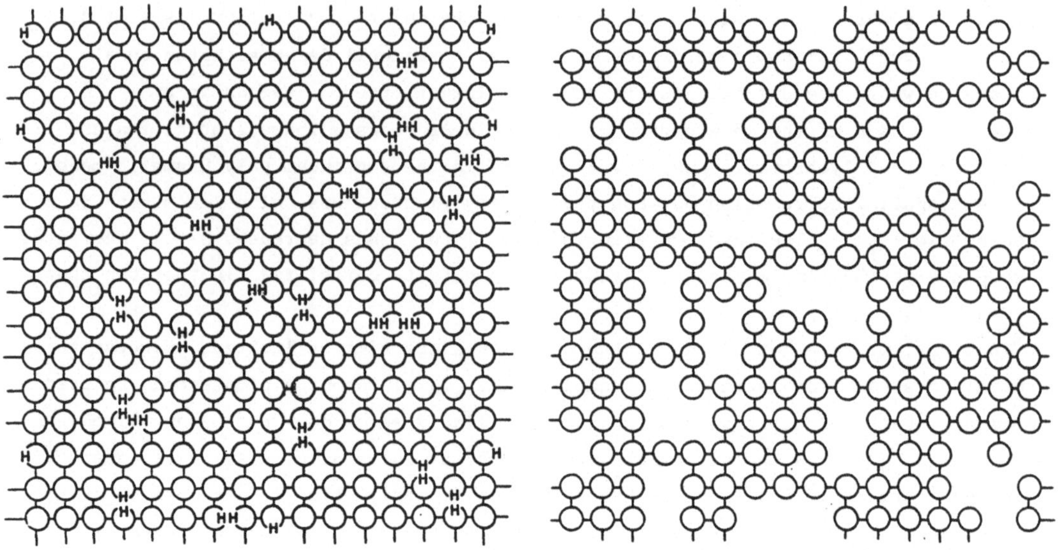

(a) FIG.23 (after Ref. 32) (b)

on the system of 'disconnected Si patches,' Brodsky succeeds in predicting the results of several hitherto inexplicable experiments. Who knows--given the energy crisis facing every country represented at this meeting--if the 'least optimistic' possibility is not sufficient justification for continuing our research program in polychromatic correlated-site percolation?

* * * * *


[1] R. J. Glauber, J. Math. Phys. 4, 294 (1963).

[2] T. D. Lee and C. N. Yang, Phys. Rev. 87, 410 (1952); the identification of the upper marginal dimensionality $d_+ = 6$ and the first expansions in $(d_+ - d)$ were carried out in M. E. Fisher, Phys. Rev. Lett. 40, 1610 (1978).

[3] G. Parisi and N. Sourlas, Phys. Rev. Lett. 46, 871 (1981); T. C. Lubensky and A. J. McKane, J. Phys. A 14, L157 (1981).

[4] S. Kirkpatrick, AIP Conf. Proc. 40, 99 (1978).

[5] G. Shlifer, W. Klein, P. J. Reynolds, and H. E. Stanley, J. Phys. A 12, L169 (1979).

[6] The distinction between the Hausdorff-Besicovitch or 'fractal' dimensionality $D = 2y_h - d = \gamma/\nu$ of the entire system and the fractal dimensionality $D^\dagger = y_h = \beta\delta/\nu$ of the incipient infinite cluster was made clear by remarks of D. Stauffer, G. Harrison, G. Bishop, and W. Klein; it is described in H. E. Stanley, J. Phys. A 10, L211 (1977). The fractal dimension was first related to critical exponents in H. E. Stanley, R. J. Birgeneau, P. J. Reynolds, and J. F. Nicoll, J. Phys. C 9, L553 (1976). Recently, critical properties have been calculated very elegantly for various fractal lattices (see talk by A. Aharony in this conference; also

Y. Gefen, B. Mandelbrot, and A. Aharony, Phys. Rev. Lett. 45, 855 (1980). A description of many objects with non-integer Hausdorff-Besicovitch dimensionality may be found in B. B. Mandelbrot, Fractals: Form, Chance, and Dimension (W. H. Freeman, San Francisco, 1977).

[7] R. J. Birgeneau and H. E. Stanley, unpublished.

[8] R. Pike and H. E. Stanley, J. Phys. A 14, L169 (1981).

[9] A. Coniglio, unpublished.

[10] A. Coniglio, Phys. Rev. Lett. 46, 250 (1981).

[11] See, e.g., G. Deutcher, this conference; M. Lagües, R. Ober and C. Taupin, J. Phys. (Paris) 39, L487 (1978).

[12] P. J. Flory, J. Am. Chem. Soc. 63, 3083, 3091, 3096 (1941); W. H. Stockmayer, J. Chem. Phys. 11, 45 (1943).

[13] M. E. Fisher and J. W. Essam, J. Math. Phys. 2, 609 (1961); H. L. Frisch and J. M. Hammersley, J. Soc. Ind. Appl. Math. 11, 894 (1963); D. Stauffer, Chem. Soc. Faraday Trans. 72, 1354 (1976); P. G. de Gennes, J. Phys. (Paris) 36, 1049 (1976).

[14] S. W. Haan and R. Zwanzig, J. Phys. A 10, 1547 (1977); T. Vicsek and J. Kertész, J. Phys. A 14, L31 (1981); E. T. Gawlinski and H. E. Stanley, "Continuum percolation in two dimensions: Monte Carlo tests of scaling and universality for non-interacting discs", preprint submitted to J. Phys. A Lett; A. Geiger and H. E. Stanley, "Continuum percolation in three dimensions: Molecular dynamics tests of universality for interacting ST2 particles", preprint.

[15] A. Coniglio, H. E. Stanley and W. Klein, Phys. Rev. Lett. 42, 518 (1979); "Solvent effects on polymer gels: A statistical mechanical model" (preprint submitted to Phys. Rev. B).

[16] T. Tanaka, G. Swislow and I Ohmine, Phys. Rev. Lett. 42, 1556 (1979).

[17] T. W. Barrett, "Site-Bond Correlated-Percolation theory of polymer gelation applied to a polymer with preferential binding at two disparate sites", preprint (submitted to Phys. Rev. Lett.). The CSK model has also been extended by A. E. Gonzales and S. Muto [J. Chem. Phys. 73, 4668 (1980)] to explain the phenomenon of crossing gelation curves observed by P. Ruiz-Azuara, T. Tanaka, A. Coniglio, H. E. Stanley and W. Klein, "Dependence of the gelation curve on solvent composition: Crossing points", preprint submitted to J. Chem. Phys.

[18] F. Delyon, B. Souillard, and D. Stauffer, "Percolative phase transition without appearance of an infinite network", preprint submitted to J. Phys. A Lett. See also M. Aizenman, F. Delyon, and B. Souillard, J. Stat. Phys. 23, 267 (1980).

[19] R. B. Griffiths, Phys. Rev. Lett. 23, 17 (1969).

[20] T. C. Lubensky and A. J. McKane, this Conference (poster session).

[21] A. Coniglio and W. Klein, J. Phys. A 13, 2775 (1980).

[22] H. Müller-Krumbhaar, Phys. Lett. A 50, 2708 (1974).

[23] J. Roussenq, Thèse, Université de Provence, 1980.

[24] D. Stauffer, J. Phys. (Paris) 42, L99 (1981).

[25] H. E. Stanley, J. Phys. A 12, L329 (1979). A more detailed description and development of the percolation picture, together with extensive comparison to experimental data, is in H. E. Stanley and J. Teixeira, J. Chem. Phys. 73, 3404 (1980). Detailed Monte Carlo calculations have been carried out by R. L. Blumberg, G. Shlifer and H. E. Stanley, J. Phys. A 13, L147 (1980), while renormalization group calculations have been performed by A. Gonzales and P. J. Reynolds, Phys. Lett. 80A, 357 (1980). Detailed network analysis of computer water is presented in A. Geiger, H. E. Stanley and R. L. Blumberg, "Connectivity studies of liquid water: Hydrogen bond networks and four-coordinated water molecules", preprint. An elementary summary of the overall research thus far is presented in H. E. Stanley, J. Teixeira, A. Geiger, and R. L. Blumberg, Physica 106A, 260 (1981).

[26] E.g., using no adjustable parameters, one finds agreement between the predictions of the binomial theorem (6), based on the assumption of random bond breaking, and the computer simulations of Geiger (Ref. 25) and also M. Mezei and D. L. Beveridge, J. Chem. Phys. 74, 622 (1981); W. L. Jorgensen, Chem. Phys. Lett. 70, 326 (1980); C. Pengali, M. Rao and B. J. Berne, Mol. Phys. 40, 661 (1980).

[27] L. Bosio, J. Teixeira, and H. E. Stanley, Phys. Rev. Lett. 46, 597 (1981); see also the earlier work of R. W. Hendricks, P. G. Mardon and L. B. Shaffer, J. Chem. Phys. 61, 319 (1974).

[28] G. D'Arrigo, G. Maisano, F. Mallamace, P. Migliardo and F. Wanderlingh, "Raman scattering and structure of normal and supercooled water", preprint scheduled for J. Chem. Phys. 75 (1981); see also R. Bansil, J. L. Taafe, and J. Wiafe-Akenten, "Raman spectroscopy of supercooled water", preprint submitted to J. Chem. Phys.

[29] In this context the word "continuum" means that the sites are not in two discrete states (as in bichromatic percolation), nor in 5 discrete states (as in polychromatic discrete percolation), but rather in a continuum of possible states. Since the term "polychromatic continuum percolation" is easily confused with the term continuum percolation used to describe systems whose elements are not constrained to the vertices of a lattice, we prefer the simpler and more descriptive designation "rainbow percolation".

[30] D. Bertolini, M. Cassettari, and G. Salvetti, "The dielectric relaxation time of supercooled water down to -21°C", preprint.

[31] C. A. Angell, "Supercooled Water", preprint.

[32] M. H. Brodsky, Solid State Commun. 36, 55 (1980).

[33] Recent references on directed percolation include J. Kertész and T. Vicsek, J. Phys. C 13, L343 (1980); D. Dhar and M. Barma, J. Phys. A 14, L1 (1981); S. P. Obukhov, Physica 101A, 145 (1980); J. L. Cardy and R. L. Suger, J. Phys. A 13, L423 (1980); W. Kinzel and J. Yeomans, preprint; E. Domany and W. Kinzel, preprint; E. Domany and W. Kinzel, preprint; P. J. Reynolds, preprint; S. Redner, preprint; S. Redner and A. C. Brown, preprint.

[34] Recent work on restricted valence percolation includes S. G. Whittington, K. M. Middlemiss, G. M. Torrie and D. S. Gaunt, J. Phys. A 13, 3707 (1980); R. Cherry and C. Domb, J. Phys. A 13, 1325 (1980); D. S. Gaunt, J. L. Martin, G. Ord, G. M. Torrie and S. G. Whittington, J. Phys. A 13, 1791 (1980).

[35] Random-site random-bond percolation is analyzed by series expansions in P. Agrawal, S. Redner, P. J. Reynolds and H. E. Stanley, J. Phys. A 12, 2073 (1979) and by renormalization group in H. Nakanishi and P. J. Reynolds, Phys. Lett. 71A, 252 (1979).

[36] The s-state hierarchy was proposed in R.B.Potts,Proc.Camb.Phil.Soc.48,106 (1952).

[37] The n-vector hierarchy was proposed in H.E.Stanley,Phys.Rev.Lett.20,589 (1968).

# DISORDERED MAGNETIC SYSTEMS

# THE PRESENT EXPERIMENTAL SITUATION IN SPIN-GLASSES

J.A. Mydosh
Kamerlingh Onnes Laboratorium
der Rijks-Universiteit
Leiden, The Netherlands

## Abstract

A survey of the salient features is presented for some recent experiments on archetypal spin-glass systems. Here we include four dynamical, and mainly zero field techniques, namely the frequency dependence of the ac susceptibility, neutron scattering, muon spin rotation and relaxation, and ultrasonic propagation. We also consider some field dependent investigations such as M(H), $\chi$(H) and the ESR all of which require an associated anisotropy. A critical review of the different interpretations is given for these experiments and a percolation model is introduced and offered as a possible way to resolve the various and conflicting interpretations.

## I. INTRODUCTION

Now that the spin-glass problem has settled a bit between the large conferences [1], we can perhaps look back at some of the major and recent experimental developments and their significance to the understanding of the spin-glass behavior. Three basic characteristics of a spin-glass are well-known and apply to many different systems, namely the sharp peak in the low-field susceptibility at $T_f$ (the freezing temperature), the broad, almost featureless, temperature dependence of the specific heat, and the onset below $T_f$ of irreversibilities, remanences, and time dependences with the application of an external, static field. These "traditional" results have spurred a great deal of further and more sophisticated experimentation, and it is some of the latter investigations that we wish to survey and discuss in the present paper.

Due to the large number of experimental studies on a great variety of spin-glass systems we must limit our considerations to a few arbitrary choices which will be biased by the author's own interests. So here we will only consider the so-called archetypal spin-glass systems such as CuMn, AuFe, etc. Then we shall examine the frequency dependence of the susceptibility $\chi(\nu)$ around $T_f$. This brings us into the realm of dynamical effects and three additional experimental techniques which give special insights into the freezing phenomenon are discussed. These are neutron scattering, muon spin rotation and relaxation, and ultra-sonic measurements. Finally we will review some static-field dependent properties, and conclude with a brief treatment of the ESR situation and anisotropy effects in spin-glasses.

The present purpose is to use these experiments in order to develop a simple, "experimentalist's" model of the spin-glass behavior. This then might serve as a starting point for more elaborate calculations and the construction of a complete spin-glass theory. We shall deliberately omit a comparison of the Edwards-Anderson model with these experimental results.

## II. FREQUENCY DEPENDENT SUSCEPTIBILITY

From ac susceptibility measurements one is able to study the reversible susceptibility $\chi(T)$ and accurately determine the freezing temperature [2]. A static susceptibility technique with a sufficiently large dc field H will distort the temperature dependence of $\chi$ near $T_f$, and introduce irreversibilities and time dependences for $T<T_f$ [3]. Thus by using a small, oscillating ($\nu$=2Hz to 10kHz), driving field h<1Oe, the fully reversible susceptibility is conveniently obtained, $\chi\equiv(\frac{\partial M}{\partial H})_{H\to 0}=(\frac{\partial M}{\partial h})_{h\neq 0}$. The equality holds due to the long time scale of the irreversible behavior and the slow time dependences in small fields, so that a frequency of a few hertz will not cause deviations from the reversible susceptibility.

Very recently Malozemoff and Imry [4] have applied SQUID magnetometry to detect the long-time dependence of the spin-glass freezing. By cooling a CuMn (5 at.%) sample in a constant field H=5 Oe to a temperature around $T_f$, they very accurately determined the magnetization M at this T. After waiting very long times, 6min<t<2400min, at this H and T, they observed no variation in M. Lowering the temperature in fine steps and waiting the above time periods allowed them to trace out the spin-glass "cusp" which was completely independent of time. They claim these results are evidence for an equilibrium phase transition. This experiment is different from the ac-$\chi$ measurements described below in two aspects. Firstly, the static field cooling (certainly in large H-fields) prepares the sample in the infinite time state. Consequently there is no time dependence as long as the field-cooling field is kept constant for any $T<T_f$. Secondly, the effective times of measurement $\tau_m \simeq 1/\nu$ are $\tau_m \lesssim 10^{-1}$sec in the ac-$\chi$ and $\tau_m \gtrsim 10^3$sec in the SQUID-M. It is plausible that the time or frequency dependence of M and $\chi$ is significantly different between these two regimes separated by a factor of $10^4$. Indeed, very low $\nu$ ac-$\chi$ measurements on an unusual $Pr(P)_y$ spin glass [5] exhibit no frequency dependences for $\nu<10^{-1}$Hz while definite variations of $T_f(\nu)$ are observed for $\nu>10$Hz.

A natural consequence of the ac technique is then the availability of frequency as a parameter with which to probe the time scale of the freezing process. However, two words of caution are necessary. For metals the limited skin-depth and the large eddy currents at high frequencies make it difficult to obtain the physical quantities which are the in-phase $\chi'$ (dispersion) and out-of-phase $\chi''$ (absorption) susceptibilities. This means that powdered or thin film metallic samples must be used and that $\chi'$ and $\chi''$ (calibrated values) must both be measured simultaneously before reliable data may be obtained.

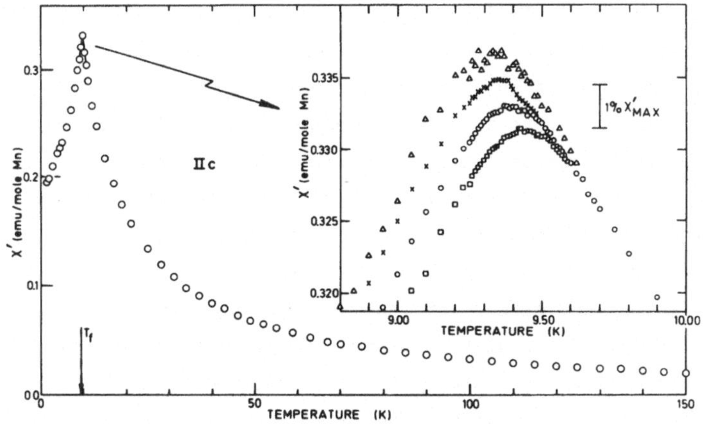


Fig.1. ac susceptibility as a function of T for CuMn (1 at.%). Measuring frequencies:
□ , 1.33 kHz
○ , 234 Hz
X , 10.4 Hz
Δ , 2.6 Hz.
After Mulder et al. [6].

In Fig. 1 we show the frequency dependence of $\chi'(T)$ for a CuMn 1 at.% alloy [6]. The significant feature of this measurement is the small, but clear, frequency shift of $\chi(T)$ around $T_f$. A measure of the way this shift affects $T_f$ is given by $\Delta T_f/T_f(\Delta \log\nu)$ which is equal to 0.005 for concentrations of Mn in Cu between 1 and 6 at.%. Also there appears a convergence of the various frequency curves as the temperature is further lowered below $T_f$ [6]. The small $T_f$ and the convergence at low temperatures are two properties which make up a type I behavior. A contrasting behavior (type II) is shown in Fig. 2 where $\chi'(\nu)$ with T as a parameter is given for PdMn (5 at.%) [7].

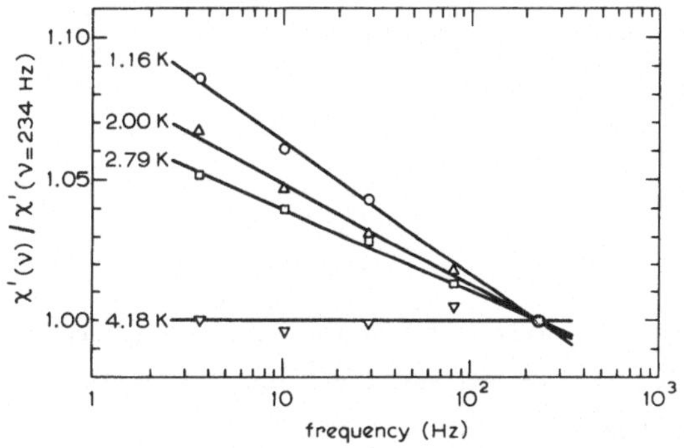


Fig. 2. ac susceptibility as a function of ν with T as a parameter for PdMn (5 at.%).
The data for $T_f(\nu=234\ \text{Hz})=3.19$ K overlap the T=2.00 K line.
After Mulder et al. [7].

Here the frequency dependence is much larger with $\Delta T_f/T_f(\Delta\log\nu)=0.015$ and there is no convergence of the different frequency curves for $T \ll T_f$. For both behavior types unphysical values of $\nu_o$ and $E_a$ occur when an Arrhenius law $\nu=\nu_o\exp(-E_a/k_B T_f)$ is used to describe the data. The use of a Fulcher law [8] $\nu=\nu_o\exp[-E_a/k_B(T_f-T_0)]$ is also pointless without a physical interpretation for $T_0$. See also the discussion in Ref.[4].

In the literature there are additional spin-glass systems exhibiting both type I and type II behavior [7]. Some typical examples of type I (small $\Delta T_f(\nu)$ and convergent $\chi(\nu)$ as $T\to 0$) would be CuMn, AgMn and AuMn. For type II (larger $\Delta T_f(\nu)$ and non-convergent $\chi(\nu)$ as $T\to 0$) we have PdMn, AuFe and AuCo. What then is the distinguishing factor? Originally it was suggested by Zibold [9] and expanded by Hardiman [10] that ferromagnetic clusters or nearest neighbor ferromagnetic interactions are responsible for the frequency shifts. Furthermore as is known from the insulating and superparamagnetic systems, the large moment clusters cause a strong frequency dependence in the ac susceptibility. So we would conclude here that the net amount of ferromagnetic exchange determines the frequency dependence. This does not necessarily have to be nearest neighbor ferromagnetic exchange as is demonstrated in the case of PdMn [7]. And the net exchange resulting from averaging the competing ferro and antiferromagnetic interactions is further affected by the short range chemical ordering of the particular alloy. The effect of various heat treatments on the sharpness of $\chi(T)$ near $T_f$ has been recently demonstrated [6]. For developing a model of the spin-glass freezing, one must use data on samples which are as random as possible and which exhibit sharp cusp-like behavior in $\chi(T)$. In addition as has been previously pointed out [11] the amplitude of $\chi$ at $T_f$ is also dependent on the amount of ferromagnetic clusters. Since these clusters will always be present - even for completely random systems - as long as there is some ferromagnetic exchange, they compose an intrinsic part of the spin glass problem. But there is more to this problem than simply ferromagnetic clusters, and the combination of local with extended properties governs the freezing process. Clearly a superparamagnetic model is insufficient to describe the spin-glass freezing. From the opposite extreme a uniform phase transition theory based on the Edwards-Anderson model is equally inadequate. The necessity of bringing clusters into the model also involves the concepts of percolation and dynamics, and we now consider some faster, dynamical experimental techniques and their interpretation.

## III. DYNAMICAL EFFECT EXPERIMENTS-ZERO FIELD

### a) Neutron Scattering.

In a recent article Murani [12] has collected a series of neutron scattering data on the CuMn spin-glass and proposed a model for the spin-glass freezing. The neutron scattering experiments include various techniques with different resolution times $\tau_R$ for observing the "central peak" and then analyzing it in terms of elastic and quasi-elastic contributions. All magnetic scattering events with relaxation times longer than $10^{-11}$s or $10^{-9}$s (the two instrumental resolution times) appear as "elastic" scattering. A plot of this elastic scattering as a function of temperature for a given $\vec{q}$ (momentum transfer)-value is shown in Fig. 3. As T is reduced, there is an increase of elastic scattering whose exact T-dependence is a function of the instrumental resolution time $\tau_R$. The increase is sharper for $\tau_R=10^{-9}$s than for $\tau_R=10^{-11}$s, since only those scattering processes with relaxation times slower or longer than $10^{-9}$s (or $10^{-11}$s) will be resolved as elastic. This evolution of elastic scattering suggests a change in the distribution of spin relaxation times from a sharply peaked,

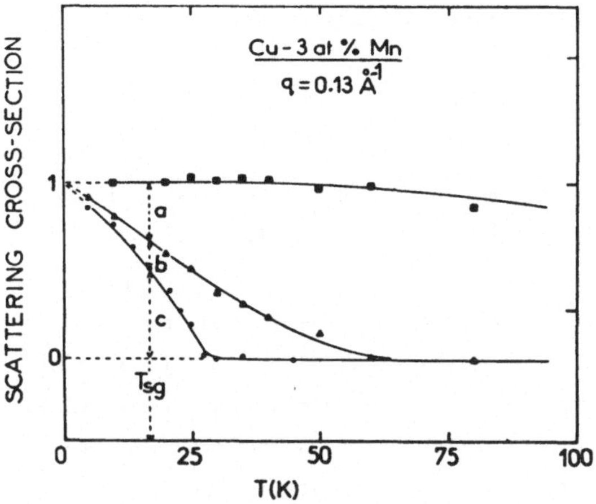


Fig. 3. Elastic scattering cross-sections with time resolutions $10^{-9}$s ● and $10^{-11}$s ▲ ; and the total quasi-static cross-section ■. After Murani [12].

Korringa, paramagnetic relaxation time at high T to a broadened and shifted distribution function of relaxation times as $T \to T_f$.

The second category of neutron scattering experiments involves the spin echo technique [13] which permits a direct observation of the time dependent spin correlation function $\xi(q,t)$ in the $10^{-12}$ to $10^{-8}$s region. In Fig. 4 we show this normalized function plotted against log of the time at different temperatures. Clearly the temperature dependence of $\xi(q,t)$ reflects the evolution of slower and slower relaxation processes, for, $\xi(q,t)$ changes from a sharply decaying exponential in t at high T, to a power law $(1-t)^n$, then to a log t near $T_f$, and finally to t independent at

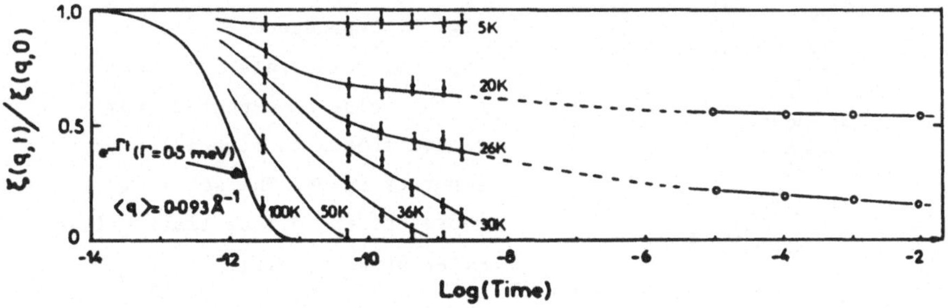


Fig. 4. Measured time dependent spin correlation function $\xi(q,t)$ for CuMn (5 at.%) at various temperatures ($T_f$=27 K). After Murani [12].

very low T. Murani has then combined these results with $\xi(0,t)$ obtained from the relatively slow ($10^{-5}$-$10^{-2}$s) measurements of $\chi(\nu)$. A nice extrapolated connection exists (see Fig. 4) between the two measurements in the two different time regimes. In these data there is a rather sudden temperature development (within 10 K) of very long time correlations. Based upon these measurements Murani [12] cultivates a model in terms of a wide spectral distribution of relaxation times evolving continuously with decreasing temperature, but which is devoid of any critical behavior. Thus according to his model the spin-glass transition is basically a freezing phenomenon occurring in a continuous manner.

b) Muon Spin Rotation and Relaxation.

Since a number of years the positive muon spin rotation (μSR) technique [14] has been employed to study the spin-glass freezing. Originally the depolarization of polarized muons was used to determine the distribution of local fields in a spin-glass. This method requires the application of a field transverse to the initial muon polarization direction and has a characteristic time of measurement $\approx 10^{-7}$s. At $T_f$ a rather abrupt change in the depolarization rate was observed. In these experiments dynamical broadening could not be distinguished from inhomogeneous static broadening. Nevertheless the depolarization was attributed to a distribution of static fields whose rapid onset was interpreted in terms of an Edwards-Anderson phase transition.

Recently there has been a resurgence of interest in the muon technique to study the spin-glasses. Additional and more detailed experiments have been performed with the $\mu^+$spin rotation method on CuMn [15] and AgMn [16]. Here the necessityof an external field smears out the temperature dependence of the depolarization rate Λ. The results on CuMn (slow cooled) [15] show a gradual change in Λ beginning at temperatures far above $T_f$. In the region around $T_f$, Λ rapidly increases, but even with finely resolved data taken at small temperature intervals, there is no discontinuity in the

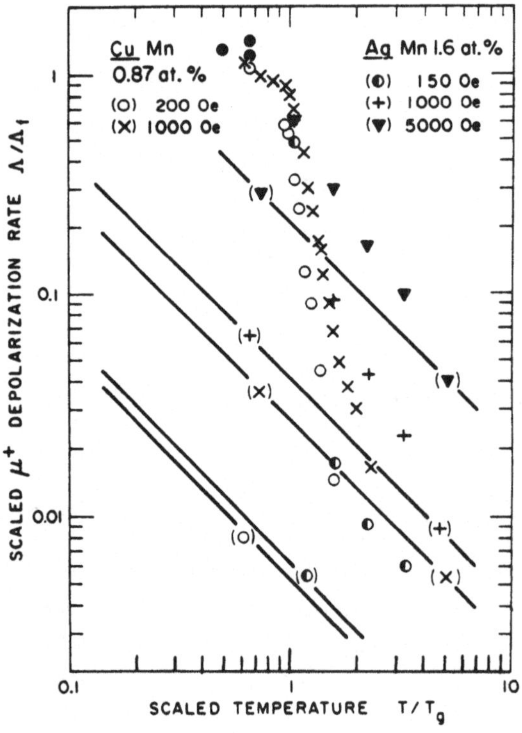


Fig. 5. Dependence of the depolarization rate Λ normalized to the value $\Lambda_f$ expected from a random frozen configuration of impurity spins. The solid lines represent the dilute limit values. After Brown et al.[16].

slope of $\Lambda$. This lead the authors [15] to conclude that the freezing of magnetic moments in a spin glass represents a continuous process and is describable in terms of an Arrhenius law. The experiments on AgMn [16], similar to those mentioned above, exhibited the same general behavior: rapid increase in $\Lambda$ as $T_f$ is approached from above and a field smearing of $\Lambda$ in the region around $T_f$. See Fig. 5 where both sets of data (Ref. 15 and Ref. 16) have been collected. There is always more polarization than that expected from the dilute, non-interacting impurity limit, and at $T \lesssim T_f$ the measured depolarization rate roughly corresponds to that of a random configuration of frozen spins. The data for AgMn do not obey the Arrhenius law found for CuMn and this has lead the authors of Ref.16 to interpret their results in terms of a static inhomogeneous broadening in the muon hyperfine field. Further they claim that the absence of long-lives muon precessions indicates a cooperative effect in the freezing of spins. This development of frozen ($>10^{-7}$s) spins is enhanced by the applied field. Their model [16] is then a cooperative and uniform (non-cluster or non-percolative) spin-glass transition.

A second and different muon technique has also been brought to bear on the spin-glass problem [17]. This method allows one to determine the longitudinal relaxation function $G_z(t)$ even in zero applied field and to discriminate between dynamic and static effects. The characteristic feature of zero-field (longitudinal) µSR is its response to the slow modulations of local fields, $G_z(t) \approx \frac{1}{3} \exp(-\frac{2}{2} t/\tau_c)$, where $\tau_c$ is the correlation time of an impurity spin. Very recently such investigations were made on AuFe and CuMn spin glasses [18]. Experimental resolution was sufficient to permit $\tau_c$ to be determined between $10^{-5}$ and $10^{-10}$s. The data of $\tau_c$ versus T for the above two alloys are given in Fig. 6. One can clearly see a large change $\approx 10^4$ in the

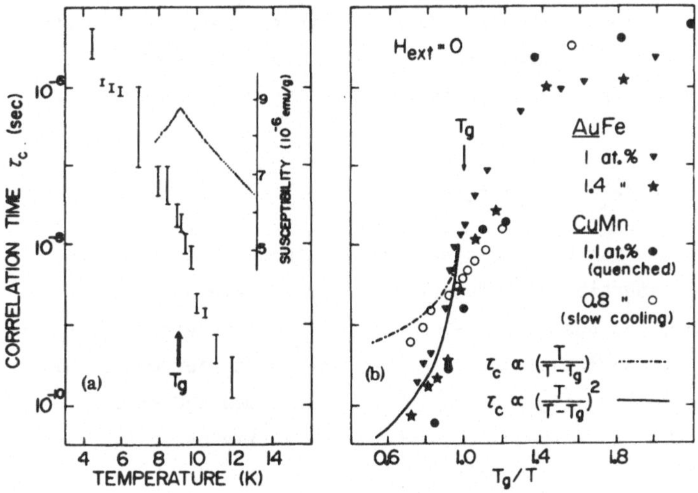


Fig. 6. Spin correlation time $\tau_c$ plotted against T (left) and $T_f/T$ (right). After Uemura et al.[18].

correlation time within a small T-interval. The authors [18] have fitted their data above $T_f$ with a function $\tau_c \propto [T/(T-T_f)]^2$. Note, however, in Fig. 6 the tail at high temperatures and lack of a constant $\tau_c$ at $T<T_f/2$. In addition a strong dependence on the heat treatment was observed. More homogeneous (quenched) samples exhibited much sharper changes in $\tau_c$ than slow cooled ones. The heat treatment is an important and complicated factor and the above effect is fully consistent with that observed in the ac-susceptibility [6]. The authors [18] cautiously interpret their results as disagreeing with the superparamagnetic cluster model of Tholence and Tournier [3]. They further state that presently it is too premature to attempt a direct comparison with the Edwards-Anderson phase transition theory. They also mention the possibility of a spatial distribution of $\tau_c$ and call for a great deal of further experimentation.

Very recently the same authors [19] have extended their muon spin relaxation measurements in CuMn to include a finite longitudinal magnetic field. The results of this investigation support the previous interpretation and allow a clearer separation between dynamic and static effects. A sharp disjunction exists between the dynamic, fast relaxing, paramagnetic region for $T>T_f$ and the almost static (most fluctuations are slower or longer than $10^{-5}$s) local fields for $T\approx\frac{1}{2}T_f$. At $T_f$ the measurements suggest the fast dynamical component to be coexisting with the static component. This coexistence is not a quasi-macroscopic cluster structure, but a more microscopic, spatial distribution of relaxation times with the possible onset of an order parameter. However, the effects of the external field (up to $\approx$600 Oe) on the spin dynamics and freezing has not yet been considered.

From the above discussion it seems that both the depolarization rate $\Lambda(T)$ and the spin correlation time $\tau_c(T)$ exhibit similar qualitative behavior, namely a rapid upturn around $T_f$. This behavior disagrees with the analysis of the neutron scattering measurements discussed in the previous section where a slow development of long relaxation times evolves. The problem here is with the quantitative meaning of "fast" and "slow", "rapid" and "gradual" for a random, nonuniform system. Thus the various authors are able to offer different interpretations of their experiments when guided by the two extreme theoretical models, that of superparamagnetic blocking and that of a mean field phase transition.

c) Ultra-Sonic Attentuation and Velocity Changes.

Another zero-field measurement, which is most useful for probing the spin-glass freezing, is the ultra-sonic propagation as a function of temperature and frequency. This technique, a dynamical one, depends upon the strength of the magnetostrictive coupling. There is sufficient evidence that adequate coupling exists for measurable effects, yet very few ultra-sonic investigations have been performed and none involving large variations of the sound frequency $\nu$ which can easily be changed throughout the MHz range ($10^{-6}$-$10^{-9}$s).

An ultra-sonic velocity change experiment at 30 MHz has been carried out by Hawkins et al.[20] on a CuMn (5 at.%) single crystal. Here the authors spanned $T_f$ with very accurate $\Delta v/v$ measurements as is shown in Fig.7. After subtracting the

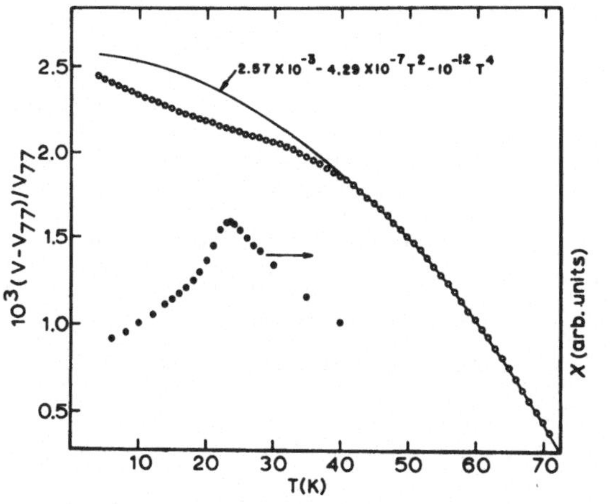


Fig. 7. Change in the ultrasonic velocity versus T for CuMn (5 at.%). The ac susceptibility is shown for comparison. After Hawkins et al. [20].

extrapolated background, they obtained the magnetic part of $[\Delta v/v]_m$. As can be seen from Fig. 7, this portion is rather smooth and exhibits a gentle minimum somewhat below $T_f$ as determined from the peak in the ac-$\chi$. From their analysis Hawkins et al. attempted to extract the specific heat critical exponent $\alpha$, since $[\frac{\Delta v(T)}{v_o}]_m = A\int_0^T C_m dT + BTC_m$, where $C_m$ is the magnetic specific heat and A and B are temperature independent constants. The result for $\alpha$ was a large negative value -1.9 which is consistent with the freezing anomaly being represented in the specific heat as a broad maximum. However, the magnetic parts $C_m$ and $[\Delta v/v]_m$, without additional analysis, both possess a broad maximum or minimum, but for $C_m$ the maximum is usually 20-30% above $T_f$ and for $[\Delta v/v]_m$ the minimum is ≈30% below $T_f$. This puzzle of no real, observable anomaly in the specific heat or ultra-sonic propagation at $T_f$ remains today unresolved and offers strong evidence against a "usual" type of phase transition.

Very recently a theoretical work by Hertz et al. [21] has appeared which calculates via the time dependent Ginzburg-Landau model the sound propagation in a spin-glass. $[\Delta v/v]_m$ is predicted to vary linearly with $T-T_f$ and the attenuation diverges as $(T-T_f)^{-1}$, both with T approaching $T_f$ from above. In addition at $T_f$, $v(T_f) \propto \nu^{\frac{1}{2}}$ while the attenuation is proportional to $\nu^{-\frac{1}{2}}$. These predictions with distinctly different anomalies in the velocity and the attenuation call for further and more systematic ultra-sonic experimentation.

## IV. FIELD DEPENDENCE OF THE MAGNETIZATION, $T \leq T_f$

A series of recent magnetization experiments have mapped out the M(H) behavior for the CuMn spin-glass at and below $T_f$. Let us now consider three of these experiments. Knitter and Kouvel [22] have employed a field stepping and cycling technique to remove the time dependence from the isothermal magnetization in an external field. The time independent M(H) curve exhibits a distinct "S"-shape character (Fig. 8) which begins at $T_f$ and becomes more pronounced for further decreasing T. This behavior is

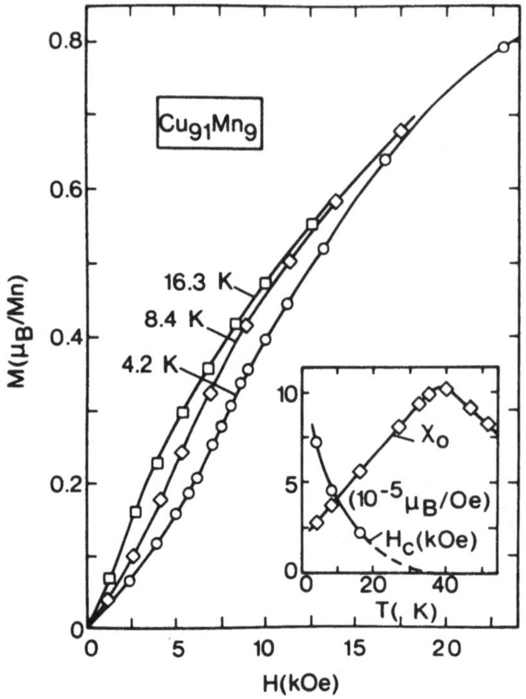


Fig. 8. Time independent M vs H for CuMn (9 at.%). Insert shows $H_c$=$H_t$ (threshold field) and $\chi_o$ (initial susceptibility versus T. After Knitter and Kouvel [22].

indicative of a field-induced transition to a state of higher magnetization above a certain threshold field $H_t$ - the point of inflection in M(H). The value of this field increases as the temperature is lower and it seems to separate the boundary between reversible, time independent behavior for a zero-field cooled sample at $H<H_t$, and the irreversibilities, remanences and time dependences for $H>H_t$. In addition once the external field has been larger than $H_t$, the high-M state does not return isothermally to the initial low-M state upon subsequent removal of the field. A peak is found in the susceptibility $\chi=\Delta M/\Delta H|_{H\to 0}$ as determined from the initial slope of the virgin magnetization curve, see Fig. 8. The temperature of the peak agrees with the value observed in the ac-$\chi$ cusp and marks the point at which $H_t$ extrapolates to zero in Fig. 8. Thus a definite connection exists between the cusp in $\chi$ and the field induced transition in $M(H_t)$.

Similar studies of M(t, H, T) for $T<T_f$ on CuMn have been carried out by Emmerich and Schwink [23]. Again the M(H) curves at fixed time exhibit the characteristic S-shape field dependence. The relaxation of the magnetization, $\Delta M=M(t_2,H,T)-M(t_1,H,T)$, shows a H and T dependent maximum. As $T\to 0$ a zero point magnetization appears which corresponds to the magnetization at t=0 for all temperatures. In order to describe these behaviors Emmerich and Schwink [23] have assumed the classical Néel model with Arrhenius-activation of magnetic clusters in an uniaxial anisotropy field with magnetic potential energy $U= -\mu H\cos(\phi-\alpha)-\mu H_a\cos^2\alpha$, where H is the applied field, $H_a$ is the uniaxial anisotropy field assumed to be randomly distributed over all

directions, and magnetic cluster moments exist far below $T_f$ with equal strength $\mu$. $\phi$ is the angle between H and $H_a$, $\alpha$ between $\mu$ and $H_a$ and $(\phi-\alpha)$ between $\mu$ and H. The relaxation processes may be calculated using a distribution of energy barriers f(E) (where $E=U_{max}-U_{min}$) and $\tau=\tau_o \exp(E/kT)$. f(E) depends strongly on $H/H_a$ in determining its position and shape. For a suitable choice of model parameters good agreement exists with the measured properties as a function of the three variables t, H and T. Also a connection is found between the "S"-point of inflection in M(H) at $H_t(T)$ and the maximum relaxation rate in $\Delta M(T)$, they both occur at the same T. This means that at the field-induced high M transition, the relaxation of M is fastest as would be expected for the onset of a new state. Thus this extended Néel superparamagnetic model seems to work quite well at low temperatures $T \ll T_f$ and can explain most of the experimentally observed M(t, H, T) behavior [23].

An effect which the above model cannot describe is the magnetic hysteresis of the CuMn spin-glass. A resurgence of interest in this effect has lately come from the systematic investigations of Monod et al. [24]. They have measured the low T ($<0.3\ T_f$), field-cooled hysteresis of CuMn as is shown in Fig. 9. Here the sample is cooled in a few kOe field to T=1.3 K, then the field is reduced to zero and we arrive at pt.A in Fig. 9. The remanence magnetization is time dependent ($\propto \log t$), but by quickly reversing the applied field (negative values) M can still be measured without any great change in slope ( pt.B'). At a certain reverse field (pt.B) there is a sudden (<0.1s) jump in M to negative values at pt.C. The reversal of remanent magnetization (H=0) is approximately equal to the initial value of the remanence. Thus almost the entire remanence has been reversed in a very short time and at a very sharp value of the negative field. The same is true for the positive field reversal pts.E→F in Fig. 9, except for the somewhat reduced M values due to the ever present time decay of $|M|$. In addition to the sharp, step-like reversal, a series of smaller discrete jumps may be contained in the magnetization transition. The resulting hysteresis loop is a rhombus with a width of a few hundred Oe, displaced from the

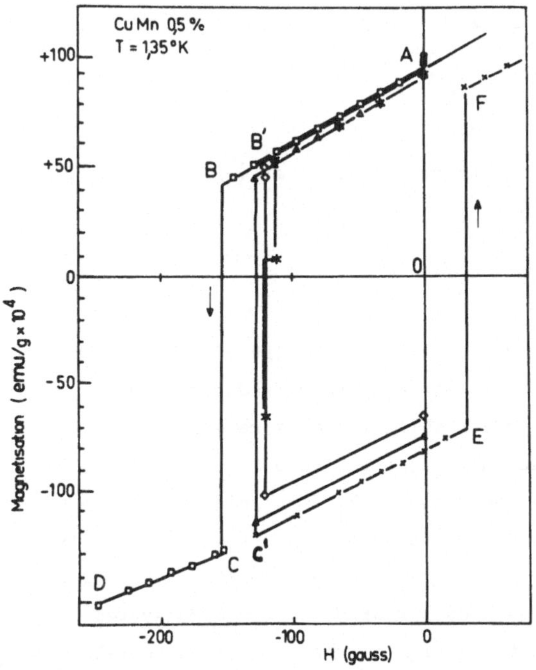


Fig. 9. Hysteresis for field-cooled CuMn with saturated remanence magnetization. Different symbols are used for different runs. After Monod et al. [24].

origin by about 100 Oe. Monod et al.[24] have also studied the c, T, t, transverse field and metallurgical dependences of the magnetization loop, and they have concluded the existence of a well-defined anisotropy field in the field cooled spin-glass state. The sharp, macroscopically coherent reversal of the magnetization further indicates a cooperative behavior among a large number of frozen spins. A puzzling fact here is that the above effects in CuMn do not exist for other archetypal spin-glasses such as AuFe and AuMn. Furthermore these effects disappear when small amounts of non-magnetic impurities with large spin-orbit coupling are added to the CuMn alloys.

The effect of an external, static field H||h on the ac susceptibility has been discussed by Mulder et al.[6]. As has long been known the application of a small H field reduces $\chi$ such that the cusp becomes rounded. This effect is most dramatic at $T_f$ with $\chi$ at all other temperatures being less sensitive to H. Fig. 10 illustrates

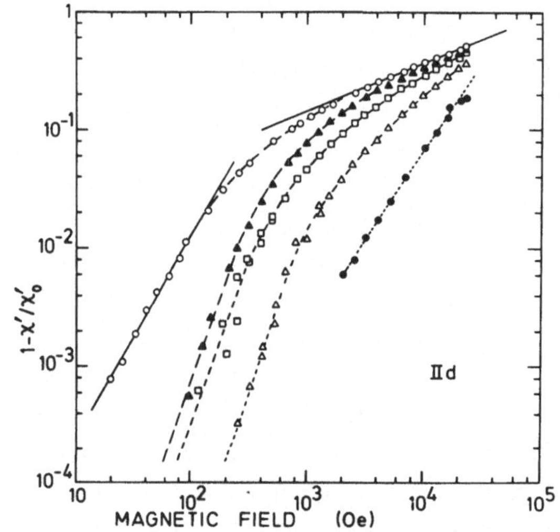


Fig. 10. Normalized changes in ac-$\chi$ as a function of applied magnetic field at several temperatures for CuMn (2 at.%).
● T = 31 K = 2 $T_f$
▲ T = 17 K
○ T = 15.5 K = $T_f$
□ T = 14 K
△ T = 12 K
After Mulder et al [6].

this dependence for a CuMn (2 at.%) sample. For sufficiently large fields >10 kOe, there is a rather flat maximum in $\chi(T)$ which has its cusp value reduced by about 30%. The external field does not seem to cause irreversibilities or time dependences in the ac-$\chi$. Also it does not matter for the ac susceptibility when this H-field is applied, above or below $T_f$. At temperatures above about $2T_f$ the field dependence of $\chi$ is negligible in fields less than 10kOe, and the high temperature susceptibility may be studied in order to trace the evolution of the freezing process. A very recent series of such measurements (up to 300K) have been performed on the CuMn system (1 to 6 at.% Mn) by Morgownik and Mydosh [25]. In Fig. 11 the inverse susceptibility is plotted against the temperature for five Mn concentration. Note the Curie-Weiss behavior at high temperatures with the paramagnetic temperature $\Theta$ positive (or ferromagnetic) and proportional to the concentration. Below about $5T_f$ there are deviations, which increase with decreasing temperature, in the susceptibility from the Curie-Weiss law. A cluster model [26] has been developed to analyse the onset of

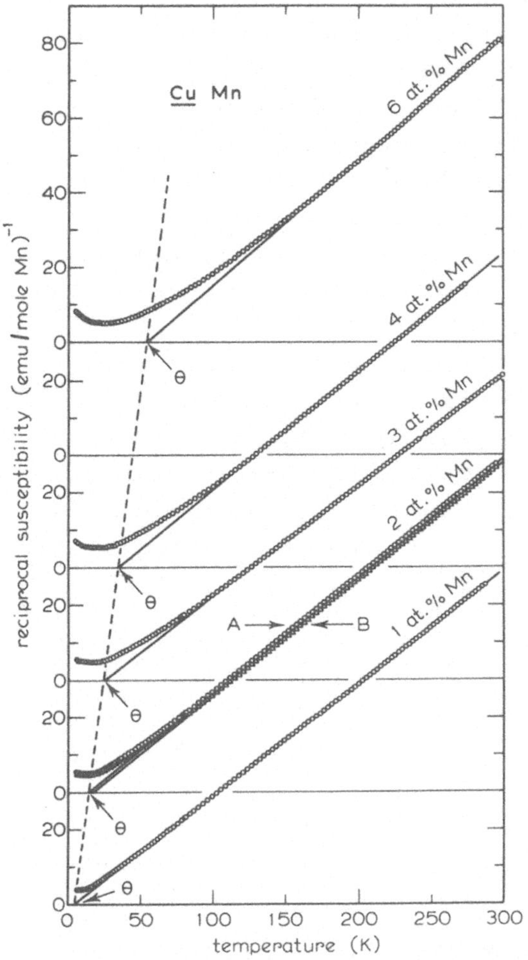


Fig. 11. Inverse susceptibility of CuMn as a function of T. "A" (slow cooled) and "B" (quenched) refer to the different heat treatments of the 2 at.% sample. After Morgownik and Mydosh [25].

these deviations. The canonical ensemble is calculated for all possible states of the first eleven configurations in which magnetic impurities can be grouped in an fcc lattice. For $T>2T_f$ this model can explain the deviations, the Θ(c) dependence and the effective Bohr magneton number with ferromagnetic first and second nearest neighbor exchanges of 50 and 20 K, respectively. This is a surprising result since the usual RKKY interaction in CuMn gives an antiferromagnetic interaction for at least one of these neighbors. In addition the very large field 400kOe magnetization measurements of Smit et al.[27] below $T_f$ showed a 50% incomplete saturation of the moments in CuMn, and when analyzed via a molecular field treatment there was no need for ferromagnetic exchange. Thus a large difference must exist in the magnetic state above and below $T_f$. In between something dramatically has happened and this is probably related to the appearance of a random anisotropy.

## V. ESR AND ANISOTROPY

Since a few years there has been a renewal of activity in studying the electron spin resonance of noble metal-Mn impurity alloys with special interest on spin-glass effects. In 1978 Salamon and Herman [28] measured the ESR linewidth ΔH as a function of temperature for a Cu-25 at.% Mn alloy. They reported a divergence in ΔH as $(T_f-T)^{-1}$ and introduced a model in which the exchange narrowed dipole broadening is drastically modified due to the critical slowing down of the Mn spins as $T_f$ is approached from above. This result is in contrast with Dahlberg et al.[29] who observed a much smoother increase in ΔH and a shift in the field for resonance (not a g-shift) through the freezing temperature in a few tenths at.% Mn in Ag. They interpreted this gradual behavior as the onset of an internal field which adds to the applied field and shifts the resonance. Monod and Berthier [30] have field-cooled two CuMn alloys (1.4 and 5 at.% Mn) and measured the zero field or remanent magnetization ESR. The resonance frequency $\omega$ in this state is describable by a linear relation $\omega=\gamma(H+H_a)$, see Fig. 12, where $H_a$ is an anisotropy field of order 400-500 Oe. A direct connection is made with the anisotropy field obtained from the displaced magnetization loops in Ref. 24. Thus the field cooled CuMn spin glass exhibits a ferromagnetic-like

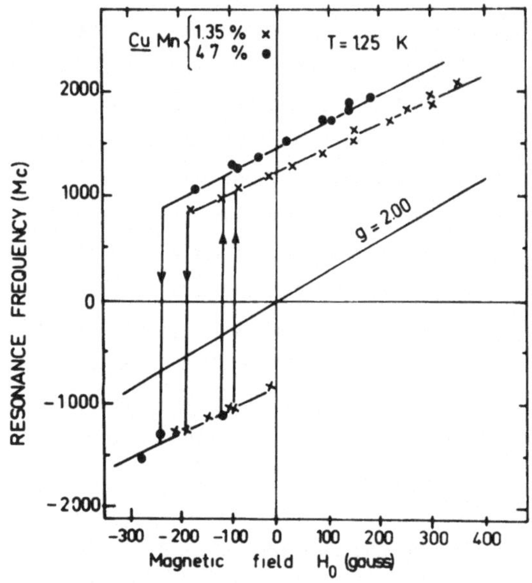


Fig. 12. ESR resonance frequency around zero applied field for two CuMn alloys after field cooling. The negative resonance frequency means an upside down resonance line shape. After Monod and Berthier [30].

behavior which allows zero field ESR and also NMR [31]. A single-valued anisotropy field of magnitude ≈ 500 Oe and non-dipolar in origin is necessary to maintain the remanent magnetization in the direction of the initial applied field. Further the experiments by Prejean et al. [32] on CuMn doped with non-magnetic impurities and AuFe have demonstrated that the hysteresis behavior and the ESR-ΔH [33] are proportional to the concentration of non-magnetic impurities with the coefficient depending upon the strength of the spin-orbit interaction. It is this interaction which is responsible for the anisotropy field in the spin-glass state. A microscopic mechanism to produce the anisotropy field and its associated energy has been proposed by Fert and Levy [34]. They showed that the anisotropy can be accounted for by a Dzyaloshinsky-Moriya type of interaction between the Mn spins arising from spin-orbit scattering of the conduction electrons by non-magnetic impurities. They calculated the magnitude of this interaction to be rather large 10-20% of the RKKY interaction for strong spin-orbit coupled impurities. It is then the field cooled, frozen state in CuMn without appreciable spin-orbit coupling which converts the random orientations of the weak, local anisotropy into a small, but well-defined, uniaxial anisotropy. In AuFe the strong spin-orbit interaction (and Fe orbital moment character) produces a large anisotropy which results in a broad symmetric hysteresis loop characteristic of a multi-domain ferromagnet. Yet in small fields the ac susceptibility exhibits quite similar "freezing" behavior for both CuMn and AuFe spin glasses. This uniaxial anisotropy only clearly manifests itself in field cooling and large field experiments.

Further progress in understanding the ESR and anisotropy properties of the CuMn spin-glass was made by Schultz et al.[35]. They also observed a linear relation in the field for resonance (see Fig. 12) with the zero external field corresponding to an internal or anisotropy field. Here the spectrum was taken by increasing the field at $T \ll T_f$ after zero field cooling. A model for the free energy and resulting equations of motion was suggested incorporating magnetic remanence, anisotropy and Zeeman energy. The resonance solution yields two coupled resonance frequencies for the zero field-cooled state (small or no remanence), $\omega^{\pm}/\gamma = \pm\frac{1}{2}H + H_a$ where $H_a = K(T)/\chi$ and K(T) is a temperature dependent anisotropy parameter. Schultz et al.[35] report the observation of this second mode $\omega^-$ by reducing the temperature and adding Ni to the CuMn sample in order to increase the anisotropy. From the normal $\omega^+$-mode, the temperature dependence of K may be determined, $K(T) = K(0)\,[1 - \frac{2}{3}\,T/T_f]$ and $K(0) = ac_{Mn}^2 + bc_{Mn}c_{Ni}$, where $a \ll b$. Good agreement for K(T) is found with the low $c_{Mn}$ measurements of Ref.30. Field cooled resonance data were also taken as a function of the angle between the applied field for resonance and the field cooling direction. Despite a marked anisotropy in the field for resonance, the simultaneously measured magnetization is fully isotropic, i.e. remains constant and simply rotates with the applied field over 360°. The authors [35] conclude the need for a memory direction which exists only below $T_f$ and play a significant role in determining the anisotropy of the spin-glass state.

In another ESR study involving the AuMn spin-glass Vaknin et al.[36] report a very large anisotropy field probably associated with the spin- orbit interaction of Ref.34. A comparison is drawn with the measurements and model of Schultz et al.[35], and a temperature dependent anisotropy parameter is also obtained. As before K(T) is approximately zero above $T_f$ and varies linearly with T for $T<T_f$. There was no indication of the second $\omega^-$-mode, although deviations from Dysonian lineshapes were found at low temperatures. Thus a general conclusion from these ESR investigations is the necessity of a temperature dependent anisotropy field which appears at $T_f$ and which significantly effects the low $T(\ll T_f)$, large field properties of all spin-glasses. In certain systems AuFe or AuMn this anisotropy field is much larger than in others CuMn or AgMn.

## VI. CONCLUSIONS AND MODEL

Based upon this brief survey one can easily see that the present experimental situation in spin-glasses is confused, but hopefully some basic trends seem to be emerging from the vast amount of available data and their varied interpretations. We would like to focus now upon two concepts which have been used time and again in the previous Sections. These are the relaxation time (or the time dependent spin correlation function) which has emanated from the dynamic measurements in zero field, and the anisotropy which clearly appears in field cooling and large field experiments.

The relaxation time $\tau_c$ observed in all the experiments in Sec.II and Sec.III definitely undergoes a change of behavior as the temperature is decreased. The same is true of the time dependent spin correlation function $\xi(q,t)$. Furthermore, in most of these dynamical experiments one makes the implicit assumption that the system is mainly uniform and can generally be described by a single, averaged $\tau_c$ or $\xi(q,t)$. Only Murani [12] has used a distribution of relaxation times, but this is based upon a rather tenuous separation of elastic from quasi-elastic scattering at a very fast ($<10^{-9}$s) measuring time. The above separation could certainly affect the form of his distribution function. Nevertheless, there is also a clear change of behavior at the freezing temperature $T_f$. Now we propose to divide the system up into dynamical, evolving magnetic clusters which develop out of the high temperature background of a paramagnetic collection of spins. There exists compelling experimental evidence [25, 37] that for $T \gg T_f$ short range magnetic interactions are present, and these interactions must be taken into account as the temperature is reduced. We suggest a competition occurs between the short range exchange, $J(r_{ij})$ (where $r_{ij}$ is the separation between two spins) and the disordering effect of the temperature, $k_BT$. When $J(r_{ij})>k_BT$ for a given group of spin, a cluster is formed. As T is decreased, this cluster will grow in size and shape based upon the distribution of other spins in its immediate neighborhood. As an illustration of this configuration, see the computer simulation given in Fig.13. We feel these clusters, with effective moments

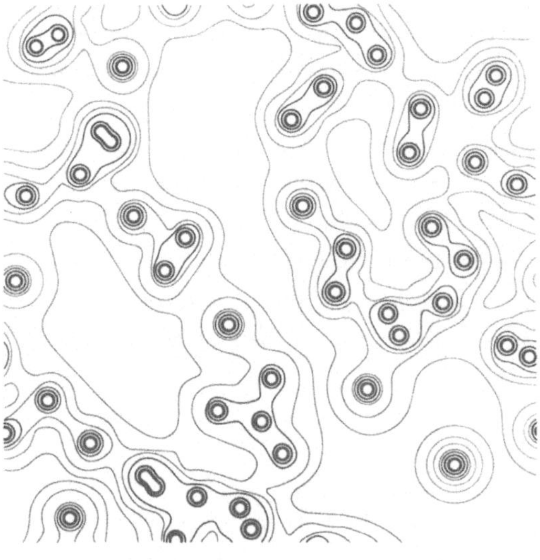

Fig. 13. Computer simulation for a random lattice of points (spins) with the contours and their overlapping representing the different ranges of spin-spin interactions.
After Verbeek et al.[39].

depending upon the dominant sign (ferro or antiferromagnetic) of the exchange, are the basic building blocks for the frozen spin-glass state. Hence in the dynamical experiments one averages a wide distribution of relaxation times from the single paramagnetic free spin to some large, slowly responding clusters where concentration fluctuations have produced many nearest neighbors. As the temperature is lowered towards $T_f$, the size and density of these clusters increase until they begin to make contact with each other. Then there is a "viscous rubbing" which affects the response to outside probes due to the competition for loose spins at the periphery of two touching clusters. Such effects cause a further broadening and shift in the distribution of relaxation times. Finally at $T_f$, we believe a percolation takes place which generates an infinite cluster of frozen spins made up of these, randomly frozen in orientation, much smaller clusters. The spins inside the infinite cluster are correlated, but they have their direction governed by the shape or exchange anisotropy of the smaller clusters. Thus the small clusters retain their original identity, however, they are no longer able to react to an external field since they are embedded in this infinite cluster. For a percolation transition the order parameter is represented by the fraction of spins belonging to the infinite cluster. Therefore, at $T \leq T_f$ there are many free, small clusters which are not part of the infinite cluster and which behave superparamagnetically. These latter clusters contribute a fast relaxation component and cause a wide distribution of relaxation times to be also observed below $T_f$. As T is lowered well below $T_f$, more of the free clusters will join the infinite group of clusters or become blocked according to a lack of thermal activation. In addition within the infinite group of clusters various types of excitations can further cause superparamagnetic-like behavior. Here the full Néel theory may be brought to bear with thermally activated processes over various magnitude energy barriers determining the low temperature properties. Indeed, this concept does nicely describe the remanence, irreversibilities and long time relaxations. It is the vast spectrum of relaxation times and excitations over a wide temperature range which makes the spin-glass behavior so difficult to measure and so confusing to interprete. Such non-simple experimental behavior forces us to an inhomogeneous model with a percolation-type of transition. It should be further noted here that in our percolation model the freezing temperature is a function of the time or frequency of measurement. Since the infinite cluster is a dynamical entity consisting of weak and strong links with many fluctuations and excitations, it will appear percolated according to the particular time window in which it is viewed. Thus with a very rapid measurement time the slowly occurring breaks in the $\infty$-cluster will not be observed and $T_f$ will be discerned at a higher T than by slower measurements. We can represent this schematically according to Fig. 14. In the limit of very long times of observation, a single $T_{f0}$ will be measured without any time dependences indicating that an equilibrium state has been reached. This is the experimental result of Malozemoff and Imry [4], and it casts doubt on non-equilibrium theories of the spin-glass transition.

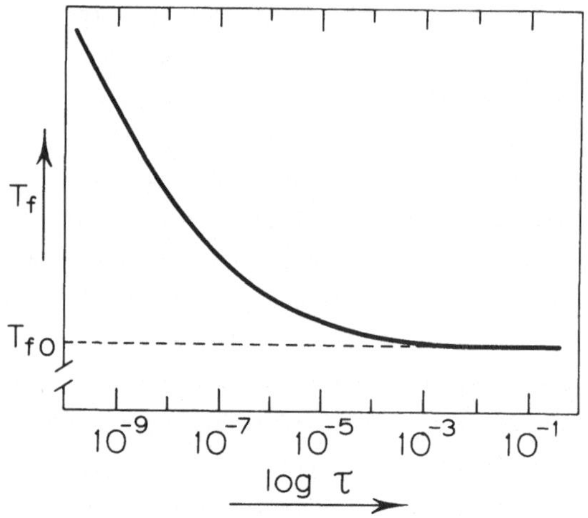


Fig. 14. Schematic representation of the freezing temperature as determined by different measurement techniques having different "time constants" of measurement.

We now turn to the second concept that of the anisotropy. In the zero field spin-glass, the varied geometries of the small clusters generate a random, local, shape anisotropy. Such an anisotropy has already been shown to exist in a number of amorphous ferromagnets [38]. This is the essential cause for the random freezing of the small clusters within the infinite cluster. What is also needed here is some antiferromagnetic exchange to prevent the formation of a ferromagnetic infinite cluster, as occurs in the giant-moment, ferromagnetic alloys, e.g. PdFe [39]. The antiferromagnetic interactions create the frustrated bonds which preserve the identity of the small clusters, even when they are touching each other. With the inclination towards a ferromagnetic percolation thus obstructed, the random, local anisotropy fixes the orientation and so the frozen spin-glass state is formed with no net moment. However, when a sufficiently large external field is applied ($H>H_t$), a small, net-macroscopic component of the anisotropy is induced in the infinite cluster along the field direction, i.e. the uniaxial anisotropy appears. This in turn produces the remanences and irreversibilities discussed in Sec.IV and Sec.V. A similar statement may be made concerning the field cooling. So as the spin-glass freezes, this net anisotropy or preferrred axis is built into the formation of the infinite cluster, and it remains so even after the external field is removed. This is why Monod et al. [24] in their magnetization reversal experiments observe a cooperative spin-flip, the entire remanent component of the infinite cluster reverses itself about this single, preferred anisotropy axis. Since a large number of excitations and fluctuations are present, the remanence will slowly decay and the memory effect of the anisotropy will gradually be lost. As these excitations are diminished by reducing the temperature, the strength of the anisotropy increases and thereby produces a T-dependent anisotropy parameter which experimentally [35] was found equal to $K(T)= K(0)\,[1 - \frac{2}{3}(T/T_f)]$. This secures the spin glass state and probably causes the ferromagnetic to spin glass transitions observed in so many systems at low temperatures

[40]. What is yet missing from this phenomenological model is the direct determination or observation of the zero field, random anisotropy; also how it develops into a small, but well-defined, uniaxial anisotropy with the application of an external field. Unfortunately our picture is qualitative, no theory exists along these lines to give quantitative support and guidance. Perhaps this will now occur as theoreticians seek a middle ground between the Edwards-Anderson model and the Néel superparamagnetism.

I wish to acknowledge the Nederlandse Stichting voor Fundamenteel Onderzoek der Materie (FOM) for their support of spin-glass research in Leiden.

REFERENCES

1. See for example J. Magn. Magn. Mater. 15-18, 99-209 (1980), J. Appl. Phys. 50, 7308-7373 (1980), and J. Appl. Phys. 52, March (1981).
2. V. Cannella and J.A. Mydosh, Phys. Rev. B 6, 4220 (1972).
3. J.L. Tholence and R. Tournier, J. de Phys. 35, C4-229 (1974).
4. A.P. Malozemoff and Y. Imry, to be published in Phys. Rev. Lett. 1981.
5. M. Guyot, S. Foner, S.K. Hasanain, R.P. Guertin and K. Westerholt, Phys. Lett. 79A, 339 (1980).
6. C.A.M. Mulder, A.J. van Duyneveldt and J.A. Mydosh, Phys. Rev. B23, 1384 (1981).
7. C.A.M. Mulder, A.J. van Duyneveldt, H.W.M. van der Linden, B.H. Verbeek, J.C.M. van Dongen, G.J. Nieuwenhuys and J.A. Mydosh, Phys. Lett. 83A, (1981).
8. J.L. Tholence, Solid State Commun. 35, 113 (1980).
9. G. Zibold, J. Phys. F8, L229 (1978).
10. M. Hardiman, Bull. Am. Phys. Soc. 25, 176 (1980) and to be published.
11. See for example P.A. Beck in Liquid and Amorphous Metals edited by E. Lüscher and H. Coufal (Sijthoff and Noordhoff, Alphen a/d Rijn, 1980) NATO-ASI Series E. Vol. 36.
12. A.P. Murani, J. Magn. Magn. Mater. 22, 271 (1981).
13. F. Mezei and A.P. Murani, J. Magn. Magn. Mater. 14, 211 (1980).
14. D.E. Murnick, A.T. Fiory and W.J. Kossler, Phys. Rev. Lett. 36, 100 (1976).
15. K. Emmerich and Ch. Schwink, Proceedings of the 2nd International Meeting on Muon Spin Rotation 1980, to be published in Hyperfine Interactions, 1981.
16. J.A. Brown, R.H. Heffner, T.A. Kitchens, M. Leon, C.E. Olsen, M.E. Schillaci, S.A. Dodds and D.E. MacLaughlin, to be published in J. Appl. Phys. 1981.
17. T. Yamazaki, Hyperfine Interactions 6, 115 (1979).
18. Y.J. Uemura, T. Yamazaki, R.S. Hayano, R. Nakai and C.Y. Huang, Phys. Rev. Lett. 45, 583 (1980). See also Y.J. Uemura, Proceedings of the 2nd International Meeting on Muon Spin Rotation 1980, to be published in Hyperfine Interactions, 1981.
19. Y.J. Uemura, K. Nishiyama, T. Yamazaki and R. Nakai, to be published in Solid State Commun. 1981.
20. G.F. Hawkins, R.L. Thomas and A.M. de Graaf, J. Appl. Phys. 50, 1709 (1979).
21. J.A. Hertz, A. Khurana and R.A. Klemm, Phys. Rev. Lett. 46, 496 (1981).
22. R.W. Knitter and J.S. Kouvel, J. Magn. Magn. Mater. 21, L316 (1980).
23. K. Emmerich and Ch. Schwink, Solid State Commun. 31, 705 (1979), and K. Emmerich, G. Felten and Ch. Schwink, J. Magn. Magn. Mater. 15-18, 173 (1980).
24. P. Monod, J.J. Prejean and B. Tissier, J. Appl. Phys. 50, 7324 (1979).
25. A.F.J. Morgownik and J.A. Mydosh, to be published in Phys. Rev. B 1981.
26. A.F.J. Morgownik and J.A. Mydosh, to be published in Proceedings of the 16th International Conference on Low Temperature Physics (LT-16), 1981.
27. J.J. Smit, G.J. Nieuwenhuys and L.J. de Jongh, Solid State Commun. 31, 265 (1979).
28. M.B. Salamon and R.M. Herman, Phys. Rev. Lett. 41, 1506 (1978).
29. E.D. Dahlberg, M. Hardiman, R. Orbach and J. Souletie, Phys. Rev. Lett. 42, 401 (1979), and R. Orbach, J. Magn. Magn. Mater. 15-18, 706 (1980).
30. P. Monod and Y. Berthier, J. Magn. Magn. Mater. 15-18, 149 (1980).
31. H. Alloul, Phys. Rev. Lett. 42, 603 (1979).
32. J.J. Prejean, M.J. Joliclerc and P. Monod, J. de Phys. 41, 427 (1980).

33. K. Okuda and M. Date, J. Phys. Soc. Japan, 27, 839 (1969).
34. A. Fert and P.M. Levy, Phys. Rev. Lett. 44, 1538 (1980).
35. S. Schultz, E.M. Gullikson, D.R. Fredkin and M. Tovar, Phys. Rev. Lett. 45, 1508 (1980).
36. D. Vaknin, D. Davidov, G.J. Nieuwenhuys, F.R. Hoekstra, G.E. Barberis and J.A. Mydosh, to be published in Proceedings of the 16th International Conference on Low Temperature Physics (LT-16), 1981.
37. J.A. Mydosh, J. Magn. Magn. Mater. 7, 237 (1978) and in Liquid and Amorphous Metals edited by E. Lüscher and H. Coufal (Sijthoff and Noordhoff, Alphen a/d Rijn, 1980) NATO-ASI Series E. Vol. 36.
38. R. Harris, M. Plischke and M.J. Zuckerman, Phys. Rev. Lett. 31, 160 (1973).
39. B.H. Verbeek, G.J. Nieuwenhuys, J.A. Mydosh, C. van Dijk and B.D. Rainford, Phys. Rev. B22, 5426 (1980).
40. See G.J. Nieuwenhuys, B.H. Verbeek and J.A. Mydosh, J. Appl. Phys. 50, 1685 (1979), and M.B. Salamon, K.V. Rao and Y. Yeshurun, to be published in J. Appl. Phys. 1981.

# MEAN FIELD THEORY FOR SPIN GLASSES

G. Parisi
INFN - Laboratori Nazionali di Frascati, Frascati, Italy, and
Istituto di Fisica della Facoltà di Ingegneria dell'Università di Roma, Italy

In these recent years our understanding of spin glasses has strongly improved. In this talk I will describe the results which have been obtained in the framework of broken replica symmetry[1-13] for Ising spins.

Generally speaking in order to build a theory to describe a certain kind of phase transitions, we must go through five steps :

I) We first identify the physical characteristics which distinguish the two phases of the system (e. g. spontaneous magnetization).

II) We introduce an order parameter which quantifies the differences between the two phases (e. g. the value of the spontaneous magnetization).

III) We write an effective free energy as function of the order parameter ; the value of the order parameter is fixed minimizing the free energy (mean field approximation) (a typical result is $m = \mathrm{th}(\beta J h)$).

IV) We verify that in the long range limit the mean field approximation gives the exact answer (e. g. for ferromagnetic systems this result can be rigorously proved).

V) We finally compute the corrections to the mean field approximation due to the finite range of the interaction ; at this stage we can apply the whole machinery of diagrammatic expansion, renormalization group.

We shall now see how this program can be implemented for Ising spin glasses.

## 1. - The Glassy phase.

The main characteristics of spin glasses is the presence of irreversible effects and of very large times of approach to equilibrium. These effects show up dramatically if we compare the alternate magnetic susceptibility ($\chi_{LR}$) with the field cooled susceptibility ($\chi_e = \frac{dM_e}{dH}$)[14-16]. It is an experimental fact that:

$$\Delta\chi = \chi_e - \chi_{LR} > 0 \tag{1.1}$$

and remanence is present in the region

$$|H| < H_c(T) \tag{1.2}$$

where the critical magnetic field $H_c(T)$ goes to zero for $T \to T_c$.

This behaviour is due to the presence of many different states in which the system may stay. These states can be characterized at the microscopical level by the value of the magnetization at each point $m_i^{[s]}$: the index i runs over the points of the lattice and s labels the different states.

When $|H| > H_c(T)$ there is only a stable state of the system; when $|H| < H_c(T)$ there are many states which are stable for small perturbations, however most of these states will be metastable states and only a small fraction will be asymptotically stable. When we change the magnetic field at fixed temperature by a very small amount, some of the previously asymptotically stable states will become metastable states and the stable will become metastable. However if the hopping time from one state to an other one is much larger than the time scale at which the magnetization M is measured, hopping can be neglected: the system remains locked in the vicinity of one state and linear response theory can be applied to compute $\chi_{LR}$. When we measure $\chi_e$, we hope to be in a asymptotically stable state: the magnetization is the equilibrium one ($M_e(T,H)$) and $\chi_e = \frac{\partial}{\partial H} M_e(T,H)$.

In other words when we change H at fixed T a spin glass system should undergo a sequence of microtransitions: in a finite volume, when we hop from one state to an other state, the magnetization jumps discontinuously[9,17,18,19]; when the volume goes to infinity the distance in H between two transitions and the discontinuity in M should go together to zero: $M_e(T,H)$ should be finally a continuous curve, it is clear however that the derivative with respect to H and the infinite volume limit do not commute. Mathematically speaking the zeros of the partition function become dense near the real H axis for all values of H in the glassy region $|H| \leq H_c(T)$.

## 2. - The order parameter.

The meaning of the local magnetizations $m_i^{[s]}$ introduced in the previous section, is very clear: we study the evolution in time (real or computer time) of the system and we define:

$$m_i = \frac{1}{t} \int_0^t d\tau\, \sigma_i(\tau) \tag{2.1}$$

where t is a large but not too large observation time, e.g. in a D-dimensional ferromagnetic system of size L, t must satisfy the conditions:

$$\tau_m \ll t \ll \tau_m \exp(L^{D-1}) \tag{2.2}$$

where $\tau_m$ is the microscopic relaxation time, e.g. one Montecarlo step.

When we change the initial condition we may obtain different results for the magnetizations; by exploring different initial conditions the full set of $m_i^{[s]}$ for different s can be obtained.

Unfortunately it is not very clear how to implement this natural definition in the framework of equilibrium statistical mechanics where all the states receive a weight $\exp(-\beta H)$. Indeed the existence of two (or more) asymptotically stable states breaks the ergodicity hypothesis (i.e. independence from the initial conditions) which is at the heart of the Gibbs-Boltzmann approach.

Which is the standard solution in the ferromagnetic case? When H is different from zero the system may stay only in one possible state; the difficulties arise at H = 0: two states of opposite magnetization are present below the critical temperature. We can define the spontaneous magnetization at H = 0 by adding to the Hamiltonian a small term $\Sigma_i \varepsilon h_i \sigma_i$ where $h_i$ are the values of an external magnetic field which may be space dependent: we define the h dependent spontaneous magnetization as

$$m_i^{[h]} = \lim_{\varepsilon \to 0^+} m_i(\varepsilon h) \tag{2.3}$$

where $m_i(\varepsilon h)$ is the magnetization in presence of the perturbation.

If in a ferromagnetic case we consider a staggered h field, $m_i^{[h]} = 0$; on the contrary if h is translational invariant there are only two possibilites $m_i = \pm m$, m being the spontaneous magnetization; s takes only two values.

The value of the spontaneous magnetization can be obtained also by considering the correlation function $\langle \sigma_i \sigma_k \rangle$ at large distance $|i-k|$. In each of the possible states of the system we must have

$$\overline{\sigma_i \sigma_k} \to m_i^{[s]} m_k^{[s]}, \qquad |i-k| \to \infty . \tag{2.4}$$

We obtain that the thermodynamic average is given by:

$$\langle \sigma_i \sigma_k \rangle \to \Sigma_s \, p_s \, m_i^{[s]} m_k^{[s]} \tag{2.5}$$

$p_s$ being the weight of the $s^{th}$ state in the thermodynamic average. At zero external field

$$p_1 = p_2 = \frac{1}{2}, \qquad m^{[1]} = m^{[2]} = m, \qquad \langle \sigma_i \sigma_k \rangle \to m^2 . \tag{2.6}$$

The spontaneous breaking of the symmetry shows up as a violation of the clustering decomposition: the spontaneous magnetization can be computed using the classical theorems on the decomposition of a non clustering state into a set of clustering pure states.

I am describing these two procedures in detail because they cannot be applied to spin glasses: in the infinite volume limit the number S of possible microscopic states $m_i^{[s]}$ may become infinite and quantities like

$$\langle \sigma_i \sigma_k \rangle , \qquad \langle \sigma_i \sigma_k \rangle^2 \tag{2.7}$$

may go to zero exponentially without giving precise informations on the values of the $m_i^{[s]}$; often the onset of spin glass order is characterized by a non zero value of the order parameter $q_{EA}^{[s]}$ defined by

$$q_{EA}^{[s]} = \lim_{V \to \infty} \frac{\Sigma_i (m_i^{[s]})^2}{V} \tag{2.8}$$

V being the volume of the system. It should be clear that $q_{EA}$ cannot be extracted from the knowledge of the spin correlation functions at large distance.

We face the same problem if we try to compute the $m_i^{[s]}$ by adding a magnetic field; from the point of view of the spin glass a constant magnetic field will look like a staggered magnetic field; for a generic choice of the $h_i$ we will not be able to split different states and also in presence of a magnetic field we have:

$$m_i^{[h]} = p_s \, m_i^{[s]} \tag{2.9}$$

where an infinite number of $p_s$ may be different from zero also at finite h. However if we take the $h_i$ proportional to one of the possible magnetizations, $h_i \propto m_i^{[s]}$ only the state labelled by s will contribute to the sum in eq. (2.9). In this way we obtain the bootstrap equation:

$$m_i^{[h]} = m_i^{[s]}, \qquad h_i \propto m_i^{[s]}. \tag{2.10}$$

For example in the ferromagnetic case previously described, the magnetic field is taken to be point independent as well as the magnetization.

In a spin glass we don't know the direction in which the magnetization points: eq. (2.10) seems to be useless and we face a dead end.

The way out can be found by introducing two real identical weakly coupled replicas of the same system[1-4]; the global Hamiltonian is:

$$H_2 = \sum_{i,k} J_{i,k}(\sigma_i^1 \sigma_k^1 + \sigma_i^2 \sigma_k^2) - 2\varepsilon \, \Sigma_i \, \sigma_i^1 \sigma_i^2 \tag{2.11}$$

where the $J_{i,k}$ are the random quenched coupling among the spins and the index $a = 1, 2$ labels the two replicas of the system; when $\varepsilon = 0$ the replicas are decoupled. For positive $\varepsilon$ each of the two replicas acts as an external magnetic field on the other one and both must be locked in the same state. In the limit $\varepsilon \to 0$ we find:

$$\Sigma_s p_s (m_i^{[s]})^2 = \langle \sigma_i^1 \sigma_i^2 \rangle . \tag{2.12}$$

If we assume that the quantity $q_{EA}^{[s]}$ is not s-dependent in the infinite volume limit and we introduce the quenched free energy density $F_2$ of the two coupled replicas we obtain:

$$\begin{aligned} q_{EA} &= - \frac{d}{d\varepsilon} F_2(\varepsilon) \Big|_{\varepsilon = 0} , \\ F_2(\varepsilon) &= \lim_{V \to \infty} - \frac{1}{2\beta V} \ln \left[ \Sigma_{\{\sigma\}} \exp(-\beta H_2(\varepsilon) \right] . \end{aligned} \tag{2.13}$$

The volume V is the total number of spins in each replica. This apparently baroque construction is the only way at our disposal to compute $q_{EA}$ in the framework of equilibrium statistical mechanics when an infinite number of possible states is present.

It is convenient to generalize this construction by introducing r replicas:

$$\begin{aligned} H_r(\varepsilon) &= \sum_{i,k} J_{i,k} ( \sum_1^r{}_a \sigma_i^a \sigma_k^a ) - \varepsilon \Sigma_i ( \sum_{a,b} \sigma_i^a \sigma_i^b - r) , \\ F_r(\varepsilon) &= - \frac{1}{\beta r V} \ln \left[ \Sigma_{\{\sigma\}} \exp(-\beta H_r(\varepsilon) \right] . \end{aligned} \tag{2.14}$$

The same arguments give:

$$Q(r) = - \frac{d}{d\varepsilon} F_r(\varepsilon) \Big|_{\varepsilon = 0} = (r - 1) q_{EA} . \tag{2.15}$$

In this way we define the function Q(r) for integer values of r; the definition can be extended to non integer values in a natural way:

$$\begin{aligned} F_r(\varepsilon) &= - \frac{1}{\beta r V} \ln \int dh_i \exp(- \frac{\beta}{4} \Sigma_i h_i^2) \left[ Z( \varepsilon^{1/2} h_i) \right]^r , \\ Z( \varepsilon^{1/2} h_i) &= \Sigma_{\{\sigma\}} \exp \left[ - \beta H( \varepsilon^{1/2} h_i) \right] , \end{aligned} \tag{2.16}$$

$$H(\varepsilon^{1/2} h_i) = \sum_{i,k} J_{i,k} \sigma^i \sigma^k + \Sigma_i \left[ \varepsilon - \varepsilon^{1/2} h_i \sigma_i \right] ,$$

$$Q(r) = - \frac{d}{d\varepsilon} F_r(\varepsilon) \Big|_{\varepsilon = 0} . \tag{2.16}$$

It easy to check that for integer $r$ eq. (2.16) coincide with eq. (2.14).

Let us see which informations are contained in the function $Q(r)$ : if linear response theory is assumed to be valid we get :

$$Q(r) = -1 + \chi_{LR}/\beta + r q_{EA} . \tag{2.17}$$

From the relation $Q(1) = 0$, we obtain the Fischer result :

$$\chi_{LR} = \beta(1 - q_{EA}) \tag{2.18}$$

which is at zero magnetic field a consequence of the linear response theory. On the other hand under reasonable hypotheses we can show that :

$$\chi_e = \beta(1 - Q(0)) . \tag{2.19}$$

We stay in the glassy phase if and only if the function $Q(r)$ is not a linear function of $r$ ; it is natural to study the function $q(r) = \frac{d}{dr} Q(r)$. As far as the function $Q(r)$ is linear on the integers for $r \geq 1$, it is reasonable to assume that

$$q(r) = q_{EA} , \qquad r \geq r_c , \qquad r_c \leq 1 . \tag{2.20}$$

The final results in terms of the function $q(r)$ are :

$$\chi_e = \beta \int_0^1 dr\, (1 - q(r)) ,$$

$$\chi_{LR} = \beta(1 - q(1)) ,$$

$$\Delta\chi = \beta \int_0^1 dr\, (q(1) - q(r)) , \tag{2.21}$$

$$q(1) = q_{EA} .$$

At low temperatures $q(r,T)$ becomes a function of only $r/T = y$ [9,11]

$$q(r,T) \simeq q_s(r/T) \tag{2.22}$$

and it can be computed from the formula:

$$q_s(y) = -\frac{1}{V}\frac{d}{dy}\frac{1}{y}\frac{d}{d\varepsilon} \ln \int dh_i \exp - G(\varepsilon, h_i) ,$$

$$G(\varepsilon, h_i) = \Sigma_i \frac{h_i^2}{4} + y\, U(\varepsilon^{1/2}, h_i) \tag{2.23}$$

where $U(\varepsilon^{1/2}, h_i)$ is the zero temperature internal energy of the Hamiltonian H of eq. (2.16); this last expression can be easily computed using standard Montecarlo techniques. The shape of the function q(r) contains many informations on the breaking of the linear response theory: it cannot be a constant inside the glassy phase. We propose to take it as order parameter for a spin glass: in this way the macroscopic state of a spin glass is characterized by a function defined on the interval 0-1 [7-9]. We will see in the next Section that q(r) can be computed in the replica formalism and that q(r) is not a constant only if the replica symmetry is broken. The spontaneous breaking of the replica symmetry should be identified with the onset of irreversible behaviour.

## 3. - The mean field approximation.

In the replica theory the expectation value of the free energy is:

$$F = -\frac{1}{\beta V} \frac{d}{dn} Z^n \Big|_{n=0} \tag{3.1}$$

and the $n^{th}$ power of the partition function Z is written as[2]

$$Z^n = \sum_{\{\sigma_i^a\}} \exp\left[-\beta \sum_{i,k} \sum_{1,a}^{n} \sigma_i^a \sigma_k^a J_{i,k}\right]. \tag{3.2}$$

It is usual to introduce as order parameter the expectation value of the product of two spins in different replicas

$$Q_i^{a,b} = \langle \sigma_i^a \sigma_i^b \rangle . \tag{3.3}$$

In the mean field approximation the free energy is written as function of Q; e.g.

$$F = -\frac{d}{dn}\left[\frac{\beta}{2} \sum_1^n {}_{a,b} Q_{a,b}^2 + \frac{1}{\beta} \ln\left(\Sigma_{\{a\}} \exp(\sigma^a \sigma^b Q_{a,b})\right)\right]. \tag{3.4}$$

The free energy is obviously invariant under the group of permutations $P_n$ which exchange the labels of the different replicas. The value of $Q^{a,b}$ is given by a stable stationary point of the free energy. It was realized by Thouless and de Almeida[3] that the replica symmetric solution (all $Q^{a,b}$ equal) is unstable for $|H| < H_r(T)$ and in this region the replica symmetry is broken. It was suggested in ref. 4 that the actual pattern of symmetry breaking is

$$P_n \rightarrow P_{n/m} \otimes (P_m)^{n/m} \tag{3.5}$$

which in the $n \rightarrow 0$ limit becomes

$$P_0 \rightarrow P_0 \otimes (P_m)^0 \ . \tag{3.6}$$

$P_0$ is the group of permutations of zero elements and $(P_m)^0$ denotes the product of $P_m$ with itself zero times ($\ln P_m$ would be more appropriate).

It was shown in ref. 7 that eq. (3.6) may be generalized to

$$P_0 \rightarrow (P_{m_1})^0 \otimes (P_{m_r})^0 \otimes \cdots\cdots \otimes P_0 \equiv G_{\{m\}} \ . \tag{3.7}$$

In this scheme $P_0$ always contains as unbroken subgroup $P_0$ so that we can have an infinite number of $m_i$. For integer n the $m_i$ s must also be integers however when n is non integer also the $m_i$'s may become non integers. If some technical conditions are satisfied one can associate to a matrice Q which is invariant under the group $G_{\{m\}}$ a function $q_R(x)$ on the interval 0-1.

It is very interesting to note that if we compute the function Q(s) defined in Section 2, we find

$$Q(s) = \int_0^s q_R(x)\,dx \ . \tag{3.8}$$

Eq. (3.8) implies

$$q_R(x) = q(x) \ . \tag{3.9}$$

This result can be proved by noticing that the free energy (2.16) is invariant under a group $P_0$ which is explicitely broken when $\varepsilon \neq 0$ to $P_0 \otimes (P_s)^0$. In presen̲ce of this symmetry breaking term, the matrix Q will orient itself in the direction of the external force as far as possible.

An x dependent q(x) function implies that the replica symmetry is broken together with the linear response theory.

## 4. - The infinite range model.

In the infinite range model[2] (Sherrington-Kirpatrick) the Hamiltonian is given by:

$$H = \sum_{1}^{N}{}_{i,k} J_{i,k}\,\sigma_i\,\sigma_k - H \sum_{1}^{N}{}_i\,\sigma_i \qquad (4.1)$$

H being the magnetic field, $J_{i,k}$ random gaussian variables $\langle J_{i,k}^2 \rangle = 1/N$.

In the thermodynamic limit one finds[7]

$$F(\beta, H) = \max_{q(x)} F_{ef}(\beta, H | q) \qquad (4.2)$$

where the effective free energy is given by:

$$F_{ef}(\beta, H\ q) = -\frac{\beta}{4}\left(1 + \int_0^1 q^2(x)\,dx - 2\,q(1)\right) - $$
$$- \frac{1}{\sqrt{2\pi}\,\beta} \int_{-\infty}^{+\infty} dZ \exp(-Z^2/2)\, g(0, H + Z\sqrt{q(0)}) \qquad (4.3)$$

where the function $g(x, H)$ satisfies the q dependent differential equation[7,23]

$$\frac{\partial g}{\partial x} = -\frac{1}{2}\frac{\partial q}{\partial x}\left[\frac{\partial^2 g}{\partial h^2} + x\left(\frac{\partial g}{\partial h}\right)^2\right] \qquad (4.4)$$

with the boundary condition

$$g(1, h) = \ln\left[\mathrm{ch}\,(\beta h)\right] .$$

In this way one finds results for the thermodynamic quantities which are in excellent agreement with the Montecarlo data.

One finds a line of transitions at $H = H_c(T)$ where

$$H_c(T) \simeq (T_c - T)^{3/2}, \qquad T \sim T_c ,$$
$$H_c \sim (2 \ln 1/T)^{1/2}, \qquad T \longrightarrow 0 . \qquad (4.5)$$

The transitions from a constant to a non constant function $q(x)$ (onset of irreversibility) is a third order transition at $H = 0$ while it becomes fourth order at $H \neq 0$. We find also

$$\Delta\chi\Big|_{H=0} \propto (T - T_c)^2 , \qquad \Delta\chi\Big|_{T\neq 0} \propto (H_c^{(T)} - H) , \qquad (4.6)$$

$$\chi = 1 - h^{4/3} + O(h^2) , \qquad T < T_c .$$

The results of the analytic approach can be recovered with very high approximation from the **PaT** hypothesis[10]:

$$\frac{\partial S}{\partial H} = \frac{\partial M}{\partial T} = 0 , \qquad \frac{\partial q_{EA}}{\partial H} = 0 , \qquad \text{for } |H| < H_c(T) \tag{4.7}$$

This hypothesis is exact in the random energy model of Derrida[20] where the entropy of the glassy phase is zero.

According to eq. (4.7) the glassy phase is characterized by temperature independent magnetization. Eq. (4.7) seems to suggest that the system has at its disposal many equivalent states and, when a magnetic field is added, it goes from one state to an other equivalent state having the same entropy.

A more detailed analysis of the predictions of the PaT hypothesis can be found in the original literature[10-13].

## 5. - The real case.

The method here expounded allows in principle for a systematic study of the corrections to the mean field approximation. In the three dimensional case these corrections should be very small for long range forces (as in diluted alloys) and should be more sizable for short range forces. Also in this last case both computer simulations[21] and real experiments[15] show small deviations from mean field theory, expecially for quantities like $\chi_e$.

The impressive success of mean field theory in predicting $\chi_e$ should not be considered as an undoubtable evidence for the existence of a transition: there are many cases for which the corrections to mean field destroy the transition, but they are small far from the would be transition and it is very difficult to distinguish experimentally a transition from a "quasitransition"[22].

The theoretical situation is rather promising, it would be very important to have a definitive experimental confirmation of the correctness of these ideas. A precise measurement of the critical line $H_c(T)$ and of $\Delta\chi$ near this line would be crucial.

References.


1. S.F.Edwards and P.W.Anderson, J.Phys. F5, 965 (1975); F6, 1927 (1976).
2. D.Sherrington and S.Kirkpatrick, Phys. Rev. Letters 35, 1792 (1975); Phys. Rev. B17, 4384 (1978).
3. J.R.L.de Almeida and D.J.Thouless, J.Phys. A11, 983 (1978).
4. A.Blandin, M.Gabay and T.Garel, J.Phys. C13, 403 (1980).
5. A.J.Bray and M.A.Moore, Phys. Rev. Letters 41, 1068 (1978).
6. E.Pytte and J.Rudnick, Phys. Rev. B19, 3603 (1979).
7. G.Parisi, Phys. Rev. Letters 43, 1754 (1979); Phys. Letters 73A, 531 (1979); J.Phys. A13, 1101 (1980); A13, 1887 (1980); A13, L115 (1980).
8. G.Parisi, Phil. Mag. B41, 677 (1980).
9. G.Parisi, Phys. Reports 67, 25 (1980).
10. G.Parisi and G.Toulouse, J.de Phys. 41, L361 (1980).
11. J.Vannimenus, G.Toulouse and G.Parisi, J.de Phys. 42, 565 (1981).
12. G.Toulouse and M.Gabay, J.de Phys. 42, L111 (1981).
13. G.Gabay, Thèse, Orsay (1981).
14. J.A.Mydosh, contribution to this Conference.
15. G.Toulouse, contribution to this Conference.
16. P.Monod, contribution to this Conference; see also H.Bouchiat, "Thèse de 3ème Cycle", Orsay (1981).
17. S.Kirkpatrick, contribution to this Conference; S.Kirkpatrick and P.Young, IBM preprint (1981).
18. F.T.Bantilan Jr. and R.G.Palmer, J.Phys. F11, 261 (1981).
19. S.K.Ma and M.Payne, La Jolla prerint (1981).
20. B.Derrida, Phys. Rev. Letters 45, 79 (1980); the same form for the free energy has also been obtained by N.Cabibbo (unpublished).
21. G.Parisi, in preparation.
22. G.Parisi, Phys. Letters 43A, 379 (1973).
23. B.Duplantier, J. Phys. A14, 283 (1981).

# EQUILIBRIUM MAGNETISATION OF A SPIN GLASS ABOVE AND BELOW $T_g$ : COMPARISON WITH MEAN FIELD THEORY

P.MONOD and H.BOUCHIAT
Laboratoire de Physique des Solides, Bât 510, Université Paris Sud 91405 ORSAY(France)

## I - INTRODUCTION

We present a series of measurements of static magnetisation performed on an alloy AgMn 10.6% between 4.2°K and 50°K,i.e.through the spin glass temperature of 37.4°K. The aim of these measurements is twofold: First we want to establish for this spin glass alloy the value of the equilibrium magnetisation for different magnetic fields H above and below $T_g$. This type of data should enable one to define a phase diagram in the (H,T) plane for the spin glass phase (1) and are necessary in order to check the Parisi-Toulouse projection hypothesis which deduce the low temperature properties from those at the transition temperature $T_g$(H) (2). Second we want to determine more accurately the non linear part of the magnetisation at and above $T_g$. Indeed a simple expansion of the Sherrington-Kirpatrick equations (3,4,5,6) shows that this non-linear term is expected to diverge at $T_g$ like the Edwards-Anderson parameter q (7). Thus a measurement of the non-linear magnetisation term can be used to check the applicability of mean field theory, at least above $T_g$. The need of such classical measurements stems from the fact that very few determinations of the static equilibrium magnetisation of a spin glass system as a function of field and temperature have been reported in a systematic way so far. A notable exception is the early work of Hirschkoff,Symko and Wheatley ( 8) on very dilute CuMn(<100ppm). One of the possible reasons for this situation is the use of SQUID magnetometers which often limit the range of use of magnetic fields to below a few hundred gauss. As explained by Tholence and Tournier (9 ) the spin glass temperature $T_g$ is best defined experimentally by the onset of irreversibility in the system: Whereas the magnetisation, or the apparent susceptibility M/H, does not depend on previous history or time for $T>T_g$, it appears that for $T<T_g$ the only procedure to obtain a time independent magnetisation state, which we take as the equilibrium state, is through field cooling from above $T_g$. As shown by Tholence and Tournier (9), Guy (10) and more recently Knitter and Kouvel (11) all other schemes by which the magnetic field is varied at $T<T_g$ invariably produce magnetic states decaying logarithmically apparently towards the field cooled state at infinite time. It further appears that this field cooled equilibrium magnetisation is practically temperature independent below $T_g$ down to zero temperature (12) when measured at sufficiently low fields. This last

property is in full agreement with the predictions of Parisi (13) and with the subsequent Parisi-Toulouse projection model ( 2). What becomes of this equilibrium magnetisation in a finite field is the main question here ( 14 ).

II - RESULTS AND DISCUSSION

Our measurements were done on a single crystal of AgMn 10.6% previously used for the X ray determination of the short range order (16). It was obtained after quenching in water from below the melting temperature and after spark cutting a cylinder 8 mm in height and 5 mm in diameter. The homogeneity of the Mn concentration was checked by an X ray microprobe analysis carried out on a separate piece.

The magnetisation was obtained with a vibrating sample magnetometer with an accuracy better than 1% and a sensitivity of about $10^{-4}$ emu. It was equipped with a He gas flow variable temperature cryostat. The He gas flow temperature was stabilised to better than 0.1°K with reference to a Platinum thermometer attached to the cryostat tail near the sample position. The sample temperature was continuously recorded with another Platinum thermometer located a few millimeters above the sample position in the sample holder. In this way very reproducible results could be obtained although no absolute calibration of our temperature was carried out. The main source of systematic error came from small contributions to either M or H which tend to blow up the ratio M/H as $H \to 0$.

a/ Equilibrium magnetisation. (field cooled)

Fig.1 displays the ensemble of our results for our fixed field, variable temperature measurements. In order to exhibit the Curie Weiss behaviour of the high temperature part ($T>T_g$) we have plotted the inverse of the apparent susceptibility, H/M, as a function of T, showing a Curie Weiss temperature of 13 ± 0.5°K for this alloy. The field dependence of this apparent susceptibility is remarkable : Indeed a very flat (<1%) "plateau" appears below $T_g$ for fields less than ∿ 1000 gauss and a small "cusp" like minimum (5% deep at most) is present right near $T_g$ for the lowest field used (26 gauss). At higher fields the transition between the Curie Weiss regime and the spin glass plateau is progressive and a smooth rounding is obtained. None of these magnetisation data are time dependent provided that they are obtained with field cooling through $T_g$. Once the lowest temperature was obtained (4.2°K) the magnetisation upon subsequent warming was observed to be identical to that measured during cooling (provided that the field was kept fixed at all times). This fact further stresses the equilibrium nature of the magnetisation. These observations allow to link the low field results obtained by various authors ( 9 )(10) (11) (12) to the high field ones, especially those of Hirschkoff, Symko and Wheatley (8 ) which have been recently reanalysed by Thompson and Thompson (17 ).

A qualitative fit of these results can be obtained for the region $T>T_g$ with the prediction of a simple Sherrington-Kirkpatrick (S.K.) set of equations including external field (18). This qualitative agreement is important as the S.K.

limit is that of the molecular field approximation. In particular the puzzling field sensitivity of the susceptibility near $T_g$ appears to be simply related to the field induced E.A. parameter q which is zero in zero field for $T>T_g$ but whose susceptibility becomes very large and diverges at $T_g$.

For $T<T_g$ it is tempting to define the occurence of the spin glass phase by the flatness of the temperature dependance of the magnetisation following the Parisi-Toulouse model (2). Unfortunately this criterion regarding the departure from flatness is difficult to apply as can be seen from Fig.(1). None the less we have tried an estimate of what a phase diagram might look like based on this criterion: The result is shown in the insert of Fig.1 but awaits confirmation by other separate techniques (11)(19).

b/ Non linear magnetisation.

The prediction for the field dependence of the magnetisation are (in reduced units (5) where $T = T/T_g$).

$$\frac{M}{H} = \frac{1}{T} - \frac{H^2}{3T^3}\left(\frac{T^2+2}{T^2-1}\right) \qquad T > 1 \qquad (1)$$

$$\frac{M}{H} = 1 - \frac{H}{\sqrt{2}} + \frac{23}{24}H^2 \qquad T = 1 \qquad (2)$$

$$\frac{M}{H} = 1 - \left(\frac{3}{4}\right)^{2/3} H^{4/3} + \frac{7}{6}H^2 \qquad T < 1 \qquad (3)$$

It is possible to quantitatively check these field dependances by measuring M(H) at fixed T for $T>T_g$ and using the field cooled results for $T<T_g$ (20).

Our measurement of the apparent susceptibility M/H for four temperatures slightly above $T_g$ are displayed in Fig.2 for the interval 10 gauss-1000 gauss. The magnetic field was produced by a watercooled set of iron free Helmholtz coil pair, allowing a very good determination of the field by the current intensity.

Qualitative agreement with the two first relations is quite-apparent: Indeed for $T \simeq T_g$ the apparent susceptibility decreases linearly with field thus indicating a non-analytic behaviour of the magnetisation (21)(i.e.it does not remain an odd function and the susceptibility an even function of H). At a temperature slightly above $T_g$(38.9°K compared to 37.4°K) it is possible to fit an $H^2$ term in the field dependence of the susceptibility(within 0-200 gauss) although with poor accuracy. The ensemble of these results (and others not reported here) are fitted to:

$$\frac{M}{H} = \chi_o(T)\ (1-b(T)\ H^2) \qquad (4)$$

The parameter b is directly proportional to $q/H^2$ where q is the E.A.parameter (3) in presence of a field. In the insert of Fig 2 we present the variation of b versus the reduced temperature $(T-T_g)/T_g$. With our admittedly crude determination of the non linear magnetisation terms it appears that b(T) varies like $(T-T_g)^{-n}$ with $1<n<2$. For $T<T_g$ the low temperature end points of the magnetisation "plateau" of Fig 1 when plotted in a similar way as Fig.2 (where they have been omitted for sake of clarity) are consistent with the third relation although the quality of the data cannot allow a direct check of the field exponent.

In summary, the ensemble of these observations tend to give very strong support to the validity of the molecular field approach at and above $T_g$ and indicate a way to check the low temperature properties of the spin glass phase. This overall qualitative agreement advocates the existence of a phase transition at $T_g$.

We thank J.Kouvel,M.Gabay, T.Garel, J.Vannimenus and G.Toulouse for many illuminating discussions and suggestions concerning this work.

REFERENCES

(1) J.R.De ALMEIDA and D.J.THOULESS, J.Phys.A 11 983 (1978).

(2) G.PARISI and G.TOULOUSE, J.Phys.Lettres 41 361 (1980).

(3) D.SHERRINGTON and S.KIRKPATRICK, Phys.Rev.Lett.32 1792 (1975).

(4) K.WADA and H.TAKAYAMA, Prog.Theor.Phys.64 327 (1980).
M.SUZUKI, Prog.Theor.Phys.58 1151 (1977).

(5) J.VANNIMENUS, G.TOULOUSE and G.PARISI, J.Phys.42 565 (1981).

(6) G.TOULOUSE and M.GABAY, J.Phys.Lett.42 103 (1981).

(7) S.F.EDWARDS and P.W.ANDERSON, J.Phys.F 5 965 (1975).

(8) E.C.HIRSCHKOFF, O.G.SYMKO and J.C.WHEATLEY, J.Low Temp. Phys 5 155 (1975).

(9) J.L.THOLENCE and R.TOURNIER, J.Phys.Colloque 35 C4-229 (1974).

(10) C.N.GUY, J.Phys.F 7 1505 (1977).

(11) R.W.KNITTER and J.S.KOUVEL, J.Mag.Mag.Mat.21 L 316 (1980).

(12) S.NAGATA, P.H.KEESAM and H.R.HARRISON, Phys.Rev.B 19 1633 (1979).

(13) G.PARISI, Phil.Mag.B 41 677 (1980) and references in (2).

(14) The connection between our results and the well known ac susceptibility cusp measurements(15) is easy to state if not easy to use: In a typical ac susceptibility experiment what is detected is that part of the magnetisation able to respond to a small ac field(in presence of a fixed static field generally parallel to the ac field). Above $T_g$ a complete response, i.e.Curie law like, is found as expected whereas below $T_g$ only a fraction of the magnetisation rapidly decreasing with T follows the driving field thus producing a sharp cusp a $\chi_{ac}$ at $T_g$. It is noteworthy that the high temperature part of the susceptibility cusp should be identical with static magnetisation measurements (in low fields) whereas the low temperature part of it depend on off equilibrium metastable properties of the spin glass system, hence cannot be compared with the field cooled equilibrium magnetisation.

(15) V.CANNELLA and J.A.MYDOSH, Phys.Rev.B6 4220 (1972).

(16) H.BOUCHIAT, E.DARTYGE, P.MONOD and M.LAMBERT, Phys.Rev.B23 1375 (1981).

(17) J.O.THOMSON and J.R.THOMPSON, J.Phys.F 11,247 (1981).

(18) J.VANNIMENUS,private communication.

(19) H.BOUCHIAT and P.MONOD, to be published.

(20) The presence of a Curie Weiss temperature indicating a positive $J_o$ will alter only the first term of these expressions.See ref.4 for the complete expressions.

(21) A similar behaviour for a CuMn 0.57% alloy is reported by C.A.M.MULDER, A.J. VAN DUYNEVELDT and J.A.MYDOSH, Phys.Rev.B23 1384 (1981).

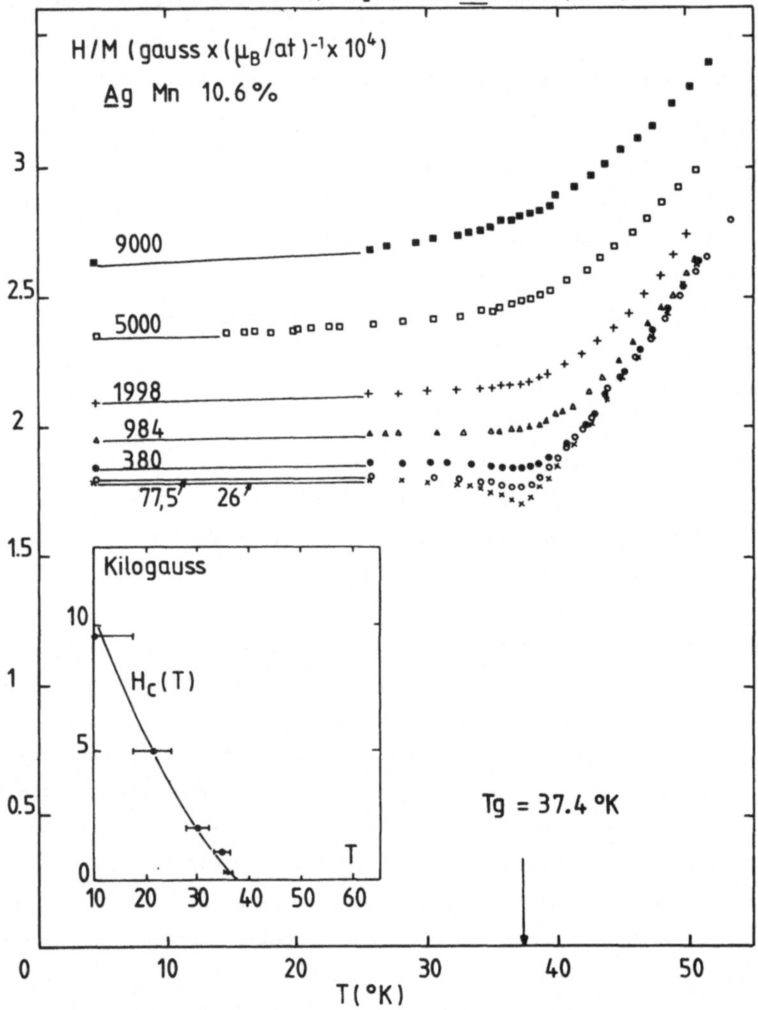


Figure 1: Inverse of the apparent susceptibility H/M for AgMn 10.6% as a function of magnetic field as indicated on each set of point symbols (in gauss). These data points have been obtained at fixed field and variable temperature i.e. by field cooling through $T_g$. As explained in the text these represent an equilibrium state of the spin glass phase. The insert shows an attempt to determine a phase diagram in the (H,T) plane. We define (arbitrarily) the boundary of the spin glass phase by the point of the M(T) curve departing by 3% from the low temperature value.

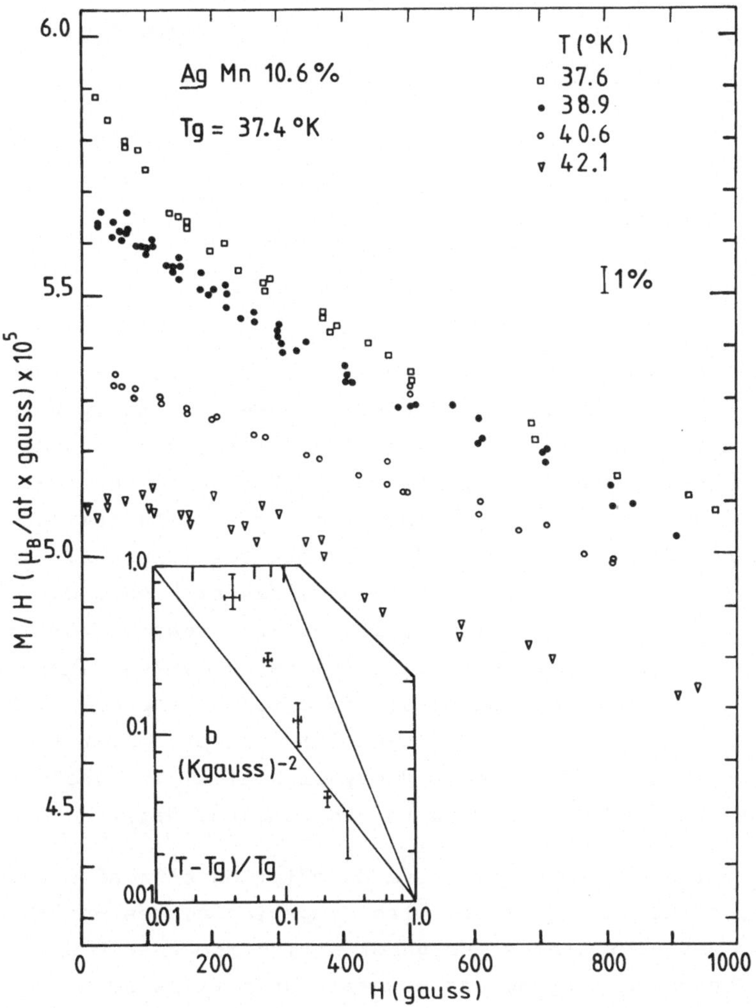


Figure 2: The apparent susceptibility M/H is plotted versus H for four temperatures above $T_g$ for AgMn 10.6% alloy on an enlarged scale. Note the linear behaviour of M/H versus H for T = 37.6°K and the subsequent rounding of the initial slope at higher temperature. By fitting these points to expression (4) we define a coefficient b for the non linear magnetisation which tend to diverge as $T \to T_g$. The insert shows a log-log plot of b versus reduced temperature. The solid lines represent a slope of 1 and 2 respectively.

# SPIN GLASS MODELS WITH SHORT-RANGE INTERACTIONS: A SHORT REVIEW OF NUMERICAL STUDIES

K. Binder and W. Kinzel

Institut für Festkörperforschung,
Kernforschungsanlage Jülich, Postfach 1913
D-5170 Jülich, West-Germany

Abstract:

Edwards-Anderson models with nearest-neighbor bonds randomly chosen by a ($\pm J$,0) or gaussian distribution are studied by various methods: Monte Carlo, transfermatrix calculations, and systematic series expansions. Evidence is presented that *thermal equilibrium phase transitions* into spin glass phases occur at *zero temperature only* at physical dimensionalities d=2,3, both for Ising, XY- and Heisenberg spin glasses. On the other hand, *dynamic* Monte Carlo studies reveal freeze-in transitions at nonzero freezing temperatures, which must be interpreted as a *nonequilibrium phenomenon* where the system remains in a "valley" of phase space for a time comparable to observation times. Specific heat, (dynamic) susceptibilities, various remanent magnetizations (and their decay with time) are shown to be in reasonable *qualitative* accord with experimental data.

As a step towards a quantitative description of real systems such as $Eu_xSr_{1-x}S$, dilution of a model with competing nearest-neighbor ferromagnetic and next nearest-neighbor antiferromagnetic exchange is studied. The phase diagram and magnetic properties such as the distribution of magnetic correlations can be understood by considering local spin configurations of magnetic atoms. *Quantitative* agreement with experimental data is obtained.

## I. Introduction

Spin glasses are magnets without conventional long-range ferro- or antiferromagnetic order, which exhibit a transition into a state where the spins are more or less frozen-in, and hence the susceptibility has a rather sharp cusp /1/. The nature of this state and of the freezing transition are heavily debated; there is no agreement about the appropriate theoretical model description /2/. This survey treats short-range interaction models containing both disorder and "frustration" /3/; we show that these models are in reasonable accord with experiment. Unfortunately, sufficient analytic methods for describing their static and dynamic properties are not available, and hence numerical methods are needed. Among the variety of interesting results /4/ this survey emphasizes recent work of the "Jülich group" /5-14/. We first discuss the ± J model (Sec. II) and the gaussian model (Sec. III), for Ising spins and various lattice dimensionalities. Sec. IV discusses XY and Heisenberg Edwards Anderson /15/ models, while Sec. V is devoted to more realistic site-disorder problems. Finally Sec. VI summarizes the conclusions.

## II. The Nearest-Neighbor ± J Edwards-Anderson Model

As a simple model of a spin glass containing both disorder and frustration, we consider the Ising square [or (hyper-) cubic] lattice [dimensionalities d from d=2 to d=8 /9/],

$$\mathcal{H} = - \sum_{<i,j>} J_{ij} S_i S_j , \qquad S_i = \pm 1 , \tag{1}$$

with bonds $J_{ij}$ randomly drawn from the distribution

$$P(J_{ij}) = p_1 \delta(J_{ij} - J) + p_2 \delta(J_{ij} + J) + (1 - p_1 - p_2)\delta(J_{ij}) . \tag{2}$$

For $p_2=0$, this problem reduces to diluting a noncompeting ferromagnet, while for $p \equiv p_1+p_2=1$ the usual ± J spin glass /3,16,17/ results.

The questions we are asking about this model are the following: (i) Is there a phase transition from a paramagnetic state (P), where all magnetic correlations are short-ranged, to a spin-glass (SG) state? These ferromagnetic correlations $\{g_F(\vec{R}) = [\langle S_i S_j \rangle_T]_{av}, \ \vec{R} \equiv \vec{r}_i - \vec{r}_j\}$ as well as antiferromagnetic ones $\{g_{AF}(\vec{R}) = \exp[i\vec{Q}\cdot\vec{R}][\langle S_i S_j \rangle_T]_{av},$

$\vec{Q}$ = reciprocal lattice vector of antiferromagnetic order} are still of short range. But long-range correlation would occur for the correlation-function

$$g_{EA}(\vec{R}) = [\langle S_i S_j\rangle_T^2]_{av} , \tag{3}$$

where (thermally averaged) spin correlations are squared, before the configurational average {with Eq. (2)} over the quenched disorder is taken $\{[\cdots]_{av}\}$.

(ii) If the answer to (i) is "yes", does this imply that there is an Edwards-Anderson [15] order parameter $q_{EA} \equiv [\langle S_i\rangle_T^2]_{av} \neq 0$? [Note $\lim_{\vec{R}\to\infty} g_{EA}(\vec{R}) = q_{EA}^2$].

(iii) Let us associate dynamics to the model by a Markovian master equation for the probability $P(\vec{X},t)$, $\vec{X} \equiv (S_1, S_1, \ldots, S_N)$ ,

$$\frac{d}{dt} P(\vec{X},t) = - \sum_{\vec{X}'} W(\vec{X}\to\vec{X}')P(\vec{X},t) + \sum_{\vec{X}'} W(\vec{X}'\to\vec{X})P(\vec{X}',t) , \tag{4}$$

where the transition probability $W(\vec{X}\to\vec{X}') = (1/2\tau)\{1-\tanh([\mathcal{H}(\vec{X}') - \mathcal{H}(\vec{X})]/2k_BT\}$ satisfies detailed balance. $\tau$ is a parameter setting the overall time-scale; Eq. (4) can physically be justified in terms of a coarse-graining, where some degrees of freedom are averaged out and act as a heat bath on the remaining ones /2/. The question then is, to what extent the model Eqs. (1,2,4) resembles static and dynamic properties of real spin glasses.

In the ground state (T=0), the answer to the first question is yes (Fig. 1a), at all $d \geq 2$. The phase boundary P-SG has been estimated /9/ from a systematic expansion of $\chi_{EA} \equiv \sum_{\vec{R}} g_{EA}(\vec{R})$ in powers of p, and looking for a divergence of $\chi_{SG}(p)$ by ratio and Padé methods. For the k-th term of such a series, one has to calculate $g_{EA}(\vec{R})$ for all "clusters" containing precisely k bonds (Fig. 1b), and weigh each configuration of bonds with its probability. For clusters which do not contain loops, $g_{EA}(\vec{R}) \equiv 1$ at T=0 trivially, while for other clusters loops may be frustrated, and then the ground state is more than 2-fold degenerate; $g_{EA}(\vec{R}) < 1$ results from averaging over the degenerate ground states. In practice, this calculation is greatly simplified by utilizing the invariance of Eq. (3) against gauge transformations /3/, $S_i$, {all bonds $J_{ij}$ at site i} $\to -S_i, \{-J_{ij}\}$.

At nonzero temperatures the behaviour of this model has been controversial for some time: High temperature series expansions of $\chi_{EA}$ for $p_1 = p_2 = \frac{1}{2}$ indicated /18/ that $\chi_{EA}$ diverges at $T_f > 0$ for $d \geq 4$ only. Unfortunately, these series are rather irregular. On the

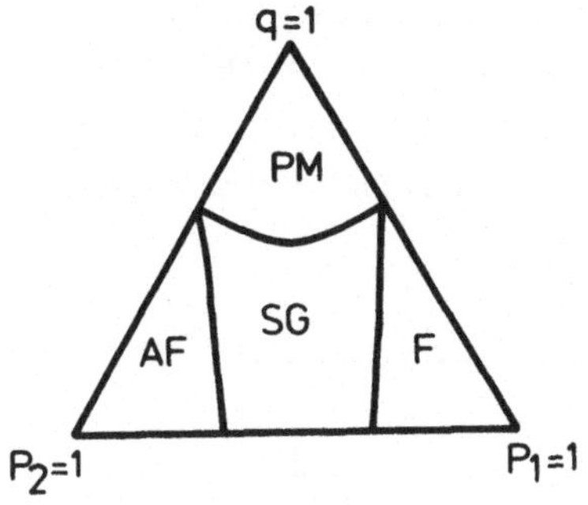


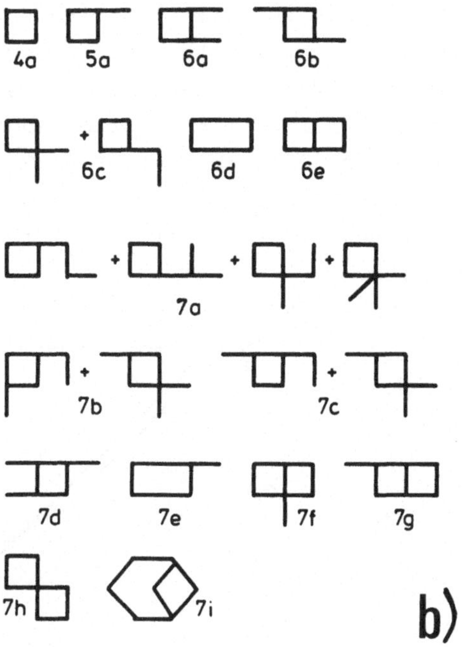


Fig. 1: a) Phase diagrams at T=0 for the diluted $\pm$ J model at the square and (hyper-) cubic lattices /9/. b) Types of clusters to be considered in the concentration expansion, including $7^{th}$ order terms /9/.

other hand, renormalization group studies showed a phase transition just below the mean field dimensionality d*=6 (expansions in $\varepsilon=6-d$ /19/). Extrapolations from d=6 to d=3 or 2 are doubtful, of course. Real space renormalization in d=2 or 3 gave ambiguous results depending on the method applied /20/. Clearly this method involves uncontrolled approximations, but it has successfully described many phase transitions in various models, particularly for d=2. Monte Carlo results suggested a nonzero $T_f$, $k_BT_f/J \simeq 1.3$ /4/. However, one should keep in mind that Monte Carlo averaging means time averaging. For instance the Edwards-Anderson-parameter is defined by /21/

$$q(t) = (1/N) \sum_{i=1}^{N} \left( \int_{o}^{t} S_i(t')dt'/t \right)^2 \qquad (5)$$

which {for symmetric bond distributions $P(J_{ij})=P(-J_{ij})$} is related to the susceptibility

$$\chi(t) = \frac{N}{k_BT} \left\{ \int_{o}^{t} \left[ \sum_j S_j(t')/N \right]^2 dt'/t - \left[ \int_{o}^{t} \left( \sum_j S_j(t')/N \right) dt'/t \right]^2 \right\} = (1-q(t)\ /k_BT) \qquad (6)$$

Thus both $q(t)$ and $\chi(t)$ are time dependent! It turns out that $q(t)$ decays very slowly with time giving by Eq. (6), a peak in $\chi(t_{obs},T)$ at some "$T_f$" for any reasonable observation time $t_{obs}$. However, a definite answer to the question whether $q(t\to\infty)$ is nonzero or not was not possible /22/. In fact, Fig. 2 anticipates results of the symmetric gaussian spin glass /7/: $q(t)$ slowly decays with time for both d=3 and d=5, although in the latter case the series /18/ do indicate a phase transition.

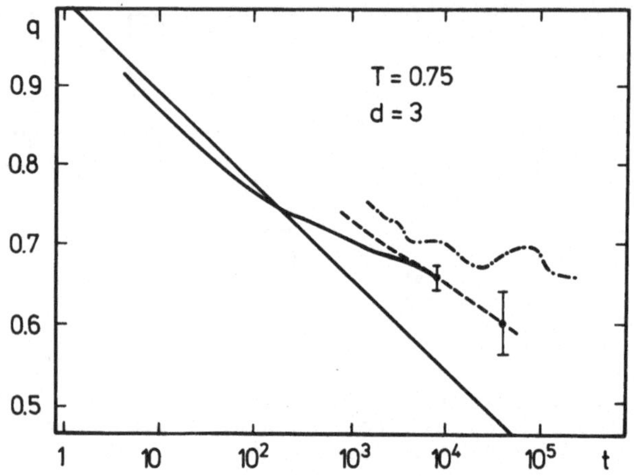


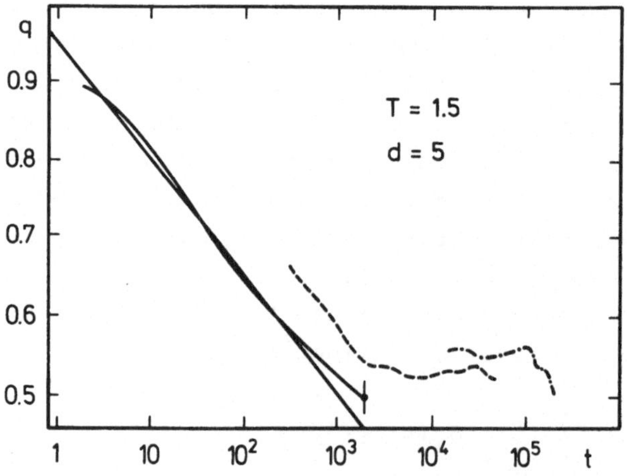


Fig. 2: Decay of $q(t)$ for the symmetric gaussian Edwards-Anderson spin glass at d=3 and d=5 /7/. Dash-dotted curve is one individual run of a lattice with periodic boundary conditions. Broken curve is an average over 13 runs, and full curve an average over 6 runs of $20^3$ ($6^5$) lattices. Full straight line shows the prediction resulting from the approximation /23/ $dq(t)/d\ell nt=-k_BT\ P_o(0)$, where $P_T(H_{eff})$ is the distribution of effective fields.

This situation has been greatly clarified by applying a recursive exact calculation of the partition function [as well as of $g_{EA}(\vec{R})$] /5,6/. There for finite lattices (up to 18x18 (d=2) or 4x4x10 (d=3), with periodic boundary conditions in the small direction(s) and free boundaries in one direction of the system) the thermal averaging is done exactly for any chosen realization of the random $\{J_{ij}\}$: only the averaging $[\cdots]_{av}$ over realizations is done numerically, which is harmless, however, since different realizations $\{J_{ij}\}$ are statistically independent, and hence $[\cdots]_{av}$ can be performed with controlled error, following standard statistical analysis. Thus the free energy (and its derivatives) were obtained, as well as $g_{EA}(\vec{R})$ and a further susceptibility $\chi_{SG} \equiv (1/k_BT) \sum_{\vec{R}} [\langle S_iS_j\rangle_o \langle S_iS_j\rangle_T]_{av}$, which measures the alignment of the spins with their ground states: if the spin orientations in the $\ell$'th of the L ground-states are $\{\phi_i^{(\ell)}\}$, we have $\chi_{SG} = (1/k_BT) \sum_{\ell} \sum_{\vec{R}} \phi_i^{(\ell)} \phi_j^{(\ell)} \langle S_iS_j\rangle_T / L$. The associate order parameter $\psi^2$ /24/ can then be estimated from the exact calculation on a finite system via $\psi^2 = k_BT\psi_{SG}/N$ (if it exists!). At T=0 we have $\psi^2 = q_{EA}^2$, of course.

Fig. 3 suggests, however, that for d=2 this order parameter does not exist: while for a Mattis /25/ spin glass - which is related via gauge transformations to the Ising ferromagnet, $\psi^2 = M^2$ where the magnetization M as well as $T_f = T_c \cong 2.27 J/k_B$ are known exactly /26/ for finite lattice calculations very clearly indicate a transition from disorder ($T > T_c$) to order ($T < T_c$), and even $T_c$ can be estimated from the inflection point of these curves to within a few percent accuracy. For the $\pm$ J spin glass, on the other hand, the decrease of $\psi^2$ with increasing N indicates that there is no order, $\chi_{SG}$ is finite at all T>0. This lack of order is also

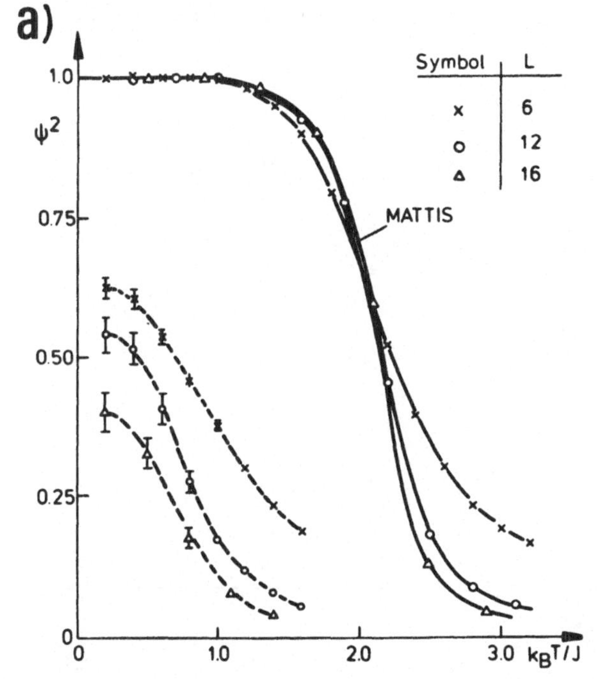


Fig. 3a: Order parameter $\psi^2$ of the Edwards-Anderson $\pm$ J model (broken curves) as compared to the Mattis model (full curves).

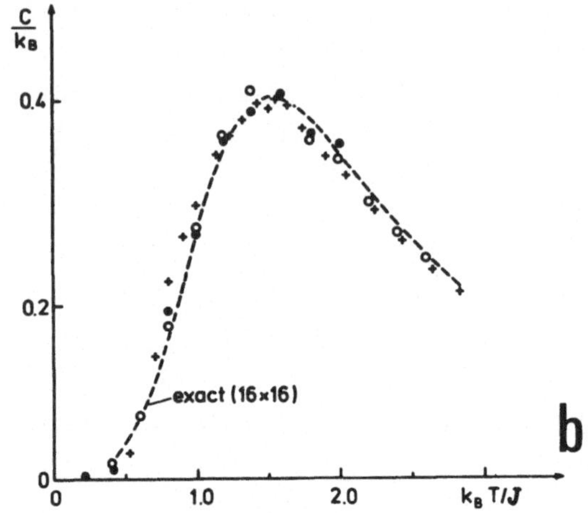


seen from $g_{EA}(\vec{R})$ decaying to zero for large $\vec{R}$, at temperatures distinctly below the "$T_f$" of the dynamic Monte Carlo studies (Fig. 4). *Thus $q_{EA}\equiv 0$ for this model at all temperatures, and there is no other nonzero order parameter whatsoever* (which would imply $q_{EA}>0$ as a "secondary" order parameter). At $T \gtrsim \frac{1}{3} T_f$ it is clear that the decay is exponential with dis-

Fig. 3b: Specific heat plotted vs. T for the $\pm$ J model /5/, for N=16x16. Monte Carlo results for runs starting either with a random spin configuration (full circles, and crosses [for N=80x80 /16/]) or with a ground state configuration (open circles) are included.

tance, $g_{EA}(R) \propto \exp(-R/\xi_{EA})$, as indicated by straight lines on the semi-log plot. The slope yields the correlation length $\xi_{EA}$. In the cases shown it is 5 (or 7) lattice spacings - thus our lattices are safely

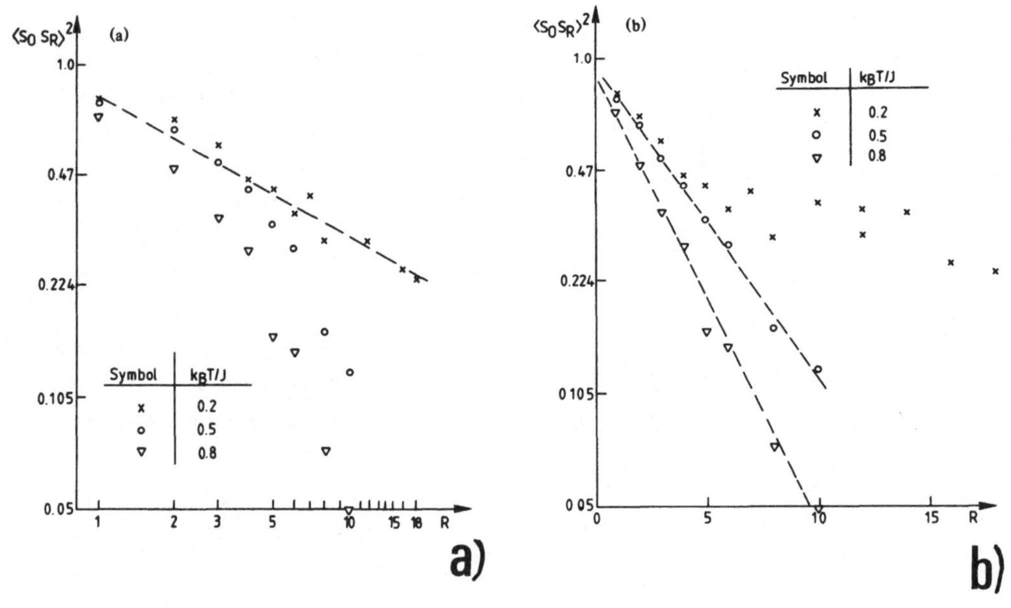


Fig. 4: a) Averaged squared correlation of the $\pm$ J model plotted vs. distance. b) Same data replotted in semi-log form /5/.

larger than $\xi_{EA}$. At very low T, this is no longer true (e.g. at T=0.2 $J/k_B$), there the data indicate a power-law decay, $g_{EA}(\vec{R}) \propto R^{-p}$, $p \cong 0.4 \pm 0.1$. We think that the latter behavior occurs for $R \to \infty$ only right at T=0, but for finite R we see it already for a range of low temperatures ("crossover").

Hence the answer to question (i) is yes - a transition occurs at T=0 to a state with power-law decay of $g_{EA}(\vec{R})$, and to question (ii) it is no, there is no nonzero order parameter. But at the same time dynamic simulations do indicate order in both q(t) and $\psi^2$ *for the same lattices* for which the exact calculation show there isn't any order, Fig. 5. Thus the freezing of spins in this model at $k_BT_f/J \approx 1.3$ is not due to a phase transition where $q_{EA}$ starts to be nonzero, but rather a dynamic nonequilibrium phenomenon.

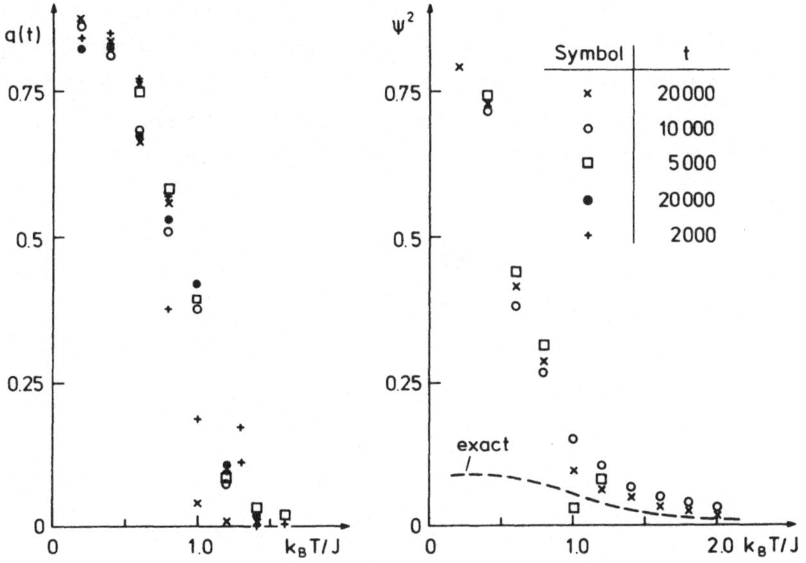


Fig. 5: q(t) [left part] and $\psi^2(t)$ [right part)] plotted vs. T for various t and lattice sizes (N=16x16 /5/ and 80x80 /16/). Broken curve is the result of the exact calculation for the same $\{J_{ij}\}$ for N=16x16 /5/.

The same conclusion emerges for d=3, where again $g_{EA}(\vec{R})$ is found to decay towards zero at temperatures distinctly below the "$T_f$" of the dynamic simulations /6/. In this case, however, it is still an open question whether the ground state has (imperfect) order /6/.

The specific heat C of this model (Fig. 3b) shows a broad Schottky-like peak, in qualitative agreement with experiment /1/. However, the nearly linear behavior $C \propto T$ of the data at low temperatures /1/ is not reproduced, of course, as the model allows for a discrete energy spectrum of excited states only. The dynamic behavior of this model /16/ turns out to be less close to experiment than that of the gaussian Edwards-Anderson model.

## III. The Nearest-Neighbor Gaussian Edwards-Anderson Model (Ising Spins)

We still consider Ising systems, Eq. (1), but use instead of Eq. (2)

$$P(J_{ij}) \propto \exp[-J_{ij}^2/2(\Delta J)^2]. \qquad (7)$$

Again the Monte Carlo work /21,24/ first was interpreted with the Edwards-Anderson transition /15/, while later doubts were raised /17/ but a clear-cut answer did not emerge /7,22/. The exact partition function calculations could clarify the situation: Again $g_{EA}(\vec{R})$ is found to *decay towards zero* exponentially with distance at T distinctly below the "$T_f$" of the dynamic simulations, Fig. 6a [$k_BT_f/\Delta J \approx 1.0$ for d=2, $k_BT_f/\Delta J \approx 1.5$ at d=3]. There is one difference, though, to the $\pm$ J model: now the ground state is only 2-fold generate, and hence

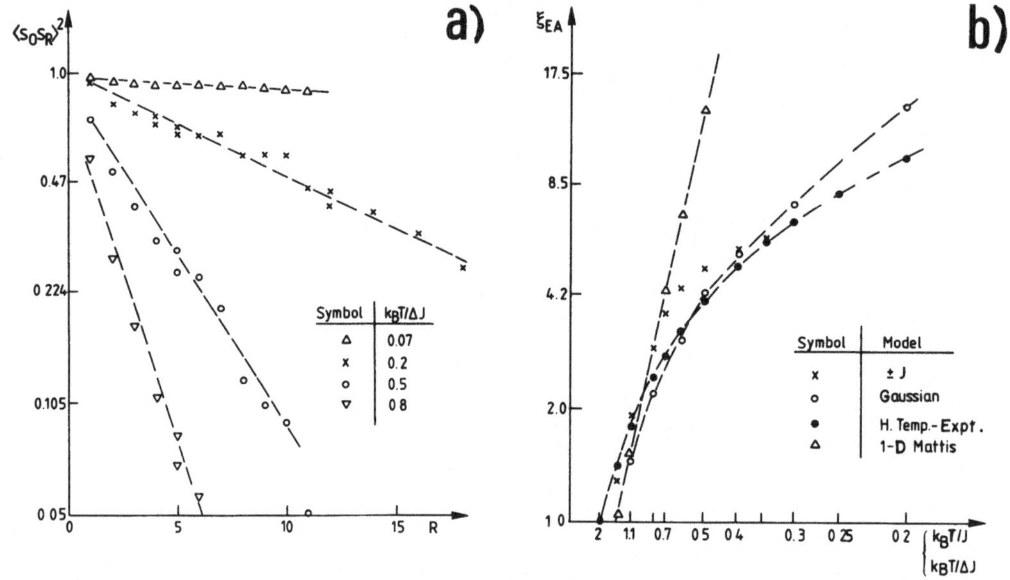


Fig. 6: a) Average squared correlation of the d=2 gaussian model plotted vs. distance. b) $\xi_{EA}$ plotted vs. 1/T for both the $\pm$ J model and the gaussian models; triangles show for comparison $\xi_{EA}$ for the d=1 Mattis model, full circles the first-term of the high temperature series expansion /5/.

$q_{EA}(T=0)=\psi(T=0)=1$. But the temperature dependence of $\xi_{EA}$ is very similar in both models (Fig. 6b). At "$T_f$", the equilibrium value of $\xi_{EA}$ is seen to be about 2 lattice spacings only - hence one cannot invoke any "critical slowing down" to explain why q(t) becomes so long-lived near $T_f$. Remember, the Edwards-Anderson /15/ phase transition hypothesis of freezing would imply that $\xi_{EA}$ diverges towards infinity at $T_f$ - this mechanism of freezing clearly is ruled out!

At this point we ask what other mechanism applies. We suggest that for $T \ll T_c$ the energy hypersurface in phase space is described by Fig. 7a: there are many equivalent "valleys", each corresponds to one of the ground-states (or low-lying excited states) $\{\phi_j^{(\ell)}\}$ introduced above. The irregular small wiggles of this energy hypersurface correspond to over-turning isolated small clusters of spins, Fig. 7b. At low T, these clusters can be interpreted as independent "two-level-systems" /27/, Fig. 8a, and a lot of experimental data can be qualitatively accounted for by thermally activated motions over the barrier (V) with

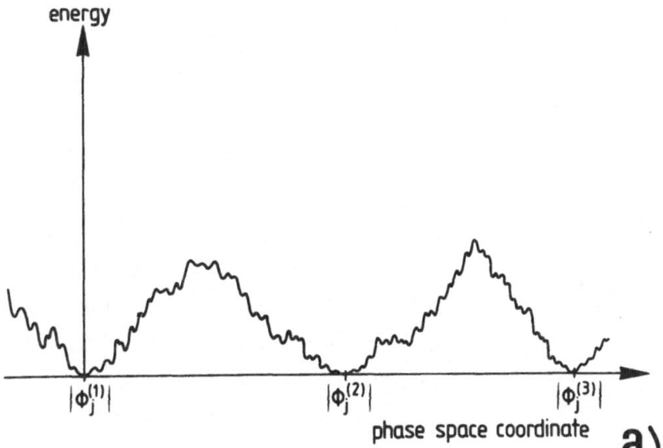


a)

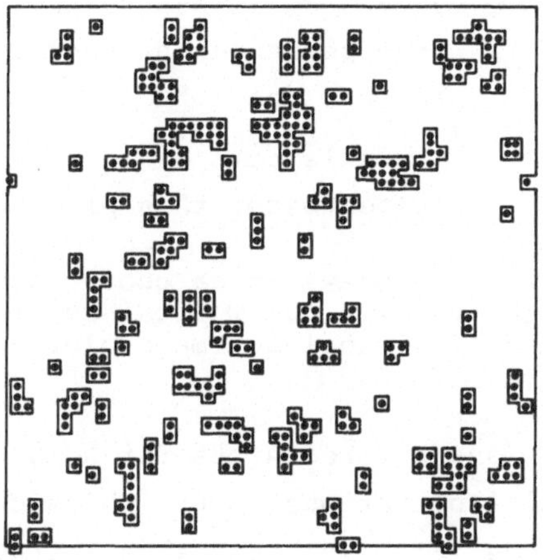

b)

Fig. 7: a) Schematic plot of energy vs. a phase space coordinate. b) Snapshot picture of a 55x55 gaussian model heated up from a "groundstate" to $k_BT/\Delta J = 0.5$. Spins which have no longer their groundstate orientation are shown as black dots. Contours indicate definition of "clusters" /21/.

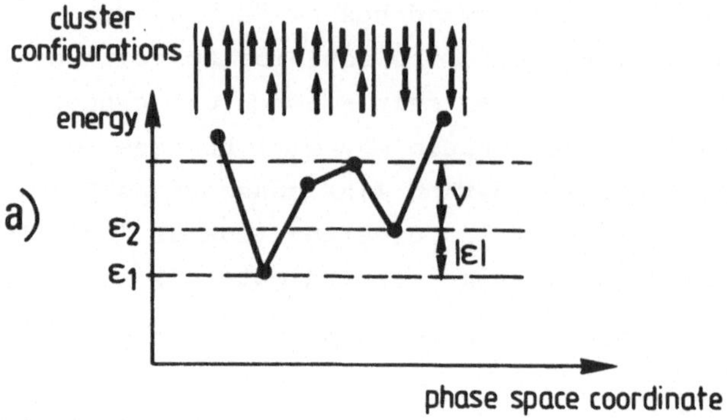


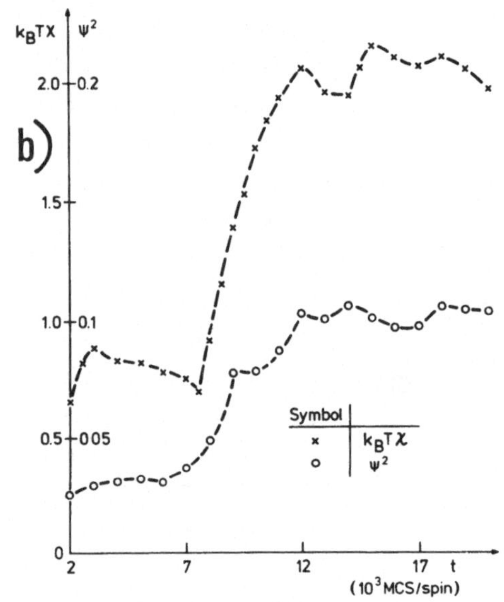


reasonable assumptions about the distribution of excitation energies $\varepsilon=\varepsilon_2-\varepsilon_1$ and barrier heights /27/. Unfortunately, much less is known about the barriers separating the various *global* minima $\{\phi_j^{(\ell)}\}$ from each other. We suggest that their heights are much larger than typical energies $(J,\Delta J)$, but stay finite in the thermodynamic limit. Thus the system stays very long in one valley, before it experiences a transition to another one.

There is some Monte Carlo evidence for this picture, Fig.

Fig. 8: a) Schematic plot of energy vs. a phase space coordinate representing various states of a "cluster" /27/. b) Typical run for a 16x16 lattice ($\pm$ J model) at $k_BT/J = 1.0$, showing time-evolution of $\chi(x)$ and $|\psi^{(1)}|^2$ /5/.

8b: often the system stays close to one ordered state for a large time ($\sim 10^4$ MCS/spin), only rather small fluctuations occur, which correspond to excitations of *small* clusters (Fig. 7b). But then a transition to another ordered state occurs, by rearranging a *large* cluster, involving to pass a fairly high saddle point in phase space. Clearly, by sampling only one (or a few) "valleys" one overestimates the order, and thus the discrepancy between dynamic simulations and equilibrium calculations (Fig. 5) arises.

The same behavior also applies for d=3: again $g_{EA}(\vec{R})_{\vec{R}\to\infty}0$ at $T<"T_f"$, Fig. 9a /6/. Although estimating $\xi_{EA}$ for d=3 is less accurate, due to the finite size of the lattices, it is worth mentioning that such small lattices do show a freezing transition in dynamic simulations, Fig. 9b, and there is surprisingly little dependence on system size.

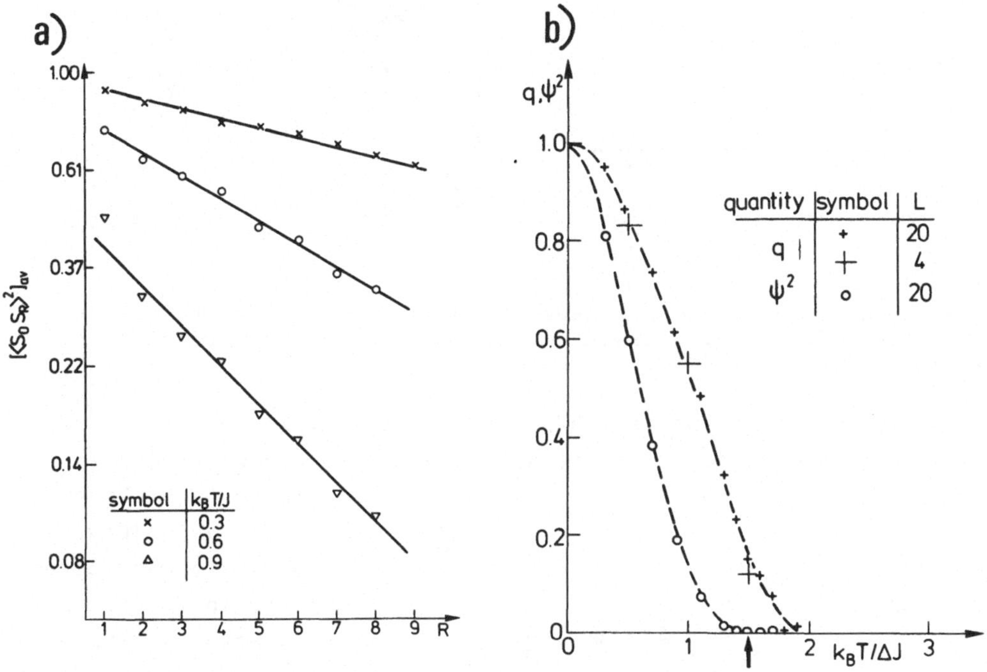


Fig. 9: a) Averaged squared correlation of the d=3 gaussian model plotted vs. distance. b) Monte Carlo estimates for order parameters of the same model, for $t_{obs}$ = 2000 MCS/spin /6/.

The gaussian model agrees surprisingly well with experiment. E.g., entropy (Fig. 10a) and specific heat (Fig. 10b) vary nearly linearly with T at low temperatures. Also the dynamic behavior of the model /11,24/ resembles experiment, particularly if the simulation imitates experimental procedures /11/: e.g., one may cool the system in an applied field B at constant cooling rate $\Delta T/\Delta t$, then switch the field off and observe the resulting "thermoremanent magnetization" (TRM) and its decay with time. This quantity differs in a characteristic way

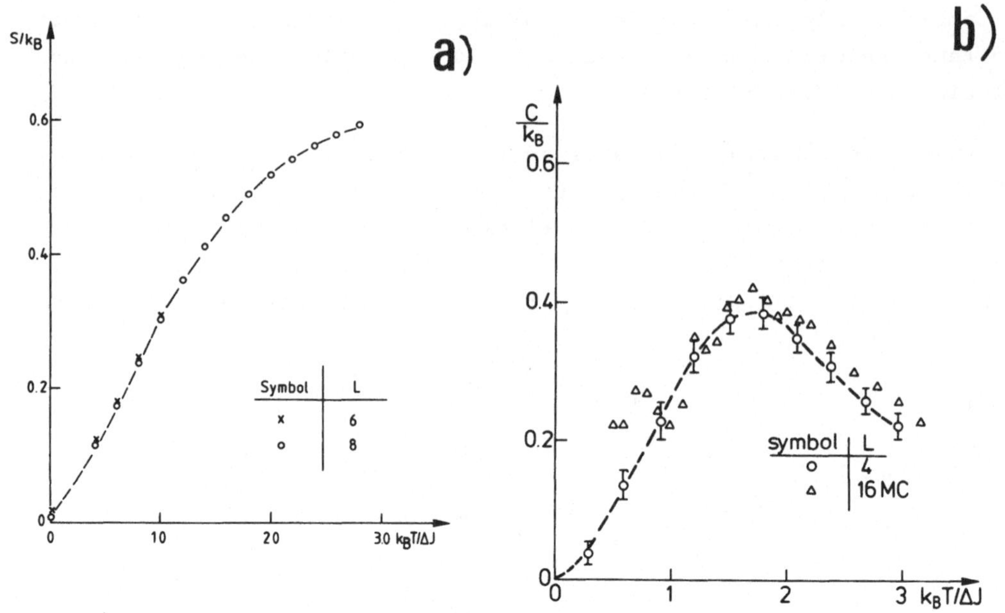


Fig. 10: a) Entropy of the d=2 gaussian model plotted vs. T /5/. b) Specific heat of the d=3 gaussian model, "exact" calculation for L=4 compared to Monte-Carlo data for L=16 /6/.

from the isothermal remanence (IRM), where a sample is cooledin zero field and then the response to the same field is measured (Fig. 11).

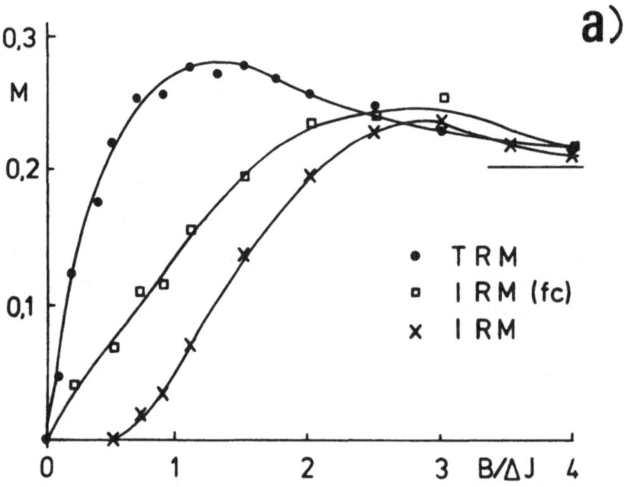


Both field- and temperature-dependence (Fig. 12a) of these magnetizations are in qualitative account with experiment, as well as their power-law decay with time (Fig. 12b), $M \propto t^{-a(T,B)}$. The apparent exponent a depends not only on the temperature and the initial value of the remanence, but also on the "magnetic history" of the sample!

Fig. 11: a) Remanent magnetization as a function of applied field for a 50x50 gaussian model /11/ IRM (fc) is obtained by some mixed cooling procedure (see /11/).

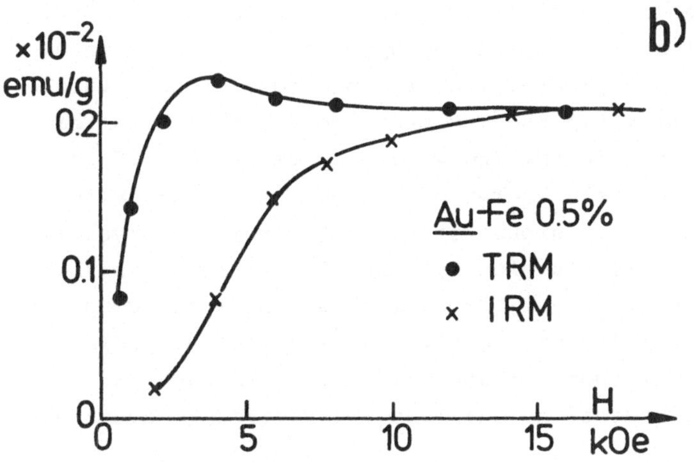


Fig. 11: b) Remanent magnetizations of $AuFe_{0.5\%}$ /28/.

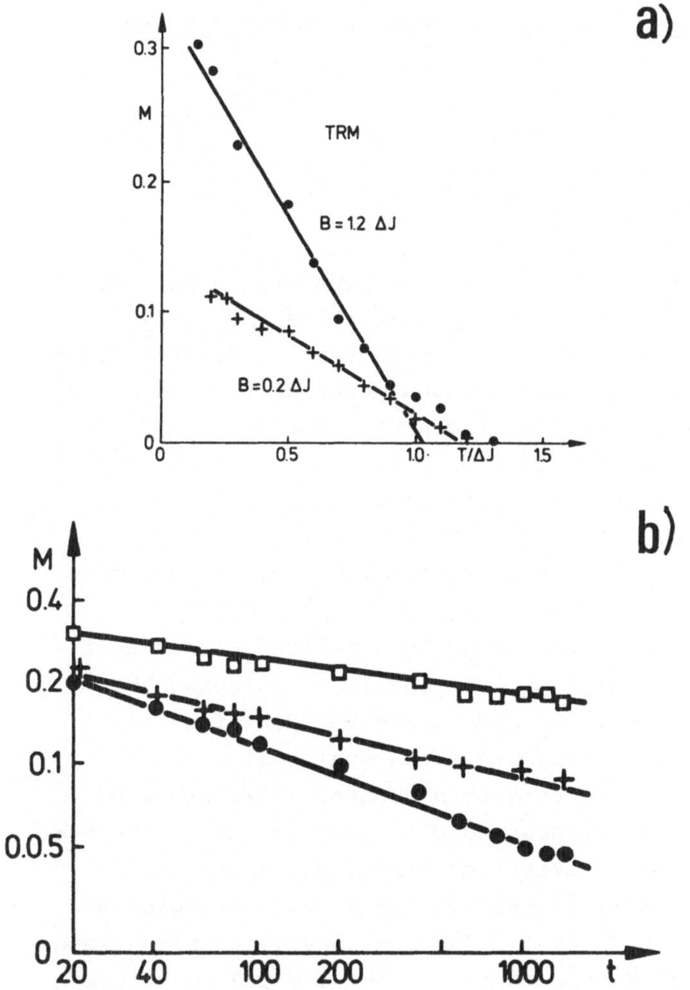


Fig. 12: a) TRM of a 50x50 gaussian model plotted vs. T for two values of the initially applied field B /11/. b) Log-log plot of remanent magnetization vs. time. The squares (TRM, $B=\Delta J$) and dots (IRM, $B=1.5\Delta J$) have the same initial energy, the dots and crosses (TRM, $B=\infty$) have the same initial magnetization ($T=0.5\Delta J$).

The picture of free energy valleys in phase space is again useful to understand the remanent magnetization. By applying a field, a local minimum becomes a global one separated from the previous ground state valleys by an energy barrier V. In the TRM case, coming from high temperatures, the system can go into the lowest valley regardless of the value of V. For the IRM, the field energy has to be larger than V in order to move the system from the ground state to the remanence valley. Since the relaxation time increases with V, this picture tells us that first, the TRM decays slower than the IRM ($a(TRM)<a(IRM)$ and second the IRM decays faster for smaller previously applied field $\partial a(T,B)/\partial B<0$ and the opposite for TRM. Both effects have been observed in computer experiments /11/, however no detailed information is available from real experiments.

## IV. The Nearest-Neighbor Gaussian Edwards-Anderson Model (XY and Heisenberg Spins)

Here we consider the classical n-vector model,

$$\mathcal{H} = - \sum_{<ij>} J_{ij} \sum_{m=1}^{n} S_i^m S_j^m \, , \quad \sum_{m=1}^{n} (S_i^m)^2 = 1 \, , \tag{8}$$

again with the distribution Eq. (7). It is straightforward to generalize the considered quantities to this case [$q_{EA}=[<\vec{S}_i>_T \cdot <\vec{S}_i>_T]_{av}$, $\psi_{\alpha\beta}^{(\ell)} = [<S_i \phi_i^{\beta(\ell)}>_T]_{av}$, $\psi^{(\ell)} = \sum_\alpha \psi_{\alpha\alpha}^{(\ell)}$ etc.]. Of course, exact partition function calculations are no longer possible, and hence only Monte Carlo work is discussed. The results /8,21,29/ can be summarized as follows:
(i) For relatively short times $t_{obs}$ = 2000 MCS/spin, one finds a freezing transition /8,21/ with order parameters $q(t_{obs}),\psi$ having a temperature dependence similar to the Ising case (Fig. 13a).
(ii) For large times, the order parameters decay with time according to a logarithmic law, similar to the Ising case (Fig. 13b) /8/.
(iii) "Two-level systems" which correspond to localized clusters of spins which have multiple equilibrium configurations (keeping the spins surrounding the clusters fixed) can again be identified /29/. These clusters are of central importance for a qualitative understanding of the dynamics of Heisenberg spin glasses /30/, since no longer unexplained anisotropy energies need be invoked, as in the Néel model /28/. [We feel that such clusters or "two-level systems", resulting from continuous distributions of interaction strength, are a natural explanation for the decay $dq/d(\ell nt) \approx$ const. in unfrustrated systems, such as generalized Mattis models in d=1 and d=2 where the

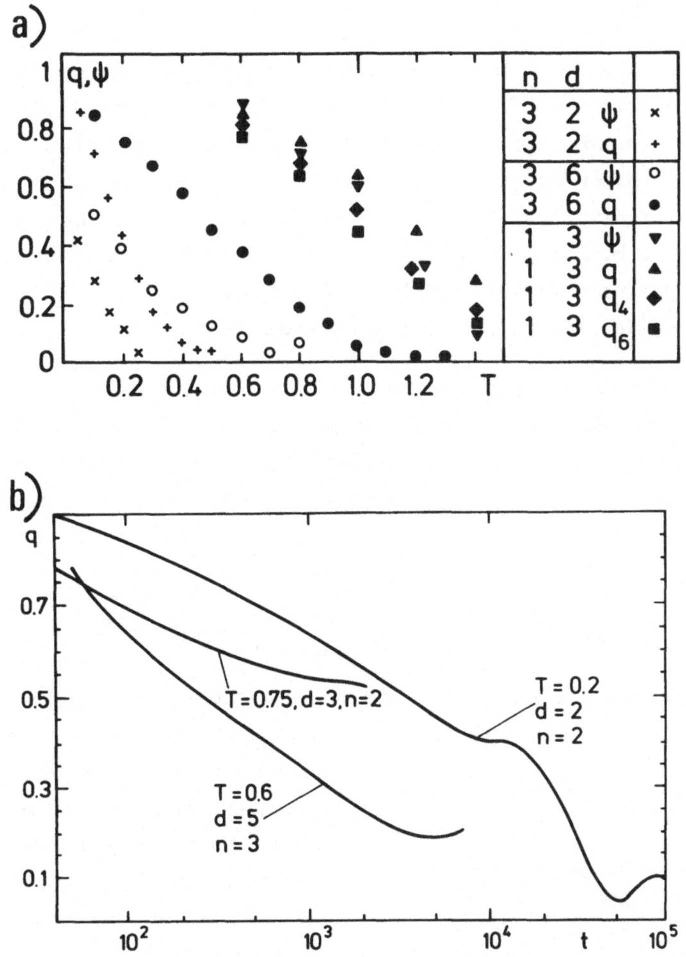


Fig. 13: a) Order parameters q and ψ plotted vs. $k_BT/\Delta J$ for several n and d /8/. b) Examples for the variation of q(t) with t /8/.

the bond strengths are chosen from Eq. (7) /31/].
(iv) At least for d=2,3, there is no nonzero $T_f$ below which $q_{EA}>0$ /8/. Again this conclusion agrees with high-temperature series /32/.

Of course, since the model is classical its low-temperature behavior is not of much interest [e.g., the specific heat behaves as /21/ $C/k_B \underset{T\to 0}{\to} (n-1)/2$]. Nevertheless it would be interesting to study the remanence also for this model in some detail, as it should be more close to experimental systems rather than the Ising spin glasses.

## V. Site-Disorder Models

It is clear that the bond disorder models studied so far cannot describe a real material quantitatively. If there were a phase transition at $T_f$, one could hope to extract universal properties (such as exponents, etc.) even from a very simplified model. As we have seen, phase transitions occur at T=0 only. However, there is not much interest in equilibrium properties which presumably are not observable, but rather in the nonequilibrium behavior near $T_f$, which is described by the Edwards-Anderson model only qualitatively.

For many aspects of spin glass systems, such as the dependence on concentration x of magnetic atoms, or magnetic short range order, the site-disorder must properly be included. As a first step, we have studied the model /12-14/ with exchange between nearest (nn) and next-nearest (nnn) neighbors,

$$\mathcal{H} = -J_1 \sum_{\langle i,j\rangle_{nn}} c_i c_j \sum_{m=1}^{n} S_i^m S_j^m - J_2 \sum_{\langle i,j\rangle_{nnn}} c_i c_j \sum_{m=1}^{n} S_i^m S_j^m, \quad J_1>0,\ J_2<0, \tag{9}$$

where the occupation variables $c_i\{=0,1\}$ are chosen at random with $[c_i]_{av}=x$. Eq. (9) is a simple model of a magnet with competing interactions, where disorder leads to spin-glass behavior. For the d=3 fcc lattice and n=3 Eq. (9) is a reasonable model of $Eu_xSr_{1-x}S$ /33/ and $Eu_xSr_{1-x}S_ySe_{1-y}$ /34/ - varying y in the latter case one can effectively change $J_2/J_1$.

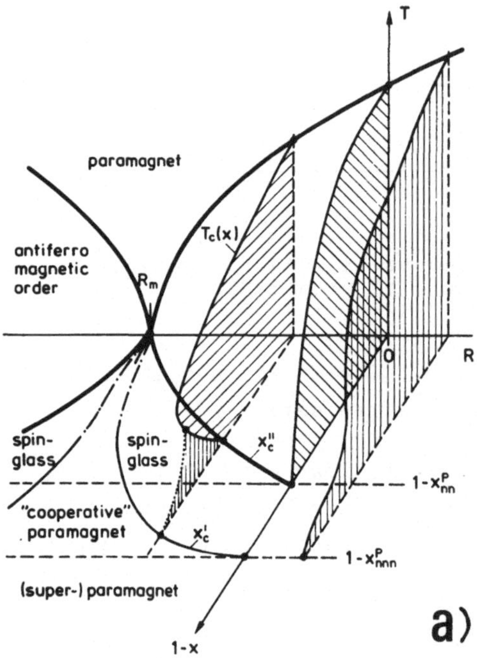


The phase diagram of the system (for n=1, d=2 or n>1, d=3, respectively) is shown in Fig. 14a, while Fig. 14b shows that quantitative agreement with experimental data is obtained. The breakdown of ferromagnetic order for x→1 when

Fig. 14: a) Schematic phase diagram of Eq. (9) as function of x, T and $R \equiv J_2/J_1$. Full curves describe $2^{nd}$-order transition, the dotted curve represents the "freeze-in" into the spin glass state /14/.

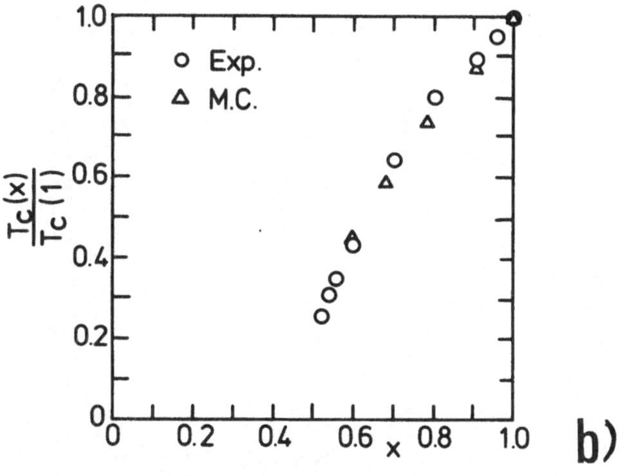


Fig. 14: b) Ferromagnetic critical temperature plotted vs. x for $Eu_xSr_{1-x}S$ /33/ as compared to Monte Carlo results for Eq. (9) with R=-1/2 /12/.

$J_2/J_1 \rightarrow -1$ has recently also been confirmed /34/. At T=O, again systematic power series in x or 1-x yielded estimates for the various phase boundaries /12/. While for noncompeting exchange the ferromagnetic phase extends up to the percolation threshold $x^p_{nnn}$ (or $x^p_{nn}$, for $J_2$=O), for R<O the groundstate exhibits both a spin-glass phase ($\xi_{EA}$ infinite) and a "cooperative paramagnet" (infinite cluster of magnetic atoms, but $\xi_{EA}$ finite). At the multicritical point where in the undiluted system the order changes from ferro- to antiferromagnetic all phase boundaries merge. By recursive exact partition function calculations Morgenstern /35/ showed that the spin glass phase of this model is similar to the $\pm$ J model: power-law decay of correlations at T=O, exponential decay at T>O.

The instability of the ferromagnetic state against dilution, when competing antiferromagnetic bonds are present, is understood qualitatively studying the spin configurations near dilution sites (Fig. 15a). It becomes energetically favorable that no longer all spins are aligned parallel. Cases a,b show that again one obtains "two-level-systems" - two equivalent possibilities of alignment. When the system is strongly diluted, more and more "clusters" of spins are no longer aligned parallel to their environment, and thus finally ferromagnetic long range order breaks down. However, even deep inside the spin glass state most nearest-neighbor pairs are nearly parallel aligned (Fig. 15b), while the probability distribution of next-nearest-neighbor correlations has about the same weight for parallel and for antiparallel orientation. In this way, one can account for the magnetic short range order seen in neutron scattering and hyperfine field measurements. In principle, even distributions such as $P(\langle\vec{S}_i \cdot \vec{S}_j\rangle)$ are accessible via the measured hyperfine field distribution. We find that frustration effects can directly be detected by enhanced variances

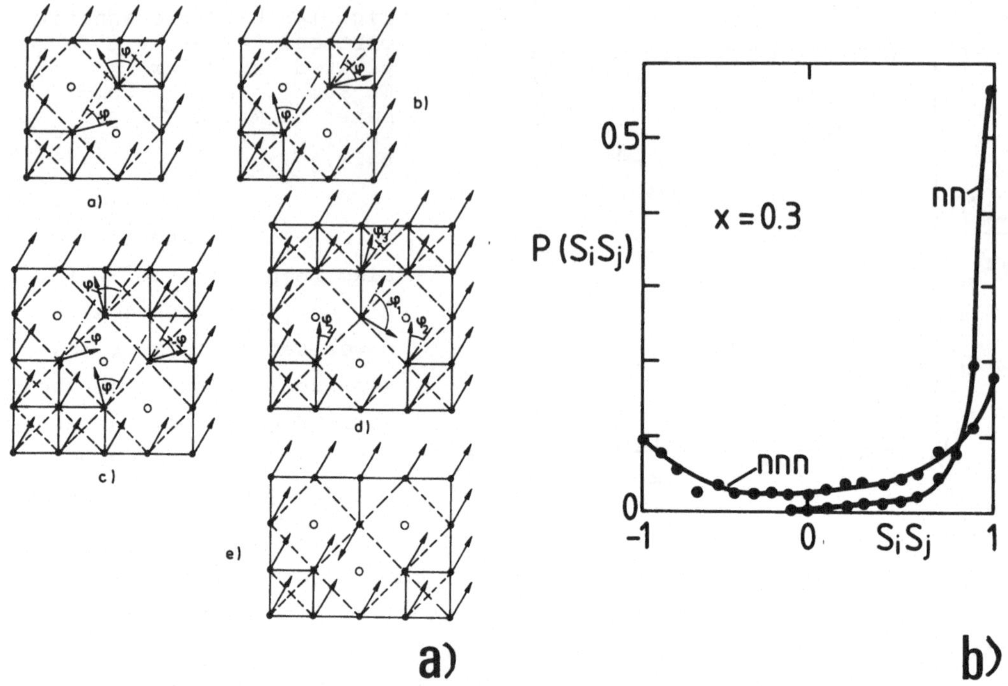


Fig. 15: a) spin configurations near clusters of nonmagnetic atoms (indicated as empty circles) in the XY-case [a)-d)] and Ising case [e)]. Nearest-neighbor exchange is indicated by full and next-nearest neighbor exchange by broken lines. b) Distribution $P(\langle\vec{S}_i\vec{S}_j\rangle_T)$ for nearest (nn) and next-nearest (nnn) neighbor sites of a classical fcc Heisenberg magnet with $J_2=-J_1/2$ at x=0.3, T=0 /14/.

of such distributions /14/. A more detailed study of local order and its distribution in spin glasses would contribute to sorting out the appropriate models for these materials.

Finally we remark that a d=1 site disorder model with long-range competing interaction seems amenable to exact solution /36/ - from which again "clusters" of strongly correlated spins emerge.

## VI. Conclusions

In this survey it was shown that spin glass models which combine both disorder and frustration (such as the nearest neighbor $\pm$ J or gaussian model, or the diluted next-nearest neighbor magnet) have a phase transition to a spin glass phase at T=0 only, while the freezing at $T_f$ is a nonequilibrium phenomenon: rather than *global* thermal equilibrium, only "local equilibrium" in a "valley" of phase space

(Fig. 7a) is achieved. A more precise description of this local equilibrium is lacking, as well as knowledge about the barriers between the "valleys" /37/. Thus we are still far from a satisfactory understanding of the freezing transition, while low-temperature excitations are, at least qualitatively, understood in terms of "two-level-systems" (clusters of spins with two or more low-lying states with respect to the surrounding frozen-in spins).

Among the models, the Edwards-Anderson gaussian model provides a surprisingly good qualitative description of most experimental facts, while the $\pm$ J model is somewhat less close to reality. The model of a diluted magnet with exchange up to next nearest neighbors has even significantly contributed to a quantitative understanding of systems such as $Eu_xSr_{1-x}S$ - a similar (but much more difficult) study of RKKY spin glasses is still lacking.

Acknowledgements: Sincere thanks are due to I. Morgenstern, D. Stauffer and A. Aharony for their fruitful collaboration on parts of the research described here.

## References

/1/ For reviews of experimental work, see e.g. J.A. Mydosh, J. Mag. Magn. Mat. 7, 237 (1978); P.A. Beck, Progr. Mat. Sci. 23, 1 (1978); A.P. Murani, J. Phys. (Paris) 39, C6-1517 (1978); J. Souletie, J. Phys. (Paris) 39, C2-3 (1978).

/2/ For reviews of theoretical work see e.g.: P.W. Anderson, J. Appl. Phys. 49, 1599 (1978); A. Blandin, J. Phys. (Paris) 39, C6-1499 (1978); K. Binder, in Ordering in Strongly-Fluctuating Condensed Matter Systems (T. Riste, ed., New York, Plenum Press 1979) p. 423 and in Fundamental Problems in Statistical Mechanics V (E.G.D. Cohen, ed., North Holland, Amsterdam 1980) p. 21.

/3/ G. Toulouse, Commun. Phys. 2, 115 (1977)

/4/ More complete reviews of somewhat less recent numerical studies of spin glasses are found in K. Binder, J. Phys. (Paris) 39, C6-1527 (1978); K. Binder and D. Stauffer, in Monte Carlo Methods in Statistical Physics (K. Binder, ed., Springer, Berlin 1979), p. 301

/5/ I. Morgenstern and K. Binder, Phys. Rev. Lett. 43, 1615 (1979); Phys. Rev. B22, 288 (1980)

/6/ I. Morgenstern and K. Binder, Z. Phys. B39, 227 (1980)
/7/ D. Stauffer and K. Binder, Z. Physik B34, 97 (1979)
/8/ D. Stauffer and K. Binder, Z. Physik B41, (1981)
/9/ A. Aharony and K. Binder, J. Phys. C13, 4091 (1980)
/10/ A. Aharony and K. Binder (unpublished)
/11/ W. Kinzel, Phys. Rev. B19, 4595 (1979)
/12/ K. Binder, W. Kinzel and D. Stauffer, Z. Physik B36, 161 (1979)
/13/ G. Eiselt, J. Kötzler, H. Maletta, D. Stauffer and K. Binder, Phys. Rev. B19, 2664 (1979)
/14/ W. Kinzel and K. Binder, Phys. Rev. B , (1981)
/15/ S.F. Edwards and P.W. Anderson, J. Phys. F5, 965 (1975)
/16/ S. Kirkpatrick, Phys. Rev. B16, 4630 (1977)
/17/ A.J. Bray and M.A. Moore, J. Phys. F7, L333 (1977); A.J. Bray, M.A. Moore and P. Reed, J. Phys. C11, 1187 (1978)
/18/ R. Fisch and A.B. Harris, Phys. Rev. Lett. 38, 785 (1977)
/19/ A.B. Harris, T.C. Lubensky and J.H. Chen, Phys. Rev. Lett. 36, 415 (1976)
/20/ W. Kinzel and K.H. Fischer, J. Phys. C11, 2115 (1978), and references therein
/21/ K. Binder, Z. Physik B26, 339 (1977)
/22/ D. Stauffer and K. Binder, Z. Physik B30, 313 (1978)
/23/ A.J. Bray and M.A. Moore, J. Phys. C12, L477 (1979)
/24/ K. Binder and K. Schröder, Phys. Rev. B14, 2142 (1976)
/25/ D.C. Mattis, Phys. Lett. 56A, 421 (1976)
/26/ C.N. Yang, Phys. Rev. 85, 809 (1952); L. Onsager, Phys. Rev. 65, 114 (1944)
/27/ C. Dasgupta, S.-K. Ma, and C.-K. Hu, Phys. Rev. B20, 3837 (1979)
/28/ J.L. Tholence and R. Tournier, J. Phys. 35, C4-229 (1974)
/29/ P. Reed, J. Phys. C12, L859 (1979)
/30/ S.-K. Ma, preprint
/31/ J.F. Fernandez and R. Medina, Phys. Rev. B19, 3561 (1979); D. Kumar and J. Stein, preprint; R. Medina, J.F. Fernandez and D. Sherrington, Phys. Rev. B21, 2915 (1980).
/32/ P. Reed, J. Phys. C11, L979 (1978)
/33/ H. Maletta and W. Felsch, Phys. Rev. B20, 1245 (1979)
/34/ K. Westerholt and H. Bach, preprint
/35/ I. Morgenstern, preprint
/36/ H. Orland, C. de Dominicis and T. Garel, Phys. Lett. 42, L73 (1981)
/37/ In this context note the interesting approach of I. Morgenstern and H. Horner (preprint) linking barrier heights and power-law decay of correlations in the $\pm$ J-model.

# GINZBURG-LANDAU SPIN-GLASS MODELS

D. Sherrington

Physics Department
Imperial College
London, U.K.

Abstract:

The use and relevance of random Ginzburg-Landau functionals as a basis for spin glass modelling is discussed.

The Ginzburg-Landau models we consider are characterized by the free energy functional

$$F[\phi] = \sum_i ((r_i/2)|\underline{\phi}_i|^2 + (u_i/8)|\underline{\phi}_i|^4) - \tfrac{1}{2} \sum_{ij} J_{ij}\, \phi_i\, \phi_j \tag{1}$$

where $\phi_i$ is an m-vector classical field associated with a lattice point i and $r_i$, $u_i$, $J_{ij}$ are quenched random parameters. $F[\phi]$ provides for statics and dynamics in the usual way, viz.

$$Z = \int (\pi_i\, d\, \underline{\phi}_i) \exp(-\beta F[\phi]) \tag{2}$$

$$\partial\underline{\phi}_i/\partial t = -\Gamma_o\, (\partial F/\partial\underline{\phi}_i) + \underline{\xi}_i\, (t) \tag{3}$$

where $\underline{\xi}_i(t)$ is a random Langevin noise satisfying

$$\langle \xi_i^\nu\, (t)\, \xi_j^\nu\, (t') \rangle = 2\Gamma_o\, \delta(t-t')\, \delta_{ij} \tag{4}$$

Let us first demonstrate a mapping of (1) onto a conventional class of random hard-spin models. Consider any site i and let

$$r_i \to -\infty,\ u_i \to +\infty,\ (-2r_i/u_i) \to s_i^2 \tag{5}$$

The $\underline{\phi}_i$-integration are then dominated by $|\underline{\phi}_i| = s_i$ and the system behaves as a classical hard-spin model with Hamiltonian

$$H = -(\tfrac{1}{2})\ \Sigma_{ij}\ J_{ij}\ \underline{S}_i\ \underline{S}_j \tag{6}$$

where the $\underline{S}_i$ are classical m-vector spins of magnitude $s_i$. The Edwards-Anderson models have all $s_i$ equal but $J_{ij}$ random. If, however, at any site $r_i$, $u_i$ are positive and either is infinitely large then $\phi_i$ is effectively quenched to zero and (6) applies with spins missing from these sites. The result is the usual diluted alloy model of which the diluted ferromagnet and RKKY spin glass models are examples.

The choices of r, u in the last paragraph are rather extreme, but it is clear that all other choices contain qualitatively similar physics, e.g. a Hubbard-Stratonovich transformation of an idealized model of a transition metal alloy[1]

$$H = \Sigma_{ij\sigma}\ t_{ij}\ a^+_{i\sigma}\ a_{j\sigma} - \Sigma_i\ (U_i/4)(n_{i\uparrow} - n_{i\downarrow})^2;\ U_i \text{ random} \tag{7}$$

yields (2) as its truncated static approximation with

$$r_i = (\chi^{(o)}_{ii} - U_i^{-1}),\ J_{ij} = \chi^{(o)}_{ij}\ ,\ u_i = \Lambda_{iiii},\ m = 1 \tag{8}$$

where $\chi^{(o)}_{ij}$ is the bare band susceptibility and $\Lambda$ the corresponding 4-point function, retained only in its local part. Henceforth, however, we shall discuss only models with non-extreme r, u and with Gaussian randomness of r or J.

The model with r, u constant but $J_{ij}$ random has been studied by Hertz and Klemm[2] perturbatively in J, u, with the assumption of a single relevant spin-glass order parameter. For the special case of infinite-range the model is exactly soluble in the same sense as the hard-spin Sherrington-Kirkpatrick model. Taking the $J_{ij}$ as quenched random parameters of mean $(J_o/N)$, standard deviation $(\Delta/N)^{\frac{1}{2}}$, replication and auxiliary variable techniques yield self-consistency equations which in the replica-symmetric approximation for m = 1 are

$$M \equiv N^{-1} \Sigma_i \langle\phi_i\rangle = \int (dz/(2\pi)^{\frac{1}{2}}) \exp(-z^2/2) \langle\phi\rangle_z \tag{9a}$$

$$p \equiv N^{-1} \Sigma_i \langle\phi_i^2\rangle = \int (dz/(2\pi)^{\frac{1}{2}}) \exp(-z^2/2) \langle\phi^2\rangle_z \tag{9b}$$

$$q \equiv N^{-1} \Sigma_i \langle\phi_i\rangle^2 = \int (dz/(2\pi)^{\frac{1}{2}}) \exp(-z^2/2) (\langle\phi\rangle_z)^2 \tag{9c}$$

where

$$\langle\phi^n\rangle_z = \int d\phi\ \phi^n A(M,p,q,z)/\int d\phi\ A(M,p,q,z) \tag{10}$$

$$A(M,p,q,z) = \exp\{(\beta J_o M+\beta\Delta q^{\frac{1}{2}}z)\phi+(-\beta r/2+(\beta\Delta)^2(p-q)/2)\phi^2-\beta u\phi^4/8)\} \tag{11}$$

The spin-glass critical condition is

$$(\beta\Delta)\ \langle\phi^2\rangle_o = 1 \tag{12}$$

where $\langle\phi^2\rangle_o$ is evaluated from (10) with M, q, z all zero and p determined self-consistently. Within the spin-glass regime the solution (9) will exhibit Almeida-Thouless[(3)] instabilities but we shall not pursue them nor the analogues of Parisi symmetry-breaking[(4)] or the Parisi-Toulouse hypothesis[(5)].

Several authors have considered a continuum analogue of (1) with r random but u, J constant and J ferromagnetic. Ma and Rudnick[(6)] argued that such a model would lead to spin-glass ordering in preference to ferromagnetism for d<4. This result is unphysical and below we indicate how the incorrect deduction arose and its probable resolution. We take m = 1 for simplicity.

In the paramagnetic regime the response $<\phi_i>$ to linear order in an applied conjugate field $h_j$ is

$$<\phi_i> = \beta G_{ij}\, h_j \tag{13}$$

where

$$G_{ij} = \int\delta\phi\ \phi_i\ \phi_j\ \exp\,(-\beta F[\phi])/\int\delta\phi\ \exp\,(-\beta F[\phi]) \tag{14}$$

We consider the application of field $h_i = h\nu_i$; $\nu_i = \pm 1$. To order $h^2$ the free energy lowering is $(-h^2/2)\,\Sigma_{ij}\ \nu_i G_{ij}\nu_j\,\beta$ . A perturbation expansion[7] in J about the local but non-uniform part of $F$ shows that a consequence of $<\phi_i^{2(n+m)}> \, > \, <\phi_i^{2n}> \, <\phi_i^{2m}>$ is that for all $J_{ij}$ non-negative the lowering is greatest if all the $\nu_i$ are equal, i.e. ferromagnetically inclined.

MR considered two paramagnetic susceptibilities, $\chi_F = \beta\Sigma_j\, \overline{G}_{ij}$, giving the average $<\phi>/h$ response to a uniform field, and $\chi_{SG} = \beta^2\,\Sigma_j\, \overline{G_{ij}^2}$ giving the average $<\phi>^2/h^2$ response to a random field. Divergence of $\chi_F$ signals ferromagnetic instability, that of $\chi_{SG}$ without $\chi_F$ a spin--glass. They expanded perturbatively about the mean Gaussian part of (1) and then averaged over the $r_i$ disorder, the choice of Gaussian disorder ensuring a Wick's theorem for the disorder averaging. Near a ferromagnetic transition G(k), the Fourier transform of $G_{ij}$, behaves as $(r + \alpha\, k^2 + 0(k^4))^{-1}$ where r is determined by summing a self-energy series. Ferromagnetism is signalled by $r \to 0$. The formal diagrammatic series for $\beta^{-2}\,\chi_{SG}$ is depicted in Fig.1. The full lines denote renormalized Green functions and the $\Gamma$ are the sum of all blocks in which (i) the $\alpha$- and $\beta$-legs are separately connected internally by G-lines but with the only connections between $\alpha$ and $\beta$-lines via $(r_i - r)$ pairings, (ii) it is not possible to split the block by cutting a single $\alpha$ and a single $\beta$-line. The use of renormalized Green functions eliminates self-energy diagrams.

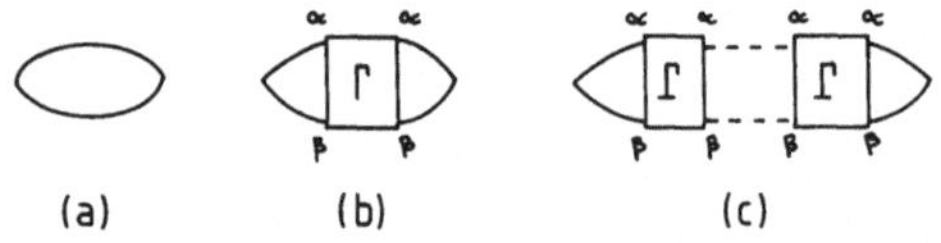


Fig.1: (a) First (b) second (c) general terms in $\beta^{-2}\chi_{SG}$ expansion.

MR retained in $\Gamma$ just the diagram with a single $(r_i - r)$ pairing connecting $\alpha$ and $\beta$-legs, i.e. the ladder approximation. This yields

$$\chi_{SG} = \beta^2 \Pi(0)/(1 - \Pi(0)\Delta) \qquad (15)$$

where $\Delta$ is the variance of $r_i$ and

$$\Pi(q) = \int d\underline{k}\; G(\underline{k})\; G(\underline{k} + \underline{q}) \qquad (16)$$

where $G(\underline{k})$ is the renormalized average Green function given above. For $d < 4$, $\Pi(0)$ is infra-red divergent in the limit $r \to 0$ and consequently the divergence of (15) occurs at positive r where $\Pi(0)\Delta = 1$, before the $r = 0$ divergence of $\chi_F$. Hence the prediction of a spin glass.

The naive attraction of the above approximation is the RPA-like summation of the apparently most divergent terms. In fact, however, corrections must remove the divergence at $r \to 0$. Fig. 2 shows the set of vertex corrections involving u which seem potentially the most important. There are also corresponding vertex corrections in which any or all of the wavy u-lines are replaced by dashed $\Delta$-lines. Summing over all of these vertex corrections effectively replaces the MR ladder rung $\Delta$ by

$$\tilde{\Delta}(q) = \Delta(1 + (3u/2 - \Delta)\Pi(q))^{-2} \qquad (17)$$

Boundedness of $F[\phi]$ ensures $(3u/2 - \Delta)$ is positive, as also is $\Pi(q)$. Thus these vertex corrections reduce the effective interaction. $\chi_{SG}$ is now given by

$$\chi_{SG} = \beta^2(\Pi(0) + \int d\underline{R} D(\underline{R})^2\, \tilde{\Delta}(\underline{R})/(1 - D(\underline{R})\,\tilde{\Delta}(\underline{R}))) \qquad (18)$$

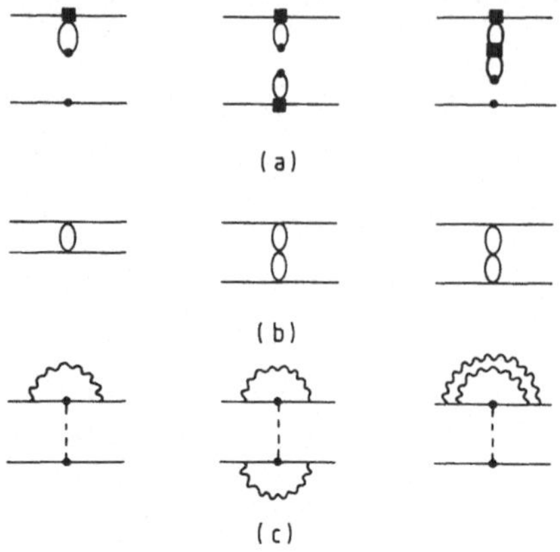


Fig. 2 -
Vertex correction involving u and one Δ-rung; (a) before disorder averaging; (b) topological consequence of disorder averaging; (c) explicit representation with u denoted by a wavy line, Δ by a dashed line.

where D(R) is the Fourier transform of $G(k)^2$ and $\tilde{\Delta}$ (R) that of $\tilde{\Delta}$ (q). As we noted Π(O) (which is the same as D(O), diverges as $r \to 0$. If Π(q) diverged in a similar fashion, the resultant vanishing of $\tilde{\Delta}$ as $\Pi^{-2}$ would overcompensate the divergence of D and leave only the bare Π(O) divergence at r = O, the onset of ferromagnetism. In fact, for $q \neq 0$ Π(q) is less divergent than Π(O) but we believe the above indicates the type of corrections needed to remove the unphysical MR divergence as well as demonstrating some of the pitfalls of the expansion around a mean Gaussian.

The introduction of oscillatory (but non-random) $J_{ij}$ into the random-r model gives rise to frustration analogous to that of the canonical RKKY spin-glasses. If the randomness is great enough, reduction of $\bar{r}$ will result in $F[\phi]$ having a multiplicity of quasi-degenerate minima with finite moments on a thermodynamically significant number of sites but no overall moment. Presumably the generalization of (18) will exibit a divergence at finite r for $d > d_c$ for sufficiently great Δ but the dominance of integrals by infrared divergences for small r, $d < 4$ suggests that for any system which is ferromagnetic in the pure limit there will be no spin-glass phase for $d < 4$.

## References


1. D. Sherrington and K. Mihill, J. Physique 35, C4-199 (1974).
2. J.A. Hertz and R.A. Klemm, Phys. Rev. B20, 316 (1979).
3. J.R.L. de Almeida and D.J. Thouless, J. Phys. A11, 983 (1978).
4. G. Parisi, J. Phys. A13, 1103 (1980).
5. G. Parisi and G. Toulouse, J. Physique Lett. 41, L-361 (1980).
6. S-K. Ma and J. Rudnick, Phys. Rev. Lett. 40, 589 (1978).
7. D. Sherrington, Phys. Rev. B22, 5553 (1980).

scale $t \gg t_c$ even below the critical point. This time scale may be determined as the average time scale of a cluster[7] in which all spins are relatively frozen except thermal excitations. This will be discussed later in more detail in connection with the frustration effect[7].

## 4. INSTABILITY AND NONLINEAR SUSCEPTIBILITY IN SPIN GLASSES

When we emphasize the equilibrium aspect of spin glasses, it is convenient to make use of a phenomenological theory[2] which assumes the existence of the spin glass order parameter q first proposed by Edwards and Anderson[1]. The free energy f(m,q) may be expanded[2] in power series of the two variables m(magnetization) and q as

$$f(m,q) = f_0 + (am^2+bm^4+\dots) + (cq^2+dq^3+\dots) + (em^2q+\dots)-hm \qquad (1)$$

From the extremization of f(m,q), we obtain

$$q = -\frac{e\chi_0^2}{2c} h^2 \text{ for } T>T_{sg} \quad \text{and} \quad q = q_0(T) + q_1(T)h^2 + \dots \text{for } T<T_{sg} \qquad (2)$$

with $\chi_0 = 1/(2a)$ and $q_1(T) = e\chi_0^2/(2c)$. From the condition of instability, we assume[2], for simplicity, that $c(T) \propto T-T_{sg}$. The uniform magnetization m(h) is also expanded as $m(h) = \chi_0 h + \chi_2 h^3 + \cdots$. Then, the nonlinear susceptibility is expressed[2] as

$$\chi_2 = -\frac{e}{2a^2} q/h^2 = \frac{e^2\chi_0^4}{c(T)} \propto \frac{-1}{T-T_{sg}} \qquad (3)$$

for $T>T_{sg}$. A similar singularity is also obtained for $T<T_{sg}$. This divergence of the nonlinear susceptibility[8-9] seems to be one of the characteristic features of the phase transition of spin glass, <u>in the time region</u> $t \ll t_c$, where the picture of equilibrium phase transition of spin glass may be valid.

## 5. RIGOROUS DEFINITION OF THE SPIN GLASS ORDER PARAMETER

Our definition of q is given[12] by

$$q = \lim_{\lambda\to 0} \lim_{N\to\infty} \left\langle \mathrm{Tr}_{\{s_j\}} P(\{s_j\},\beta',h') \sum \langle\sigma_j\rangle_{\{s_j\}} s_j \right\rangle_{av} \qquad (4)$$

where $\langle\sigma_j\rangle_{\{s_j\}}$ denotes the thermal average of the spin $\sigma_j$ in the presence of a fictitious field $\lambda s_j$ at the j th site, and P is the weight function defined by

$$P(\{s_j\},\beta',h') = \exp[-\beta'\mathcal{H}(s)+h'\sum s_j]/\mathrm{Tr}\exp[-\beta'\mathcal{H}(s)+h'\sum s_j] \qquad (5)$$

with the same Hamiltonian $\mathcal{H}(s)$ as the original one $\mathcal{H}(\sigma)$. Here, the parameter h' may be determined so that the free energy is minimized, and

# STATIC AND DYNAMIC PROPERTIES OF SPIN GLASSES

Masuo SUZUKI and Seiji MIYASHITA

Department of Physics, Faculty of Science,
University of Tokyo, Tokyo

In the present paper, we discuss several essential aspects of spin glasses[1-17].

## 1. MOTIVATION

We are interested in random systems, partly because we expect that a new type of phase transition may occur in random systems, and partly because they give us a new challenging problem of how to formulate randomness in a form of statistical mechanics. What is the randomness? It means the violation of translational invariance. There are two kinds of randomness, spatial and temporal. Here we confine our arguments into the spatial randomness.

## 2. RANDOM AVERAGE AND SPATIAL ERGODICITY

Experimental observation is performed for a sample in which the randomness is specified (for example, a set of random bonds $\{J_{ij}\}$ is fixed). It is very difficult to calculate theoretically the corresponding physical quantities for the randomness specified. Thus, we usually replace them by random averages of the corresponding quantities. How is it justified? If we observe a macrovariable which is a sum of local variables over the whole system, then the corresponding normalized macrovariable (namely intensive macrovariable) approaches the average over the randomness, according to the central limit theorem. This is the spatial ergodicity. However, we must be careful about the case in which we treat a single local variable, because of the absence of translational invariance in random systems.

## 3. OBSERVATION AND TIME SCALE

Is the phase transition of spin glass a phenomenon in equilibrium or in nonequilibrium? If a completely frozen spin appears below the transition point, then it is described as a phase transition in equilibrium. However, recent Monte Carlo simulations[3-6] suggest a possibility of a "quasi" phase transition in nonequilibrium. That is, it depends on the time scale of observation. It is expected that there exists such a characteristic time $t_c$ that a frozen spin is observed for the time scale $t \ll t_c$ and that no frozen spin is observed for the time

$\beta'$ is the so-called quenching parameter to describe the procedure of quenching, as was introduced in our previous paper[12]. At a glance, the above definition (4) of q seems more complicated than the original Edwards-Anderson order parameter $q_{EA}$ defined by $q_{EA} = \langle\langle\sigma_j\rangle^2\rangle_{av}$ . However, the meaning of $\langle\sigma_j\rangle$ is very profound in random systems which are highly degenerated at the ground states due to the frustration effect. Of course, it is identically zero from the spin inversion symmetry without an infinitesimal symmetry breaking field, and this field is very difficult to define explicitly due to the extremely high degeneracy of the ground states. An abstract formulation of this problem is given by (4) and (5). That is, the fictitious field term $\lambda\sum s_j\sigma_j$ in calculating $\langle\sigma_j\rangle_{\{s_j\}}$ plays a role of a symmetry breaking field, and the high degeneracy of the ground states can be taken into account, in principle, through the microscopic weight function $P(\{s_j\},\beta',h')$ defined by (5). If q in (4) exists in a strict sense and it is non-vanishing at certain finite temperatures, then the phase transition of spin glass is rigorously an equilibrium phase transition (i.e., $t_c\to\infty$). It is the case in the Mattis type spin glass[10], in which there is no frustration. If the spin glass transition is a quasi phase transition defined in the time region $t \ll t_c$, as was suggested by the recent Monte Carlo simulations[3-6], then a finite q does not appear in (4), and consequently we have to contruct a dynamical theory of spin glasses to describe such quasi phase transitions in the above time region.

## 6. CLUSTER THEORY OF SPIN GLASSES

Here we review briefly a cluster theory proposed by the present authors[7]. For simplicity, we consider the $\pm J$ Ising model in order to study the effect of frustration to the phase transition of spin glasses.

First we explain the concept[7] of a cluster in the above model. It is defined by a region which is surrounded by frustrated plaquettes. That is, the boundary of each cluster is composed of the equal number of "right" bonds and "wrong" bonds. Inside each cluster, there is no frustration like the Mattis type spin glass. This cluster property has been confirmed by Takase and Takayama[6], using the Monte Carlo simulation on the Ising model with the Gaussian random distribution.

Now we study the frustration effect on the spin correlation $f_{ij} \equiv \langle\langle\sigma_i\sigma_j\rangle^2\rangle_{av}$ where $\langle\ldots\rangle_{av}$ denotes the average over the randomness, as usual. If spins $\sigma_i$ and $\sigma_j$ belong to different clusters, the correlation function $f_{ij}$ vanishes, because different clusters do not correlate. If both spins belong to the same cluster, the correlation function is similar to that of the corresponding regular system. These considerations lead us to the following intuitive approximation[7] $f_{ij} \cong \langle\sigma_i\sigma_j\rangle_0^2\, C(r_{ij})$, where $\langle...\rangle_0$ is the average in the corresponding pure system and $C(r_{ij})$ is the probability that the sites i and j belong to the same cluster. In Ref. 7, we have shown that $f_{ij}$ should decay in a power law like $f_{ij} \propto r_{ij}^{-\psi}$ with $\psi \simeq 1/4$, at least in the two-dimensional Ising model. This suggests the following dynamical scaling law $f(r,t) \equiv \langle\langle\sigma(r,t)\sigma(0,0)\rangle^2\rangle_{av} \sim r^{-\psi}f(r^z/t) \sim t^{-\psi/z}g(r^z/t)$. Here $\psi$ will be calculated by the percolation theory[14]. This dynamical scaling yields a slow relaxation of spins for $T \leq T_{sg}$, which has been suggested by some experiments[16] and numerical calculations[4]. There are also exact results on frustrated regular systems[17] to show a power law decay at T=0.

## 7. DISCUSSION

There are many other important aspects, for example, (i) the relation between interaction range and spin glass from a view point of frustration (ii) the cross over effect between the Mattis type spin glass (no frustration) and the frustrated spin glass, and (iii) the time-dependent internal field and its distribution function. These will be reported in detail elsewhere.

## ACKNOWLEDGEMENTS

The present authors would like to thank Professor T. Yamazaki and Mr. Y.J. Uemura for informing their experimental results prior to publication and for their useful discussions. This study was partially financed by the Mitsubishi Foundation.

## REFERENCES

1. S.F. Edwards and P.W. Anderson, J. Phys. F:Metal Phys. 5, 965(1975). D. Sherrington and S. Kirkpatrick, Phys. Rev. Lett. 35, 1792(1975).
2. M. Suzuki, Prog. Theor. Phys. 58, 1151(1977) and unpublished note (1979, June).
3. S. Kirkpatrick, Phys. Rev. B16, 4630 (1977). A.J. Bray and M.A. Moore, J. Phys. F:Metal Phys. 7, L333(1977)
4. I. Morgenstern and K. Binder, Phys. Rev. B22, 228(1980); Z. Physik B39, 227(1980)
5. W.Kinzel, Phys. Rev. B19, 4595(1979). C. Dasgupta, S-k. Ma and G-k. Hu, Phys. Rev. B20, 3837(1979).
6. S. Takase and H. Takayama, preprint.
7. S. Miyashita and M. Suzuki, J. Phys. Soc. Japan 50(1981) in press.
8. S. Katsura, Prog. Theor. Phys. 55, 1049(1976); J. Phys. C: Solid State Phys. 9, L619(1976).
9. C. Domb, J. Phys. A: Math. Gen. 9, L17(1976).
10. D.C. Mattis, Phys. Letters 56A, 421(1976).
11. Y. Miyako, S. Chikazawa, T. Saito and Y.G. Youchunas, J. Phys. Soc. Japan 46, 1951(1979). Y. Miyako, S. Chikazawa, T. Sato and T. Saito, J. Mag. Magn. Mat. 58-18, 139(1980).

12. M. Suzuki and S. Miyashita, Physica 106A, 344(1981).
13. K. Honda and H. Nakano, Prog. Theor. Phys. 63, 1800(1980); Prog. Theor. Phys. 65, 83(1981).
14. J.W. Essam, in Phase Transition and Critical Phenomena, edited by C. Domb and M.S. Green (Academic Press, 1972), Vol.2, p. 197.
15. J.A. Mydosh, J. Mag. Magn. Mat. 7, 237(1978).
A.P. Murani, J. de Phys. 39, C6-1517(1978); Solid State Comm. 33, 433 (1980).
16. Y.J. Uemura, T. Yamazaki, R.S. Hayano, R. Nakai and C.Y. Huang, Phys. Rev. Letters 45, 583(1980).
Y.J. Uemura, K. Nishiyama, T. Yamazaki and R. Nakai, Solid State Comm. in press.
17. J. Stephenson, J. Math. Phys. 11, 413(1970).
G. Forgacs, Phys. Rev. B22, 4473(1980).


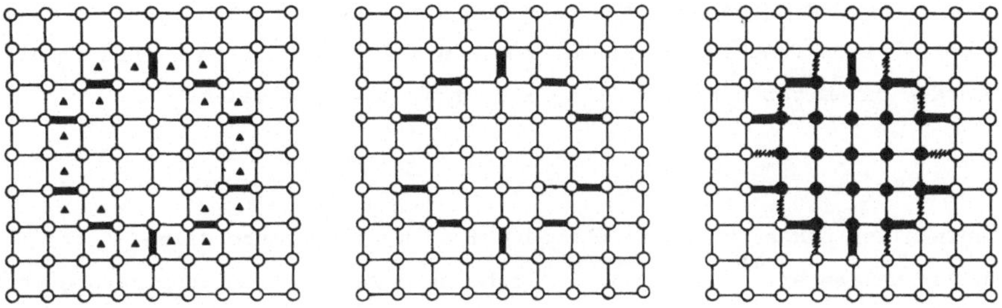

(a) (b) (c)

Fig. 1. A typical example of the cluster discussed in the paper. The bold and thin lines are antiferromagnetic bonds (-J) and ferromagnetic bonds (J), respectively. The bond ▲ denotes a frustrated plaquette. The two configurations b) and c) are degenerate in the ground state, where o and ● denote up and down spins, respectively. Bold lines in (b) are frustrated and shaded lines in (c) are frustrated. The local ground state of each non-frustrated cluster is uniquely determined except the freedom of total inversion of spins inside the cluster. The ground state energy of the whole system does not depend on the direction of the total spin of each cluster and consequently there is high degeneracy of the ground state.

# STABILIZATION OF THE ORDER PARAMETER FLUCTUATIONS IN SPIN GLASSES

J.A. Hertz
A. Khurana
M. Puoskari
NORDITA, Blegdamsvej 17,
DK-2100 Copenhagen Ø, Denmark

Abstract

The negative value of the order parameter susceptibility found in simple mean field theories of spin-glasses is related to the negative value of the kinetic coefficent of the spin response function. The zero frequency limit of the self-consistent equation for the dynamic response function has a solution which differs from the spin-spin correlation function obtained from a static calculation. This solution gives the order parameter susceptibility a finite, positive value.

The spin-glass phase, which is characterized by a non-zero value of the Edwards-Anderson order parameter [1], $q = \overline{\langle \sigma(x) \rangle^2}$, below some freezing temperature where the order parameter susceptibility diverges, has turned out to be unstable in simple mean field theories. (The bracket means the thermal average and the bar indicates the average over random configurations.) The instability, which was first discovered by de Almeida and Thouless[2] is associated with a negative value of the order parameter susceptibility $\chi_{EA}$ below $T_g$ [3]:

$$\chi_{EA} = \int d^dx \; \overline{\chi^2(x,y)} = \int d^dx \; \overline{(\langle \sigma(x)\sigma(y) \rangle - \langle \sigma(x) \rangle \langle \sigma(y) \rangle)^2} < 0 \; . \tag{1}$$

One way to overcome the instability in replica theories is to break the replica symmetry [4]. We propose an alternative approach using dynamics instead of replicas: when we calculate the dynamic reponse function at a finite frequency $\omega$ and take the limit $\omega \to 0$ we find a solution which differs from the static susceptibility. The order parameter susceptibility calculated with this new solution is positive and the "replicon mode" becomes massive.

We study a random-easy-axis model [5], which is obtained from the conventional Landau-Ginzburg ferromagnet by adding a random uniaxial anisotropy term [6,7]:

$$H = \frac{1}{2} \int d^dx \left[ r_0 |\vec{\sigma}(x,t)|^2 + \frac{u}{4m} |\vec{\sigma}(x,t)|^2 + |\nabla \cdot \vec{\sigma}(x,t)|^2 - D\sqrt{m}\, (\hat{n}(x) \cdot \vec{\sigma}(x,t))^2 \right] \tag{2}$$

Here $\vec{\sigma}(x,t)$ is an m-component classical spin density vector in d-dimensional space and $\hat{n}(x)$ is a unit vector, which points along the random easy axis. We take $\hat{n}(x)$ to be a quenched variable with a uniform distribution over a unit sphere and assume that vectors at two different points x and x' are uncorrelated. In calculating the configurational averages over the distribution of $\hat{n}(x)$, we follow Pelcovits et. al. [6] and keep only the two lowest order cumulants [8]

$$D\ \overline{n_i n_j} = \frac{D}{m}\ \delta_{ij} \tag{3a}$$

$$D^2\ \overline{n_i n_j n_k n_l} = \frac{\Delta}{m}\ (\delta_{ij}\delta_{kl} + \delta_{ik}\delta_{jl} + \delta_{il}\delta_{jk}) \tag{3b}$$

where $\Delta = D^2/(m+2)$ and the bar indicates again the angular average.

The dynamics of the system is described by a Langevin model [9,10]

$$\frac{\partial \sigma_i(x,t)}{\partial t} = -\Gamma_0\ \frac{\delta H}{\delta \sigma_i(x,t)} + \eta_i(x,t) \tag{4}$$

where $\Gamma_0$ is a bare kinetic coefficent and $\eta_i(x,t)$ is a Gaussian white noise source. We solve this equation iteratively using the standard diagrammatic expansions of the time-dependent Ginzburg-Landau model [10,11]. The result for a self-energy $\Sigma(k,\omega)$, defined via the full dynamic response function $G^{-1}(k,\omega) = G_0^{-1}(k,\omega) + \Sigma(k,\omega)$ with $G_0^{-1}(k,\omega) = r_0 + k^2 - i\omega/\Gamma_0$, is given in fig. 1a while the expansion of the order parameter q is shown in fig 1b. In these figures, a straight line represents the full response function $G(k,\omega)$ and a solid line with a circle is the dynamic correlation function $C(k,\omega) = (2/\omega)\,\mathrm{Im}\ G(k,\omega)$. Dashed lines represent random fields $n_i(x)$ and the average over the random axes is indicated by two dashed lines joined with a cross. A circle with cross-hatching represents the order parameter q. In figures 1a and 1b only the first three diagrams and the first diagram, respectively, are of order O(1), the fourth one in fig. 1a is of order $1/\sqrt{m}$ and the rest of the diagrams in both figures are of order 1/m.

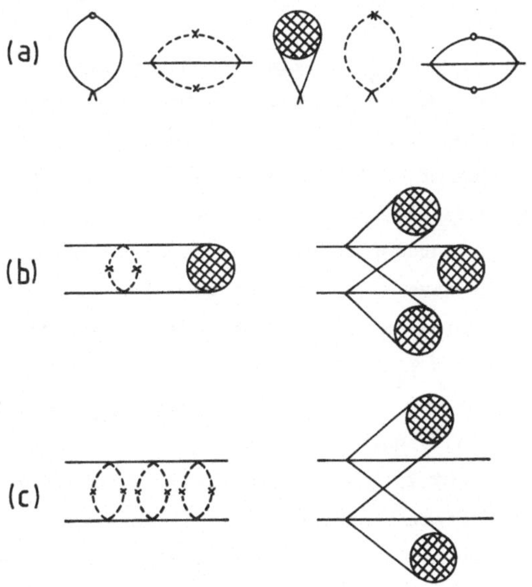


Fig. 1 Self-energy diagrams for $G(k,\omega)$ (a); ladder series for q (b) and for $\chi_{EA}$ (c).

In the limit $m\to\infty$, the self-consistent equations for the order parameter in zero external field and for the full response function are [6)]

$$q = \Delta q \int G^2(k,\omega;q) \frac{d^dk}{(2\pi)^d} \tag{5}$$

$$G^{-1}(k,\omega;q) = r + k^2 + s(\omega) \tag{6}$$

respectively. Here $r = r_0 + \Sigma(k,0)$ and $s(\omega)$ is the difference of the inverses of the dynamic and static response functions, $s(\omega) = G^{-1}(k,\omega) - G^{-1}(k,0)$. In the limit $m\to\infty$ the only frequency dependence in $\Sigma(k,\omega)$ comes from the second diagram in fig. 1a

$$s(\omega) = \frac{-i\omega}{\Gamma_0} - \Delta \int \left[ G(k,\omega) - G(k,0) \right] \frac{d^dk}{(2\pi)^d} \quad . \tag{7}$$

For the onset of a non-zero solution of the order parameter we obtain from (5)

$$\Delta\Pi(\omega=0;q) \equiv \Delta \int G^2(k,\omega=0;q) \frac{d^dk}{(2\pi)^d} = 1 \ , \tag{8}$$

which fixes the freezing temperature $T_g$ via the temperature dependence of the parameters (e.g. $r_0$) in G. It is exatly when (8) is satisfied that the order parameter susceptibility, which is given in the limit $m\to\infty$ by the first ladder sum in fig. 1c,

$$\chi_{EA} = \left. \frac{\Pi(\omega=0;q)}{1 - \Delta\Pi(\omega=0;q)} \right|_{q=0} \propto (T-T_g)^{-1} \ , \tag{9}$$

diverges. Below $T_g$ the function $\Pi$ depends on q via the inclusion of the q-dependent self-energy term in G. Thus to obtain an explicit value for $\chi_{EA}$ we also have to solve the self-consistent equation of state. In the limit $m\to\infty$, the only way in which the finite value of q affects both the equation of state and the equation for $\chi_{EA}$ is via this q-dependence of G in $\Pi$ so that to this order $\chi_{EA}$ retains its infinite value ( $\chi_{EA}^{-1} = 0$ ) through the whole spin-glass phase, as shown by Ma and Rudnick [9)]. Similar results have also been obtained for a random bond spherical model and for a generalized Sherrington-Kirkpatrick model of m-component spins in the limit $m\to\infty$, where $\chi_{EA}^{-1} = 0$ below $T_g$ as well [12)].

In higher orders in 1/m, q and $\chi_{EA}$ can still be calculated from eqs. (5) and (9), when the single rung parameter $\Delta$ is replaced by more general functions of q and G, $\Delta_q(q,G)$ and $\Delta_\chi(q,G)$ respectively, which in general are no longer equal because of possibly different combinatorial factors in figs. 1b and 1c. Furthermore, the order parameter is now non-local, $q=q(x-x')$, and in the equation of state one can no longer solve directly for $q(0)$ as before. Also the equation for $\chi_{EA}$ is non-local and no longer reduces to a simple algebraic one. However, these complications are of a technical nature; therefore for the sake of clarity we treat $\Delta_q$ and $\Delta_\chi$ as local.

We distinguish between two kinds of corrections in higher orders in 1/m. For some diagrams the combinatorial factors in $\Delta_q$ and $\Delta_\chi$ are the same so that $\Delta_q$ and $\Delta_\chi$ are still equal. If we keep only these "benign" diagrams, $\chi_{EA}^{-1}$ would still vanish below $T_g$. The instability comes from the last diagram in fig. 1b and from the corresponding susceptibility diagrams, for now the combinatorial factors in figs. 1b and 1c are different, since there are three different ways we can choose a pair of lines as external legs in the susceptibility diagram. This means that now the functions $\Delta_q$ and $\Delta_\chi$ are no longer equal but $\Delta_q < \Delta_\chi$, which means that $\chi_{EA}$ becomes negative. Since $\Delta_\chi$ is of second order in q this instability appears in order $(T_g-T)^2$ as in the random bond Ising model [13].

In the dynamics these negative order parameter fluctuations can be associated with the negative value of the kinetic coefficent

$$\Gamma^{-1} = \lim_{\omega\to 0} \Gamma^{-1}(\omega) = i \frac{\partial}{\partial\omega} G^{-1}(k,\omega) \Big|_{\omega=0} . \tag{10}$$

In the limit $m\to\infty$ the only frequency dependence in the inverse response function was in $s(\omega)$ so that $\Gamma(\omega)$ can be calculated from the equation

$$\Gamma^{-1}(\omega) - \Gamma_0^{-1} = \Delta\Gamma^{-1}(\omega) \int G(k,\omega)\, G(k,0) \frac{d^dk}{(2\pi)^d} . \tag{11}$$

At $\omega=0$ this equation reduces to $\Gamma = \Gamma_0(1 - \Delta\Pi(\omega=0;q))$ , which goes to zero like $(T-T_g)$ above the critical temperature $T_g$ and remains zero through the spin-glass phase. If we go again beyond the order O(1) we have to replace the parameter $\Delta$ in eq. (10) by a more general rung $\Delta_\chi(q,G)$. As in the order parameter susceptibility calculation, the combinatorial factor in $\Delta_\chi$ corresponding to the last diagram of fig. 1c is three times that of the corresponding diagram for $\Delta_q$. This means that the kinetic coefficient $\Gamma$ becomes negative, so $G(k,\omega)$ describes an unstable growth in response to an external perturbation, rather than a decay.

The stability of the order parameter fluctuations can be restored with the help of a non-zero solution of the self-consistent equation for the quantity $s(\omega)$ in the static limit. The self-consistent equation (11) expressed in terms of $s(\omega)$ is

$$s(\omega) = -\frac{i\omega}{\Gamma_0} + \Delta_\chi\, s(\omega) \int \frac{G^2(k,0;q)}{1 + s(\omega)\, G(k,0;q)} \frac{d^dk}{(2\pi)^d} \tag{12}$$

Expanding in $s(\omega)$ we obtain a quadratic equation in s which has in addition to the trivial solution $s(0)=0$, also a non-zero solution

$$s(0) = -\frac{(1 - \Delta_\chi\, \Pi(\omega=0;q))}{\Delta_\chi\, \Pi^{(3)}(\omega=0;q)} > 0 , \tag{13}$$

where

$$\Pi^{(3)}(\omega;q) = \int G^3(k,\omega;q) \frac{d^dk}{(2\pi)^d} .$$

We can use this new value of s(0) to calculate the quantity $\Pi(\omega;q)$ which enters the expressions for $\chi_{EA}$ and q. When we expand $\Pi$ in s(0) and use the self-consistent equation for q, we obtain in the leading order the result that $\chi_{EA}$ simply changes sign from the unstable, negative value obtained in the static calculation. Thus the order parameter susceptibility is now positive below $T_g$ (" a massive replicon" ) and $\chi_{EA}^{-1} \propto (T_g-T)^2$ when we approach $T_g$ from below.

The above analysis can be repeated for the random bond Ising model as well and we find stability restored in the same manner. In this model one can use the Hubbard-Stratonovich transformation for the partition function and employ a simple Langevin model for the dynamics of the auxiliary variables of this transformation[14).]

We have treated the order parameter q as a static quantity in that the equation of state for q (obtained from the static calculation) is not taken to be modified by the finite value of s(0). Thus the finite value of s(0) is simply a discontinuity in the response of the spin system as $\omega \to 0$. It can be looked upon as a breakdown of the linear response theory: such a possibility has been discussed earlier in connection with the static spin susceptibility [15)].

References

1. S.F. Edwards and P.W. Anderson, J. Phys. F 5, 965 (1975).
2. J.R.L. de Almeida and D.J. Thouless, J. Phys. A 11, 983 (1978).
3. A.J. Bray and M.A. Moore, J. Phys. C 12, 79 (1979) ; E. Pytte and J. Rudnick, Phys. Rev. B19, 3603 (1979).
4. A.J. Bray and M.A. Moore, Phys. Rev. Lett. 41, 1068 (1978) ; G. Parisi, Phys. Rep. 67, 25 (1980) and references therein.
5. R. Harris, M. Plischke and M.J. Zuckermann, Phys. Rev. Lett. 31, 160 (1973).
6. R.A. Pelcovits, E. Pytte and J. Rudnick, Phys. Rev. Lett. 40, 476 (1978); R.A. Pelcovits, Phys. Rev. B19, 465 (1979).
7. A. Aharony, Phys. Rev. B12, 1038 (1975).
8. A. Aharony and M.E. Fisher, Phys. Rev. B8, 3323 (1973).
9. S.K. Ma and J. Rudnick, Phys. Rev. Lett. 40, 589 (1978).
10. S.K. Ma, Modern Theory of Critical Phenomena (Addison Wesley,Reading,Mass.,1976).
11. J.A. Hertz and R. Klemm, Phys. Rev. B20, 316 (1979) ; Phys. Rev. Lett. 40, 1397 (1978).
12. J.M. Kosterlitz, D.J. Thouless and R.J. Jones, Phys. Rev. Lett. 36, 1217 (1976); J.R.L. de Almeida, R.C. Jones, J.M. Kosterlitz and D.J. Thouless, J. Phys. C 11, 1871 (1978).
13. A. Khurana and J.A. Hertz, J. Phys. C 13, 2715 (1980) ; M.V. Feigelman and A.M. Tsvelik, Sov. Phys. JETP 50, 1222 (1980).
14. A. Khurana ( NORDITA preprint, 81/10, 1981).
15. A.J. Bray and M.A. Moore, J. Phys. C 13, 419 (1980) ; D.J. Thouless, J.R.L. de Almeida and J.M. Kosterlitz, J. Phys. C 13, 3271 (1980).

ORDER AS A CONSEQUENCE OF DISORDER IN FRUSTRATED ISING MODELS

R. Bidaux, J.P. Carton, R. Conte
DPh-G/PSRM, CEN Saclay, B.P. N° 2, 91190 Gif sur Yvette, France

and J. Villain
DRF/DN, CEN Grenoble, 85X, 38041 Grenoble Cédex, France.

## 1. Frustrated Ising models with highly degenerate ground states.

An Ising model

$$\mathcal{H} = -\frac{1}{2}\sum_{ij} J_{ij} S_i S_j \qquad (S_i = \pm 1) \qquad (1)$$

is called "frustrated" if all pairs (i,j) have not their minimum energy in the ground states. Specifically, systems of interest in the present communication have a highly degenerate (but not too degenerate) ground state and show no long range order (LRO) at T=0, i.e.

$$m_k^2 = \lim_{N \to \infty} N^{-2} \mathcal{N}^{-1} \sum_{\alpha} \left(\sum_i S_i \cos k.R_i\right)_{\alpha}^2 = 0 \qquad \forall k \qquad (2)$$

N is the number of spins and $\mathcal{N}$ is the number of ground states, each of which is designated by the index $\alpha$ . Examples of such systems are

i) Danielian's model, or Ising model with antiferromagnetic interactions between nearest neighbours on a fcc lattice.[1,2,3,4] A possible ground state is the periodic structure with the unit cell displayed by Fig. 1. A set $E_z$ of $N_z$ other ground states can be obtained by multiplying all spins of the n'th (001) layer by $\varepsilon_n = \pm 1$. $N_z$ is the number of (001) layers. There is a similar set $E_x$ of $N_x$ ground states constituted by randomly stacked, antiferromagnetically ordered (100) layers. Finally there is a set $E_y$ of $N_y$ ground states. The

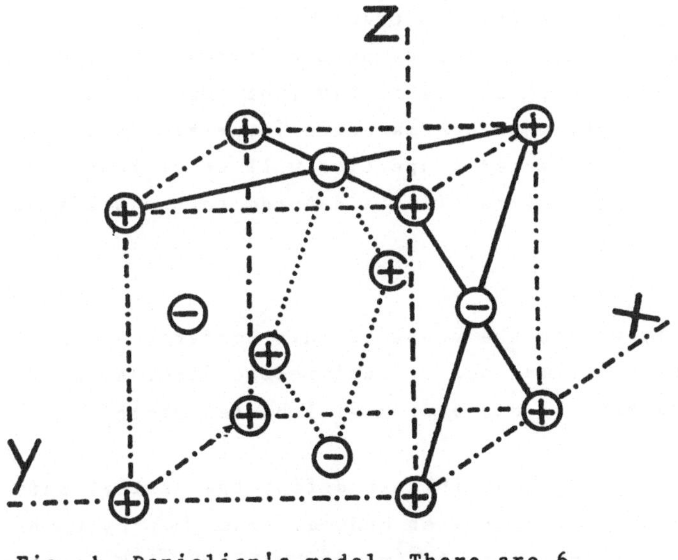


Fig. 1. Danielian's model. There are 6 low temperature structures which are periodic. The Figure shows a possible unit cell.

ground state entropy for a large cubic sample is $N^{1/3}\ln 2$. Equality (2) is easily checked.

ii) The $K_2NiF_4$ model[5,6] (in its Ising version) may be viewed as an anisotropic Danielian model. Only the set $E_z$ of ground states remains. The properties are very similar to those of the isotropic model.

iii) The two-dimensional "domino" model[7,8,9] is displayed by Fig. 2. It has two non-equivalent sites A, B and three different interactions $J_{AA}$, $J_{AB}$ and $J_{BB}$. A and B atoms form alternating chains. $J_{AA}$ is positive (ferromagnetic) and $J_{BB}$ is negative. The main interest of the domino model is to be accessible to analytic methods. In particular the partition function of the pure system can be calculated and shows a continuous transition though relation (2) is satisfied[8].

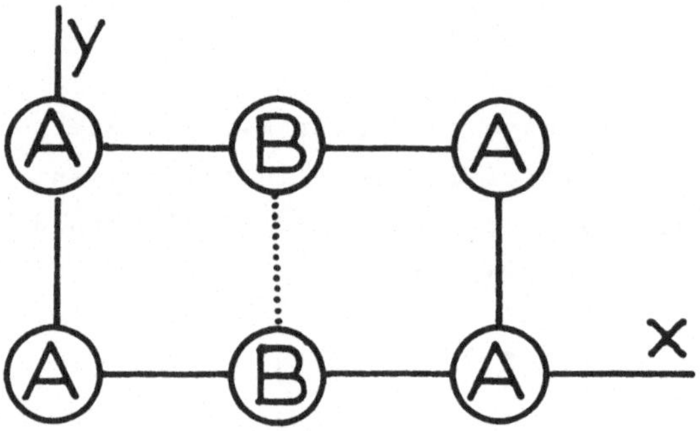

Fig. 2. Unit cell of the domino model. Full lines are ferromagnetic bonds. The dotted line is antiferromagnetic.

iv) The Ising model with antiferromagnetic interactions between nearest neighbours on the triangular lattice may also be mentioned. It is not so interesting since it undergoes no transition.

Models (i), (ii) and (iii) order at low temperature though they do not at T=0. In addition, model (iii) orders at T=0 (for appropriate values of the exchange constants) when non-magnetic impurities are randomly added. These rather unexpected properties will be reviewed in the next two Sections. The last Section will be devoted to the study of percolation in the domino model.

2. Ordering at low temperature.

This point is not really within the scope of this Conference, but a comparison between effects of thermal and stoechiometric disorder has some interest. For definiteness the discussion will be restricted to Danielian's model.

For any ground state $|\alpha\rangle$ it is possible to define the partial partition function $Z_p^{(\alpha)}$ corresponding to states deduced from $|\alpha\rangle$ by reversal of p different spins. In particular, $Z_o^{(\alpha)}=\exp(-W_o/T)$ where $W_o$ is the ground state energy. One can define

$$Z_\alpha = Z_o^{(\alpha)}+Z_1^{(\alpha)}+Z_2^{(\alpha)}+Z_3^{(\alpha)}+\ldots \qquad \text{and} \qquad F_\alpha = -T\ \ln Z_\alpha \quad (3)$$

The low temperature expansion of $Z_\alpha$ can easily be obtained, and the expansion of the $\alpha$-dependent free energy $F_\alpha$ follows. For states of set $E_z$ the result is given below as a function of the parameters $\varepsilon_n$ defined in Section 1, which define the state $|\alpha\rangle$.

$$F_\alpha/NT = W_o/NT - \exp(8J/T) - 4\exp(12J/T) - (31/2)\exp(16J/T)$$
$$+(1/2N_z)\sum_n (1-\varepsilon_n\varepsilon_{n+2}) \exp(16J/T) +\ldots\ldots \qquad (4)$$

Analogous results hold for sets $E_x$ and $E_y$. Formula (4) was essentially obtained by Slawny[10] although he did not specify the explicit dependence with respect to the parameters $\varepsilon_n$. The low temperature structure is given by the minimum of the free energy (4), which is obtained when all $\varepsilon_n$'s are equal. This defines the periodic structure with the unit cell displayed by Fig. 1. There are 5 other possible structures deduced by the symmetry operations of the cubic group.

The $\alpha$-dependence of the free energy (4) results from the following fact. Although all ground states have the same energy $W_o$, the energy of excited states depend on $\alpha$ . For instance, consider the rhombus formed by dotted lines on Fig. 1, and the excited state deduced from a ground state $|\alpha\rangle$ by reversal of the 4 corresponding spins. The excitation energy depends on $\alpha$ . It is $-16J$ in the case of the Figure, but it may be higher for some other ground states.[2,10]

The domino model is discussed in Ref. 8 in the case $J_{AA} > -J_{BB} > |J_{AB}|$. It is ferrimagnetic at low temperature and the mechanism is somewhat analogous. When T goes to zero, the A-magnetisation tends to saturation for an infinite sample, while the B magnetisation vanishes. But at T=0 the magnetisation is exactly zero.

3. Effect of stoechiometric disorder at temperature T=0.

The domino model has another surprising property for appropriate values of the exchange constants, namely

$$J_{AA} > -J_{BB} > 2|J_{AB}| \qquad (5)$$

Let non-magnetic impurities be randomly distributed on A and B sites with respective concentrations $x_A$ and $x_B$. $x_B$ is assumed to be much larger than $x_A$, but however not too large as precised below. Then the ground state becomes ferromagnetic. The mechanism is quite simple (Fig. 3b). A chains are coupled through <u>odd</u> B chains (singlets, triplets, quintuplets, etc.).

For any concentration $x_A$ of non-magnetic impurities on A sites, there is a percolation concentration $X(x_A)$ of B impurities, above which

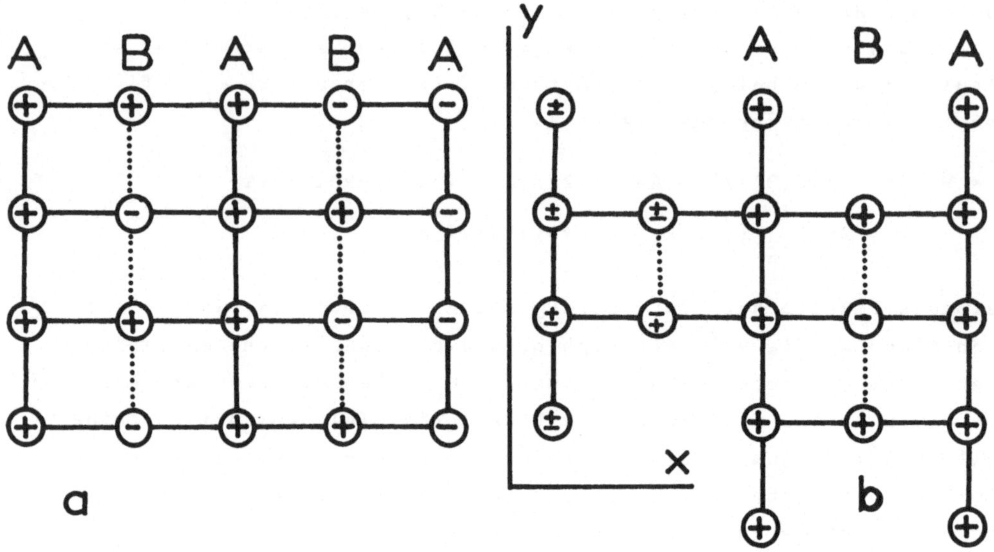


Fig. 3. (a) A particular ground state of the pure domino model. (b) Effect of non-magnetic impurities : the two right-hand A chains are ferromagnetically coupled while the left-hand chain is loose.

there is no infinite cluster. On the other hand there is a critical concentration $X'(x_A)$ above which the Ising model (1) is disordered at T=0, i.e. identity (2) is satisfied. In standard ferromagnets, $X'=X$. In the domino model, $X'<X$. The reason is that B atoms involved in even clusters are inactive (Fig. 3b, left hand part). In particular, if $x_A \ll 1$, $1-X \ll 1$, the effective B sites are the isolated atoms, the number of which is $x_B^2(1-x_B)N_B$. The critical concentration $x_B = X'(x_A)$ is reached when this is equal to $X(1-X)N_B$ (number of isolated B atoms + number of pairs at percolation). This calculation (or a more careful one[8]) yields

$$X' \simeq X - (1-X)^2 \qquad (6)$$

Restoration of conventional order by non-magnetic impurities is apparently a special feature of the domino model. In the $K_2NiF_4$-Ising model, the dominant effect is probably due to pairs of neighbouring impurities on nearest neighbour (001) layers. Assuming each layer to be antiferromagnetically ordered, these pairs can be seen to produce a random field. According to Pytte et al[11], such a random field destroys order in contradiction with the initial assumption of ordered layers. Thus non-magnetic impurities are expected to produce a disordered or spin-glass state.

4. Properties of the domino model near percolation at $T \neq 0$.

When $x_B > X'$, no LRO occurs at T=0 because paired B atoms are uneffective. At $T \neq 0$ they are effective and LRO cannot be excluded a priori. However the domino model can be shown[8] to be paramagnetic at all temperatures for concentrations $x_B > X''$, where again $X''(x_A)$ is smaller than the percolation threshold $X(x_A)$, but possibly larger than $X'(x_A)$. If $x_A \ll 1$, an upper limit of $X''$ is $X-(1-X)^2/4$.

5. Conclusion.

The main interest of the domino model is to be tractable by analytic methods. Its properties are only of interest if they have some generality. i) Temperature-induced order is probably a fairly common property. In particular it may be expected in three-dimensional systems with ground state entropy proportional to $N^{1/3}$, from low-temperature expansions. It is probably more exceptional in two-dimensional systems. ii) Order induced by stoechiometric disorder is probably exceptional in Ising models. The domino model seems to be very special. iii) The fact that $X''$ can be strictly smaller than the concentration threshold X is an interesting feature of the domino model. It would be of interest to consider this possibility in other frustrated systems.

Of course the fact that thermal, stoechiometric (or quantum) disorder can produce magnetic order in Ising models is contrary to intuition. This is the main motivation of this communication, which is quite marginal in this Conference.

REFERENCES

1. P.W. ANDERSON, Phys. Rev. 79, 350, 705 (1950)
2. A. DANIELIAN, Phys.Rev. 133, A 1345 (1964) and Refs therein.
3. M.K. PHANI, J.L. LEBOWITZ, M.H. KALOS, C.C. TSAI, Phys. Rev. Lett. 42, 577 (1979)
4. K. BINDER, J.L. LEBOWITZ, M.K. PHANI, M.H. KALOS, preprint (1981)
5. R. PLUMIER, J. Physique 24, 741 (1963) and References therein.
6. L.J. DE JONGH, A.R. MIEDEMA, Adv. Phys. 23, 1 (1974) and Refs therein.
7. G. ANDRE, R. BIDAUX, J.P. CARTON, R. CONTE, L. DE SEZE, J. Physique 40, 479 (1979)
8. J. VILLAIN, R. BIDAUX, J.P. CARTON, R. CONTE, J. Physique 41, 1263 (1980)
9. V.V. BRYKSIN, A.Yu GOL'TSEV, E.E. KUDINOV, J. Phys. C 13, 5999 (1980)
10. J. SLAWNY, J. Stat. Phys. 20, 711 (1979)
11. E. PYTTE, Y. IMRY, D. MUKAMEL, Phys. Rev. Lett. 46, 1173 (1981)

# SPIN GLASSES
## WITH SPECIAL EMPHASIS ON FRUSTRATION EFFECTS

Gérard Toulouse
Laboratoire de Physique de l'Ecole Normale Supérieure
24 rue Lhomond, 75231 Paris Cedex 05

Abstract

Some views on the historical evolution of the spin glass problem are given in introduction. Direct experimental testing of the existence of a phase transition is briefly discussed. Concepts related to frustration and developments around these themes are cursorily surveyed.

*"Vorrei e non vorrei".*
*Zerlina*

## 1. A look at the historical evolution of the problem of spin glasses

During the last twenty years, the physics of disordered systems has come to the forefront of theoretical physics, requiring the introduction of new concepts and raising well defined problems. Like the physics of polymers, percolation and localization, the physics of spin glasses has had an interesting history, with a good lot of surprises, which we now try to put into perspective.

The original materials, on which were first observed the characteristic properties of spin glasses, were dilute alloys of transition metal impurities in noble matrices. With his 1932 experiments [1], Louis Néel was hoping to get a better understanding of the magnetic properties of pure transition metals. It was only in 1956 that the flattening of the magnetic susceptibility was observed by J. Owen et al. [2], being able to go to lower temperatures. First surprise. In the categories of the time, this could only be interpreted as an antiferromagnetic transition. However, in some respects (remanent magnetization, hysteresis), the materials behaved like ferromagnets and besides it was hard to imagine a long range antiferromagnetic ordering in these alloys.

The specific heat experiments, at the end of the 50 's, showing a contribution linear in T at low temperatures and a very round maximum, led to a different type of interpretation, which essentially negated the existence of a transition. This interpretation came to dominate in the 60's and found its expression in the term "spin glass" coined by B.R. Coles at the end of the decade. So much so that the Mössbauer observations of C.E. Violet and R.J. Borg in 1966, finding a well-defined transition temperature [3], received little attention. When E.C. Hirschkoff et al., in 1971, observed a plateau in the low temperature magnetization [4], they attached little importance to this effect, felt like a nuisance because they were really interested in the Kondo effect (the isolated impurity problem) and not in interaction effects.

Rebirth of interest came with the experiments of V. Cannella and J.A. Mydosh, in 1972, finding a sharp cusp in the low-field a.c. susceptibility [5]. In 1975, S.F. Edwards and P.W. Anderson [6] reactivated the theoretical front, by opening the way to a third type of interpretation, namely showing the possibility of a phase transition of novel type.

During the 70's, it has been shown that the spin glass properties were observable in a much wider diversity of materials, than it had been contemplated before. For instance :

i) materials concentrated in magnetic atoms (and not only dilute) [7],
ii) rare earth impurities (and not only transition) [8],
iii) amorphous materials (and not only substitution alloys) [9],
iv) electric dipolar [10] or quadrupolar [11] glasses (and not only magnetic moments).

The progressive extension of the label "spin glass" to new classes of materials has considerably influenced our vision of the problem. Thus some classification or interpretation schemes have become obsolete because they were too restrictive or specific to encompass this diversity. In order to illustrate this evolution of ideas, I shall evoke a recent controversy (one after many others). In a recent review article [12], J.A. Mydosh says :

"Experimentally there does not seem to exist today a spin glass where only antiferromagnetic exchange is present. Yet two attempts with well-defined and randomly distributed, antiferromagnetically coupled impurities, In doped into CdS [13] and P doped into Si [14], have failed to find spin glass-like freezing down to temperatures of about 3 mK. However, solely antiferromagnetic spin coupling will certainly lead to 'frustrated' spins..."

But, within a year, several materials with solely antiferromagnetic spin couplings have been found to exhibit spin glass behaviour [15]...

Computer simulation has also contributed a great deal to the advancement of ideas during the last five years. The interpretations, concerning the existence of a phase transition for short range models in three dimensions, have evolved grosso modo from a positive to a negative answer [16]. There has been a kind of rotation of positions with experimentalists, many of the latter finding now evidence for a sharp phase transition and a well-defined equilibrium spin glass phase, in a new series of measurements (magnetization laws, frequency dependence of the susceptibility, magnetic hysteresis, nuclear magnetic resonance, mixed phases ferro-spin glass).

The theoretical analysis of spin glasses has also made some unexpected steps forward. It has been slowly realized that mean field theory itself was very subtle indeed for spin glasses, containing a fair share of the mystery : replica symmetry breaking, critical fields, failure of a simple linear response formula. An ensemble of detailed predictions is now offered, and may stimulate experimental investigations : magnetization plateau at low temperatures, freezing of the transverse degrees of freedom, mixed phases [17-24].

## 2. The existence of a sharp phase transition

It should be made clear that this question can be experimentally resolved, by measurements made at high temperatures $T > T_c$, i.e. without having to worry with remanence effects. The theory predicts that the magnetization behaves at high temperatures like :

$$M(H) \simeq \chi H - aH^3 + O(H^5) \quad , \quad T > T_c, \tag{1}$$

where the susceptibility $\chi$ varies continuously with temperature, and the coefficient $a(T)$ diverges, for $T \to T_c$, like

$$a(T) \simeq (T - T_c)^{-\gamma} \quad . \tag{2}$$

At $T = T_c$, the expected behaviour is :

$$M(H) \simeq \chi H - b\,H^{1+2/\delta} + \dots \tag{3}$$

This type of behaviour is suggested by the Mattis model [25-27] and the exponents are defined in such a way that the usual scaling laws still hold. In mean field theory (defined by the infinite range model), one finds $\gamma = 1$ and $\delta = 2$.

Change of analytic behaviour from (1) to (3), as $T \to T_c$, is proof of a transition and it is astonishing that comparatively so little experimental endeavor has been directed toward the determination of $\gamma$ and $\delta$. Concerning exponent $\gamma$, only one experimental group [28] has presented a result (logarithmic divergence for $a(T)$), which calls for checks. As for exponent $\gamma$, the values produced up to now [8, 29, 30] are scattered between $\delta = 1{,}3$ and $\delta = 5$. But the mere fact of finding a value different from $\delta = 1$ fits well with the observation of a rapid field rounding of the susceptibility cusp and lends support to the existence of a phase transition [31].

## 3. Frustration

Taken generally, the word frustration expresses a contradiction of the interactions. Is frustrated a material or a model which possesses no state where all interactions would be simultaneously satisfied. Because their ground states have little stability, these systems, whether regular or disordered, have a tendency to exhibit metastability effects. They are also particularly sensitive to modifications of external parameters, leading to cascades of phase transitions, in systems with commensurate-incommensurate transitions, and some strange behavior in spin glasses (e.g. magnetic linear response).

In magnetic models, where the Hamiltonian is :

$$\mathcal{H} = -\sum_{(ij)} J_{ij}\, \vec{S}_i . \vec{S}_j \tag{4}$$

it is possible to push the analysis further [32]. Consider, for specificity, a square lattice with Ising spins ($S_i = \pm 1$), nearest-neighbour interactions of fixed modulus but arbitrary sign. Clearly, this Hamiltonian is invariant under the following local transformations (acting on both spin and interactions variables) :

$$\begin{array}{lcl} S_i & \to & - S_i \\ J_{ij} \text{ (j adjacent to i)} & \to & - J_{ij} \end{array}$$

Analogy with the gauge transformations of electrodynamics suggests that the interactions $J_{ij}$ are not the good physical variables. One is therefore led to introduce the frustration function (loop function) defined by :

$$\Phi(c) = \prod_C J_{ij}$$

where the $J_{ij}$'s are taken along a contour C.

The thermodynamic properties (partition function, specific heat,...) of the system defined by (4) are gauge invariant and they depend on the interactions only via the frustration functions. This gauge invariance is broken by an external magnetic field.

It is not possible to mention all the developments that are related to the concept of frustration. But we shall present here some vistas.

i) Periodic frustrated models

Various families of models have been actively studied : "domino" models [33], ANNNI (Axial Nearest Neighbour Ising) models [34], fully frustrated models (for convenience, these will be discussed in a special entry), etc.

A variety of properties, sometimes unusual, have been discovered :

- finite transition temperatures, despite large ground state denegeracies (lack of long range order and vanishing interface energy, at zero temperature) [35],
- restoration of order with the introduction of impurity disorder [36],
- Lifshitz points, commensurate-incommensurate transitions with more or less "devilish" staircases [37].

ii) Fully frustrated models

A vast, rooted in old times, literature exists on these systems where every elementary loop (plaquette) is frustrated. Let us mention antiferromagnetic models on triangular lattices [38] or f.c.c. lattices [39], the "odd model" of Villain [40], the Ashkin-Teller model [41], and generalizations in arbitrary dimension for f.c.c. [42] or simple cubic [43] lattices.

Some general features have emerged from the study of these models :

- "overblocking effect" for space dimensions $d > 4$ (i.e., compulsory appearance in any state of a density of plaquettes with more than one unhappy bond due to geometric hindrances) leading to a ground state energy $E_0$ varying as

$$E_0 \simeq -\sqrt{z} \quad , \tag{5}$$

where z is the coordination number. This differs from the usual linear relation, normally found in non frustrated systems, but it is alike spin glass behaviour,

- Onsager-type local field corrections [44], which are not negligible in the mean field limit (again alike spin glasses),

- in dimension two, absence of a phase transition at finite temperatures, but oscillating behaviour superposed on an algebraïc (exponent 1/2) asymptotic decay for the zero temperature correlation function [45],

- in dimension three, the transition may be first order (f.c.c. case [39]) or second order (s.c. case, apparently). Fully frustrated simple cubic lattices are specially interesting, because they constitute a natural limit in the phase diagram of lattice gauge theories [46], which are actively studied by field theorists [47].

iii) Systems with finite residual entropy

Some of the previous models (but not all of them) plus some others (such as the q-state antiferromagnetic Potts model) belong to this category. It has been conjectured that these systems might exhibit a particular low temperature phase (with algebraïc decay of the correlations), in high enough dimensionalities [41].

iv) Approach to spin glasses, by dilution of periodic frustrated systems

The idea is to introduce the essential ingredients in two steps : frustration firstly, disorder secondly [48]. It is also a realistic description for part of the phase diagram of various materials, with possible appearance of mixed phases antiferromagnetic-spin glass.

v) Connections with gauge theories ; topological defects and their hydrodynamics

Here are an ensemble of approaches, which appear natural and therefore promising, but which have not yet made their junction with the mainstream, either theoretically or experimentally. We enter here some references [49-50]. En passant, we note the extension of analogous concepts into the physics of glasses [51].

vi) Random frustration ($J = \pm 1$) models, in various space dimensions

It has become a general rule, in numerical simulations as well as in theoretical analyses, to look systematically for the difference of properties between frustrated models, on one side, and non frustrated models, on the other side. And in the case of random frustrated systems, it is instructive to compare the cases of a ($\pm$ 1) and a gaussian distribution (for the bond interactions).

In particular, in dimension two, for Ising spins, there has been quite a number of studies to find the threshold of disappearance of ferromagnetism and to characterize the nature of the ground state beyond, when the concentration of negative bonds is increased. At the present time, no consensus has yet been reached : grosso modo, the controversy may be described as a debate between the Jülich [16] and Grenoble [52] view points.

Jülich, by a numerically "exact" calculation of the partition function and the correlation function of finite samples, estimates the threshold of ferromagnetism at a concentration $x_c = 0{,}12 \pm 0{,}02$ of negative bonds. Beyond this threshold, the correlation function $\overline{\langle S_0 S_R \rangle^2}$ is predicted to have a power law asymptotic decay, at zero temperature (for a gaussian distribution of interactions, this function would tend asymptotically toward a finite limit).

Grenoble, using graph theory algorithms to analyze its samples together with real space renormalization arguments, obtains the ferromagnetic threshold at $x_c \simeq 0{,}10$. Beyond this threshold, a range of concentrations is found where some ground state rigidity subsists, corresponding to a regime characterized as "random antiphase state". A second threshold at $x_c \simeq 0{,}15$ would mark the transition to a non rigid, paramagnetic, ground state.

vii) Quantum spins and frustration

The studies, heretofore mentioned, assume classical spins. It is natural to introduce the quantum nature of the spins, but it is not a simplification and few solid results have been obtained [53].

To close this incomplete review, let us mention that some interesting discussions of the concepts related to frustration may be found in various review articles [16-18]. Perhaps the most spectacular theoretical advances in the last couple of years have come from the study of the infinite range model of spin glasses [54]. It is worth noting that, for this model, in the thermodynamic limit, the gaussian or (± 1) distributions for bond interactions have exactly the same physics. The frustration ingredient is therefore quite present in what is now being considered as the mean field theory of spin glasses. There remains a hope that the concepts related to frustration will be able to give a clearer, more geometric, picture of both the mean field and fluctuation aspects in spin glasses, still so much enrobed in mystery.

## References


1 - L. Néel, Journal de Physique et le Radium, série VII, t.3 (1932) 160.
2 - J. Owen, M. Browne, W.D. Knight, C. Kittel, Phys. Rev. 102 (1956) 1501.
3 - C.E. Violet, R.J. Borg, Phys. Rev. 149 (1966) 540.
4 - E.C. Hirshkoff, O.G. Symko, J.C. Wheatley, J. Low Temp. Phys. 5 (1971) 155.
5 - V. Cannella, J.A. Mydosh, Phys. Rev. B 6 (1972) 4220.
6 - S.F. Edwards, P.W. Anderson, J. Phys. F 5 (1975) 965.
7 - G.J. Nieuwenhuys, B.H. Verbeek, J.A. Mydosh, J. Appl. Phys. 50 (1979) 1685.
8 - H. Maletta, W. Felsch, Phys. Rev. 20 (1979) 1245.
9 - J.P. Renard, J.P. Miranday, F. Varret, Solid State Comm. 35 (1980) 41.
10 - F. Borsa, U.T. Höchli, J.J. van der Klink, D. Rytz, Phys. Rev. Letters 45 (1980) 1884.
11 - N.S. Sullivan, M. Devoret, B.P. Cowan, C. Urbina, Phys. Rev. B 17 (1978) 5016.
12 - J.A. Mydosh, in Liquid and Amorphous Metals, ed. by E. Lüsscher and H. Coufal (Sijthoff and Noordhoff, 1980).
13 - R.B. Kummer, R.E. Walstedt, S. Geschwind, V. Narayanamurti, G.E. Devlin, Phys. Rev. Letters 40 (1978) 1098.
14 - K. Andres, Bull. Am. Phys. Soc. 24 (1979) 262.
15 - J. Ferré, J. Pommier, J.P. Renard, K. Knorr, J. Phys. C 13 (1980) 3697 ;
S. Nagata et al., Phys. Rev. B 22 (1980) 3331 ;
R.R. Galazka, S. Nagata, P.H. Keesom, Phys. Rev. B 22 (1980) 3344.
16 - K. Binder, in Fundamental Problems in Statistical Mechanics, ed. by E.G.D. Cohen (North-Holland, 1980).
17 - P.W. Anderson, in Ill-condensed Matter, ed. by R. Balian, R. Maynard, G. Toulouse (North-Holland 1979).
18 - S. Kirkpatrick, ibidem.
19 - G. Parisi, J. Phys. A 13 (1980) 1887 ; and references therein.
20 - G. Parisi, G. Toulouse, J. Physique Lettres 41 (1980) L 361.
21 - G. Toulouse, J. Physique Lettres 41 (1980) L 447.
22 - J. Vannimenus, G. Toulouse, G. Parisi, J. Physique 42 (1981) 565.
23 - G. Toulouse, M. Gabay, J. Physique Lettres 42 (1981) L 103 ;
M. Gabay, G. Toulouse, submitted for publication.
24 - A more detailed survey, with extensive list of references on the matters of this chapter, may be found in : G. Toulouse, Proc. of the French Physical Society meeting in Clermont-Ferrand (1981), to be published.
25 - D.C. Mattis, Phys. Letters A 56 (1976) 421.
26 - J. Chalupa, Solid State Comm. 24 (1977) 429.
27 - M. Suzuki, Progr. Theor. Phys. 58 (1977) 1151.
28 - S. Chikazawa, Y.G. Yuochunas, Y. Miyako, J. Phys. Soc. Japan 49 (1980) 1276.
29 - M. Simpson, J. Phys. F 9 (1979) 1377.
30 - C.A.M. Mulder, A.J. van Duyneveldt, J.A. Mydosh, Phys. Rev. B 23 (1981) 1384.
31 - P. Monod, H. Bouchiat, this conference.
32 - G. Toulouse, Commun. Phys. 2 (1977) 115 ; also Lecture Notes in Physics Vol 115 (Springer 1980) ;
G. Toulouse, J. Vannimenus, La Recherche (novembre 1977).
33 - G. André, R. Bidaux, J.P. Carton, R. Conte, L. de Sèze, J. Physique 40 (1979) 479.
34 - M.E. Fisher, W. Selke, Phys. Rev. Letters 44 (1980) 1502.
35 - B. Derrida, J.M. Maïllard, J. Vannimenus, S. Kirkpatrick, J. Physique Lettres 39 (1978) L 465.
36 - J. Villain, R. Bidaux, J.P. Carton, R. Conte, J. Physique 41 (1980) 1263.
37 - S. Aubry, in Solitons and Condensed Matter Physics, Solid State Science 8 (Springer 1978) ;
P. Bak, Phys. Rev. Letters 46 (1981) 791 ; J. Vannimenus, ENS preprint (1981).
38 - G. Wannier, Phys. Rev. 79 (1950) 357.
39 - M.K. Phani, J.L. Lebowitz, M.H. Kalos, C.C. Tsai, Phys. Rev. Letters 42 (1979) 577.
40 - J. Villain, J. Phys. C 10 (1977) 1717.
41 - A.N. Berker, L.P. Kadanoff, J. Phys. A 13 (1980) L 259.
42 - S. Alexander, P. Pincus, J. Phys. A 13 (1980) 263.
43 - B. Derrida, Y. Pomeau, G. Toulouse, J. Vannimenus, J. Physique 40 (1979) 617 ; 41 (1980) 213.

44 - R. Brout, H. Thomas, Physics 3 (1967) 317.
45 - M. Gabay, J. Physique Lettres 41 (1980) L 427.
46 - G. Toulouse, J. Vannimenus, Phys. Reports 67 (1980) 47.
47 - G. Bhanot, M. Creutz, Phys. Rev. B 22 (1980) 3370 ;
G. Jongeward, J. Stack, private communication.
48 - L. de Sèze, J. Phys. C 10 (1977) L 353.
49 - I. Dzyaloshinskii, G. Volovik, J. Physique 39 (1978) 693.
50 - G. Toulouse, in Recent Developments in Gauge Theories (Plenum, 1980) ; Phys. Reports 49 (1979) 267.
51 - N. Rivier, Phil. Mag. 40 (1979) 859.
52 - F. Barahona, R. Maynard, R. Rammal, J.P. Uhry, to appear in J. Phys. A.
53 - F. Fazekas, J. Phys. C 13 (1980) L 209 ;
A.J. Bray, M.A. Moore, J. Phys. C 13 (1980) L 655.
54 - G. Parisi, this conference.

# METAL-INSULATOR TRANSITIONS AND LOCALIZATION

# LOCALIZATION AND INTERACTION EFFECTS IN A TWO DIMENSIONAL ELECTRON GAS

M. Pepper
Cavendish Laboratory
Cambridge, U. K.

## ABSTRACT

A review is presented of recent work on the logarithmic corrections due to weak localization and interaction effects in the 2D electron gas in the silicon inversion layer. It is shown that very good agreement exists between experiment and theory, and that the two logarithmic corrections can be separated by a magnetic field or a combination of magnetic and electric fields. Unlike exponential localization in a band tail, the weak (power law) localization is suppressed by a magnetic field and it is suggested that this leads to the return of a discontinuity in conductance at zero temperature, i.e. a minimum metallic conductance. It is suggested that these corrections do not have any relevance to the quantized, 2D, Hall resistance.

## INTRODUCTION

The inversion (or accumulation) layer of the Si MOSFET is of interest in solid state physics for three reasons.

(1) Due to the strong surface electric field the electron wavefunction is quantized in the direction normal to the interface. This results in two dimensional transport.

(2) The concentration of carriers in the inversion layer can be altered simply by varying the applied gate voltage. The carrier concentration can be increased up to about $\sim 2.10^{13}$ $cm^{-2}$. For the (100) Si surface this corresponds to a Fermi energy of $\sim$ 100 meV.

(3) The control offered by the silicon technology allows the preparation of interfaces which are stable, free of hysteresis, and with a low density of fast interface states. Specimens can be produced which do not suffer from the contact effects often found with other semiconductors.

Other two dimensional systems have been studied, such as thin metallic films [1] and inversion or accumulation layers on other semiconductors [2]. A recently introduced system which could be very powerful in this type of work is the two dimensional electron gas at the interface between GaAs and GaAℓAs [3]. This system has not been widely used in localization studies [4], although some Hall effect measurements will be discussed in this paper.

## GENERAL FEATURES OF LOCALIZATION IN THE INVERSION LAYER

Many studies have shown that inversion layer carriers are localized at low values of carrier concentration [5]. Increasing the number of carriers results in a transition from activated to metallic conduction, an effect first found by Fang and Fowler [6] and discussed by Stern and Howard [7]. Mott [8] suggested that this was an example of an Anderson transition, and that the effect of increasing the carrier concentration is to push the Fermi level through the localized states into the extended states at, and above, the mobility edge. The device can thus be used as a model system for the investigation of Anderson localization and the associated metal-insulator transition [9]. The following points have been established.

1. When carriers are localized the mechanism of conduction is by a process of excitation to the mobility edge, passing into variable range hopping ($\sigma \propto \exp(-(T_0/T)^{1/3})$) as the temperature is decreased | 5 | .
2. The position of the mobility edge is determined by both the background disorder, i.e. charges within the Si and/or $SiO_2$, and by the electron-electron interaction. As the carrier concentration rises this increases the tendency to localization and so the mobility edge rises [10,11]. This is qualitatively similar to the impurity band of a doped semiconductor, where both correlation and disorder contribute to the localization [9].
3. In a particular specimen there is a well defined value of conductance at which the metal-insulator transition occurs. Metallic conduction is not obtained below this value. The values of minimum metallic conductance, $\sigma_{min}$, found are greater than $\sim 10^{-5}\Omega^{-1}$ . The value can be altered by sample preparation techniques, and the application of a substrate bias, which has been interpreted in terms of an increase in $\sigma_{min}$ when the potential fluctuations become long range [11].
4. The number of localized carriers (between $10^{11}$ cm$^{-2}$ and $2.10^{12}$ cm$^{-2}$) is dependent on the sample preparation techniques, i.e. the nature of the interfacial change, so offering a means of investigating the interface.

The cause of the behaviour of the Hall effect when carriers are localized is not clear. When the conductance is activated, implying excitation to the mobility edge, it is generally found that the Hall mobility is activated rather than the carrier concentration [12,13].

A point not pursued in early work was the existence of a very small decrease in the extended state conductance with decreasing temperature [14]. This was subsequently shown to be logarithmic with decreasing temperature [15], in the manner suggested by recent theories of two dimensional transport [16,17]. These will now be discussed.

DIFFUSION AND ASSOCIATED LOCALIZATION AND INTERACTION EFFECTS IN TWO DIMENSIONS

From a classical point of view Polya [18], in 1921, showed that true diffusion would not occur in one and two dimensions. According to his theorem, as the system size goes to infinity, the probability of a random walker returning to the origin tends to unity. Abrahams, Anderson, Licciardello and Ramakrishnan [16] first suggested that quantum diffusion would not occur in two dimensions. A considerable literature now exists showing that all states are localized in two dimensions [19,20,21] and Hodges [22] has investigated the correspondence between Polya's theorem and the quantum case.

Abrahams et al [16] and Gorkov et al [23] suggest that the conductance of a 2D system can be written as

$$\sigma = \sigma_0 - \frac{e^2\alpha}{\pi^2\hbar} \mathrm{Ln}(L/\ell) \tag{1}$$

where $\sigma_0$ is the normal conductance, $ne^2\tau/m$, and the logarithmic correction arises from the localization. The constant $\alpha$ is unity or one half, depending on the spin flip length, and at zero temperature L is the specimen length; at finite temperatures, L is the inelastic length $L_{IN} = (\ell_i\ell)^{\frac{1}{2}}$ where $\ell_i$ and $\ell$ are the inelastic and elastic mean free paths respectively. The complete pre-logarithmic factor is only strictly valid for $k_F\ell > 1$, being obtained by perturbation theory, where $k_F$ is the Fermi wave-vector. Recently Kaveh and Mott [24] have considered the form of the 2D wavefunction and the relation between this and the conductance. They find that, at absolute zero, there is a transition between exponential and power law localization [25], this wave-function varying as $1/r \exp(i\underline{k}.r)$. However, at finite temperatures this is converted into exponential localization, the wavefunction now decaying as $1/r \exp(-r/L) \exp(i\underline{k}.r)$.

The Hall effect in this regime was first considered by Fukuyama [26] who calculated that the Hall constant, $R_H$, is unaffected by the logarithmic correction. Thus the Hall mobility will have the same temperature dependence as the conductance, and it appears as if the logarithmic correction is reflected in the scattering time.

A significant difference between this type of localization and that in a band tail (both 3D and 2D) is in the effect of a magnetic field, B. Hikami et al [27] and Altshuler [28] have predicted a negative magneto-resistance, the conductance correction, $\delta\sigma$, being given by

$$\delta\sigma = \frac{e^2}{\pi^2\hbar}\left[\psi(\tfrac{1}{2} + \hbar/4eB\tau_{IN}D) + \ell n\left(\frac{4eB\tau}{\hbar}\right)\right] \tag{2}$$

when $\psi$ is the Digamma function, D the diffusivity, $\tau$ and $\tau_{IN}$ are the elastic and inelastic scattering times respectively, $L_{IN}^2 = D\tau_{IN}$. This negative magneto-resistance is discussed in further detail later; it is to be contrasted with the strong positive magneto-resistance found for band tail localization. Here, the shrinkage of the wavefunction is the dominant effect, resulting in a reduced tunnelling probability in hopping, and an increase in the activation energy when the conduction is by excitation to the mobility edge [29].

Another theory which predicts a logarithmic correction to the conductance was proposed by Altshuler, Aronov and Lee [17] and is based on the three dimensional work of Altshuler and Aronov [30]. It is suggested that a density of states singularity is at the Fermi surface, produced by a combination of the electron-electron interaction and impurity scattering. The conductance correction $\delta\sigma$ is given by

$$\delta\sigma = \frac{e^2}{4\hbar\pi^2}(2 - 2F)\,\ell n\,T$$

where the factor F is determined by screening in the system, and, if $k_F/K \ll 1$, where K is the inverse 2D screening length, $F \sim 1 - 4k_F/\pi K$. At large values of $2k_F/K$, F tends to zero. Thus, at low values of carrier concentration, n, this type of logarithmic correction is small, and at the higher values of n becomes comparable with that due to localization.

Kaveh and Mott [31] have obtained this result by a non-diagrammatic method and point out that localization is reflected in the diffusivity whereas interactions affect the density of states. Therefore, provided the changes are small, the two corrections add to give the final conductance correction, a conclusion also suggested by Fukuyama [32].

The temperature dependence of the conductance does not allow a clear distinction between these two mechanisms unless both F and the inelastic scattering mechanism are known with confidence. However, the predicted behaviour of the Hall effect is quite different. As previously mentioned, Fukuyama predicts that the Hall constant, $R_H$, is not altered by the localization; on the other hand, Altshuler et al [17,28] show that for the interaction mechanism $\sigma_{xy}$ does not possess a logarithmic correction, and so

$$\frac{\delta R_H}{R_H} = 2\,\frac{\delta R}{R} \qquad (3)$$

where $\delta R/R$ is the proportional change in specimen resistance. Thus in the localized regime the Hall mobility $\mu_H$ decreases with decreasing temperature, whereas $\mu_H$ increases with decreasing temperature on the interaction model. Physically, we may appreciate the interaction result by the following argument. Writing $\sigma_{xy}$ as

$$\sigma_{xy} = \sigma_{xx}\,\omega\tau$$

where $\omega$ is the cyclotron frequency and $\tau$ is the elastic scattering time. We assume the diffusivity $V_F{}^2\tau$ is unaffected, hence

$$\frac{\delta\sigma_{xy}}{\sigma_{xy}} = \frac{\delta\sigma_{xx}}{\sigma_{xx}} + \frac{\delta\omega}{\omega} + \frac{\delta\tau}{\tau} .$$

Since $\tau$ depends upon the unperturbed density of states, $\delta\tau = 0$; $V_F$ is not altered hence:

$$\frac{\delta\sigma_{xx}}{\sigma_{xx}} = \frac{\delta k_F}{k_F} , \qquad \text{as} \quad \frac{\delta\omega}{\omega} = - \frac{\delta k_F}{k_F}$$

we see that $\delta\sigma_{xy}/\sigma_{xy} = 0$ and the factor of 2 follows in equation (3). It is physically interesting that this argument, which assumes the unperturbed density of states determines the scattering time, produces the factor of 2.

The effect of a magnetic field on the interaction models has been considered very recently by Lee and Ramakrishnan [33]. They suggest that if $g\beta B > kT$, where $\beta$ is the Bohr magneton, then the factor 2 - 2F is converted into 2 - F, so producing a logarithmic correction similar to that caused by localization.

As in many other types of experiment the localization can be investigated as a function of temperature under Ohmic conditions, or with constant lattice temperature but an increase in electron temperature produced by an electric field. Anderson et al [34] proposed that a comparison of these two effects gives information on the nature of the inelastic scattering. Although this type of analysis has been performed it will not be discussed here.

EXPERIMENTAL INVESTIGATIONS INTO WEAK LOCALIZATION AND INTERACTIONS

The first evidence for the existence of the logarithmic correction was provided by measurements on thin metal films [1]. Subsequently Bishop, Tsui and Dynes [15] found the weak temperature dependence observed in the metallic region of Si MOSFET's was also logarithmic. Due to the low value obtained for $\alpha$ it was not clear if the origin of the logarithmic behaviour was localization or interaction. Negative magnetoresistance was reported by Kawaji and co-workers [35,36] but at temperatures too high for the logarithmic term to dominate the temperature dependence of conductance. The factor of two predicted by Altshuler et al [17] for the behaviour of $(\Delta R_H.R/\Delta R.R_H)$ was verified by Uren, Davies and Pepper [37] (UDP) and Bishop, Tsui and Dynes [38]. An example is shown in figure 1. UDP showed that the magnetic field required to measure this type of Hall correction suppressed the localization and enhanced the interaction effect. It was thus possible for UDP to confirm the existence of both mechanisms. The work of these authors with Kaveh will now be described.

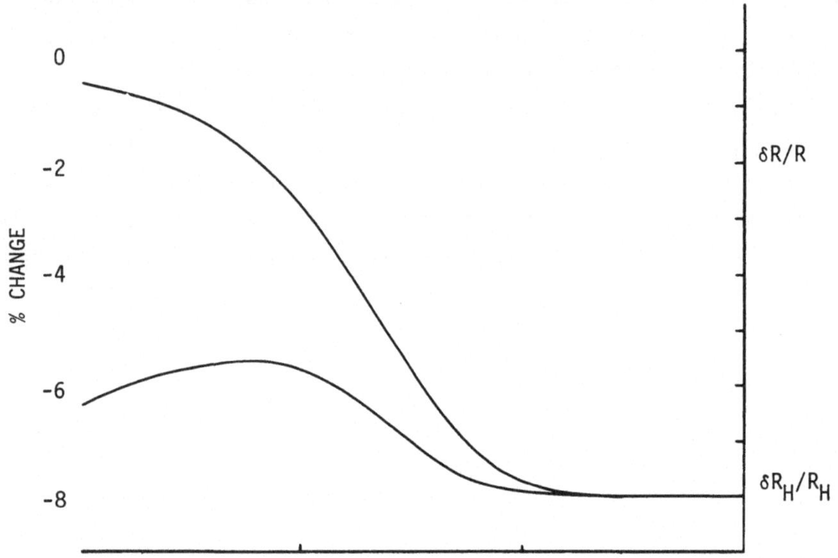


Figure 1. The percentage change in both resistance, R, and Hall constant, $R_H$, against electric field for an inversion layer with lattice temperature = 50 mK, from UDP. The rate of change of $R_H$ is twice that of R, UDP show this ratio as a function of carrier concentration.

Kaveh, Uren, Davies and Pepper [38] showed that the Digamma function dependence of the negative magneto-resistance, equation (2), can be approximated by the following formula for $\delta\sigma$,

$$\delta\sigma = \frac{e^2\alpha}{\pi^2\hbar} \text{Ln}\,(L_D/\ell) \tag{4}$$

where the new diffusion length $L_D$ is given by

$$\frac{1}{L_D^2} = \frac{1}{L_{1N}^2} + \frac{1}{L_c^2}$$

$L_c$ is the cyclotron orbit $(e\hbar/B)^{\frac{1}{2}}$. Figure 2 shows the agreement between this formula and the Digamma function, the only discrepancy is at very low values of B where equation (4) fails to capture the $B^2$ variation of $\delta\sigma$. Thus, as B increases, $L_c$ and $L_d$ decrease and the conductance increases. The negative resistance only ceases when $L_d \approx \ell$, but as, when $L_c < L_{1N}$, $\delta\sigma$ varies as $\ell n\, L_c$, the final variation with B is very slow, as is clear in figure 2. When $L_c \ll L_{1N}$, $\delta\sigma$ no longer has a temperature dependence, even though the negative magneto-resistance will still be present.

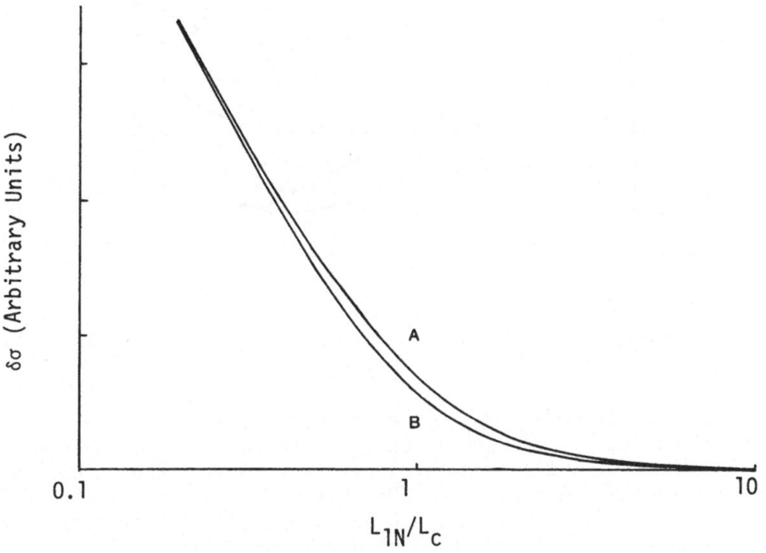


Figure 2. The calculated change in conductance, $\delta\sigma$, versus $L_{1N}/L_c$ for varying $L_{1N}$. Curve A is derived from equation (4), and B is the digamma function, equation (2). The curves are normalised to fit at high and low $L_{1N}$, as in the interpretation of experiment, this procedure removes the effects of additive constants, from Kaveh et al [40].

Figure 3 shows the resistance, R, of an inversion layer versus logarithm of electric field, E (and electron temperature, this being proportional to E), for different values of B. The increase in R for E > 1V/m is due to the change in elastic length with screening and is not important in this context.

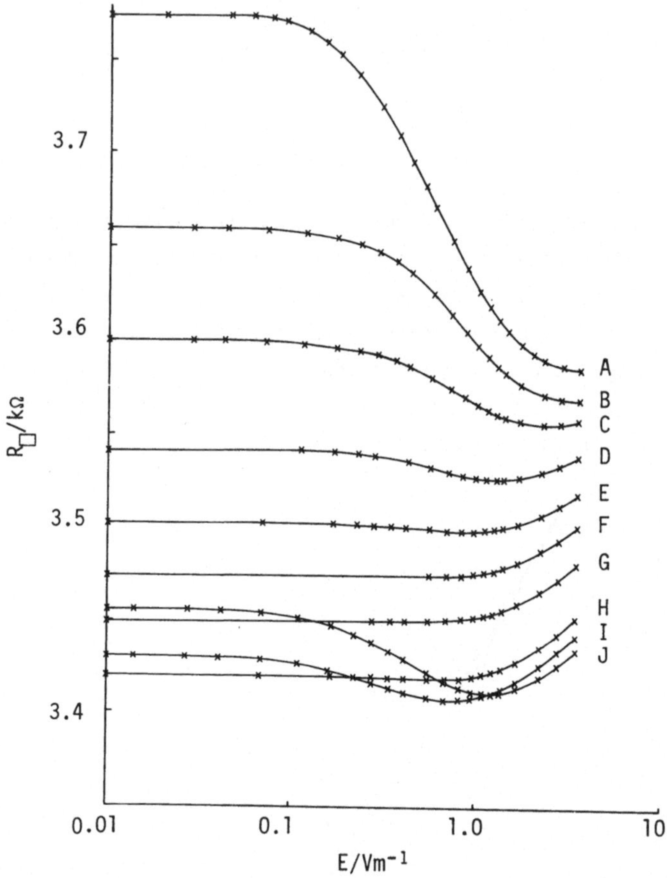


Figure 3. The sheet resistance $R_\square$ is plotted against log electric field for various values of magnetic field and a constant carrier concentration of 3.8 10" cm$^{-2}$, the lattice temperature was ∿ 50 mK. The values of magnetic field in Tesla are A = 0, B = 0.008, C = 0.021, D = 0.047, E = 0.074, F = 0.1, G = 0.15, H = 0.26, I = 0.32, J = 0.32. The elastic scattering length was 36 mm, from Davies, Uren and Pepper [39].

The following are the significant points of figure 3.

1. As B increases the specimen shows a negative magneto-resistance in the ohmic regime.
2. Increase in B results in a poorer fit to a log E law, we term this regime "quasi-logarithmic". The total change in R decreases as B is increased.
3. As B increases, the value of electric field required for a change of R increases.
4. Eventually, with further increase of B, the field (temperature) dependence of R disappears but the negative magneto-resistance is still observed.
5. Further increase in B above that required to induce the above point results in positive magneto-resistance in the ohmic regime and the return of a logarithmic correction. This second mechanism disappears at a lower electric field than the localization for comparable values of "quasi-logarithmic" slope. When the second mechanism disappears at higher electric fields the negative magneto-resistance again becomes observable, this can be observed up to high electron temperatures $\sim$ 4 K.

These points can all be satisfactorily interpreted on the localization and interaction theories. In the original UDP paper the Hall ratio of 2 was measured when this second regime was quite dominant - thus allowing identification as the interaction correction. Therefore this figure shows the gradual transition from localization to interaction behaviour. Initially, as the magnetic field increases, the cyclotron length decreases and so the length scale $L_D$ decreases giving negative magneto-resistance. As B increases it is necessary to increase the temperature of the electron gas further for the inelastic length $L_{IN}$ to become shorter than the cyclotron length $L_c$. Eventually, over the electron temperature range below $E \simeq 1\ vm^{-1}$ it is not possible for $L_{IN}$ to become shorter than $L_c$, all temperature dependence is now lost but the negative magneto-resistance is still present, this will disappear when $L_c \gtrsim \ell$. Finally the magnetic field brings in the interaction process with positive magneto-resistance in the ohmic region. This is due to both the introduction of this process and the normal magnetic change of $\ell$. The interaction mechanism is destroyed at quite low values of electron temperature, experiment indicates that the condition for its magnetically induced presence is $g\beta B \gtrsim 3kT$. Hence, when it is removed the negative magneto-resistance due to localization can be found.

This figure is discussed in detail by Davies, Uren and Pepper [39].

Although, for a certain range of B, complete separation of the localization and interaction mechanisms is found as well as metallic conduction down to 50 milliKelvin, this situation will not persist down to absolute zero. Eventually kT will become smaller than gβB and the logarithmic, interaction, correction will become clear. The localization theory indicates that all states are localized and so $\sigma = 0$ at $T = 0$ for an infinite specimen; however, it seems that the interaction correction is a perturbation and so metallic conductance should be found at $T = 0$. For finite specimens, the sample length imposes a cut off[17,31] and so $\sigma$ is finite at $T = 0$.

The work described in UDP is an example of the coexistence of interaction and localization behaviour. This will always occur when the disorder is sufficiently high, i.e. $\ell$ and $L_{IN}$ are small, and the condition $g\beta B > kT$ is achieved at values of B too small for a suppression of the temperature dependence due to localization. Figure 4 shows the "quasi-logarithmic" gradient plotted against B for this situation [37]. It is seen that the gradient initially decreases (localization) and then rises

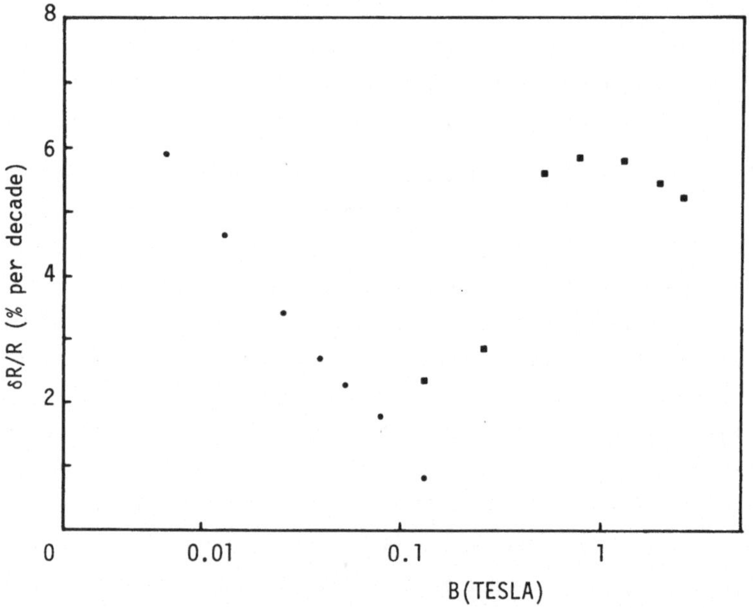


Figure 4. The "Quasi-Logarithmic" slope is plotted against B in Tesla. The slope is in the form of resistance per decade change in electric field, with lattice temperature $\sim$ 85 mK. The circular points arise from localization, and the square points are due to interactions. The elastic scattering length was 19 nm, the $\omega\zeta = 1$ condition occurs for $B \sim 5.0$ Tesla, from UDP [37].

(interactions). The gradient due to the interaction mechanism is constant then falls when $\omega\zeta$ approaches 1.

For B $\sim$ 0.1 Tesla it is possible to observe a transition from interaction plus localization to localization as the electron temperature is increased. This is shown in detail in figure 5.

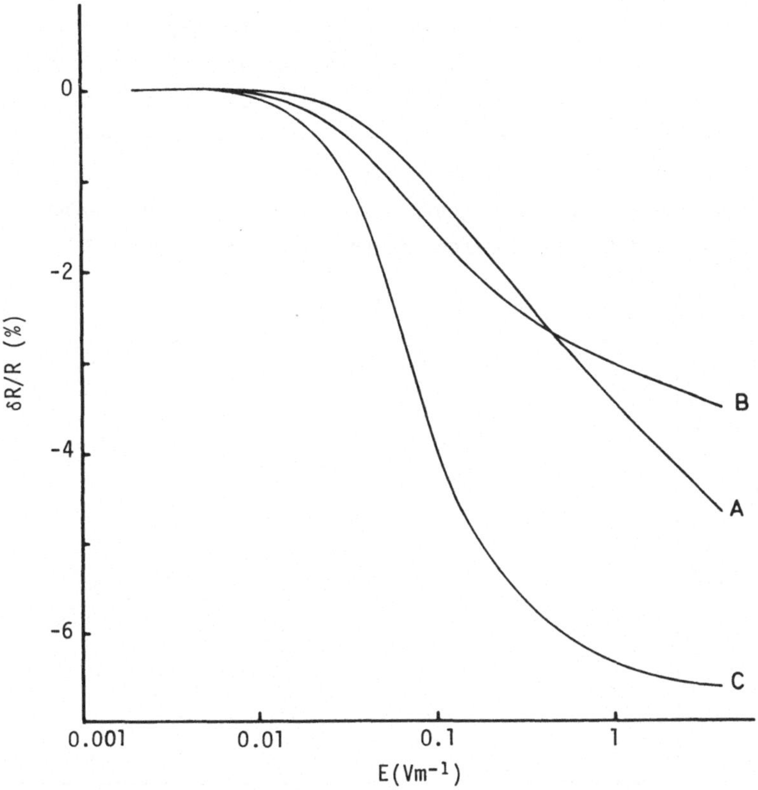


Figure 5. Percentage change in resistance against electric field for different magnetic fields and constant carrier concentration, lattice temperature $\approx$ 85 mK. A = 0.053 Tesla, B = 0.131 Tesla, C = 0.525 Tesla: A is localization, B is localization plus interaction and C is interaction, from UDP [37].

The value of the pre-logarithmic factor found by UDP for the magnetic field induced interaction regime was in reasonable agreement with (2 - F), where F was in the range 0.74 - 0.9. However for higher $\ell$ the experimental value of the pre-logarithmic factor was less than expected. This was because $\omega\zeta$ approached 1 and the effect of the interaction decreased, presumably due to a change in the form of the wavefunction, before the factor 2 - 2F was completely converted into 2 - F.

Very recently Davies, Uren and Pepper [39] have shown that the magnetic enhancement of the interaction mechanism is independent of the direction of B. This is in agreement with the prediction of Lee and Ramakrishnan [33] that the change in the pre-logarithmic term is a spin, and not an orbital, effect.

Because the magnetic field affects the two mechanisms in different ways it is fairly clear that the field will tend to separate the mechanisms completely if $\ell$ is long. Thus a smaller B results in $L_c < L_{1N}$, and suppression of the logarithmic correction due to localization, whilst $g\beta B < kT$ and the interaction mechanism is not observable, 2 - 2F being near zero. The suppression of the interaction mechanism by F being near unity is further enhanced by a large $\ell$:- for the system to possess a given conductance n decreases as $\ell$ increases, hence F increases.

Kaveh et al [40] have suggested that all the factors affecting localization, i.e. magnetic field, optical excitation, inelastic length, can be combined into a single length. Thus the conductance correction can be written in terms of a dependence on a universal length. Evidence from magneto-resistance is presented in support of this suggestion, figures 6 and 7. The magneto-resistance has also been used to investigate the quantum corrections to the electron-electron scattering rates [36,38].

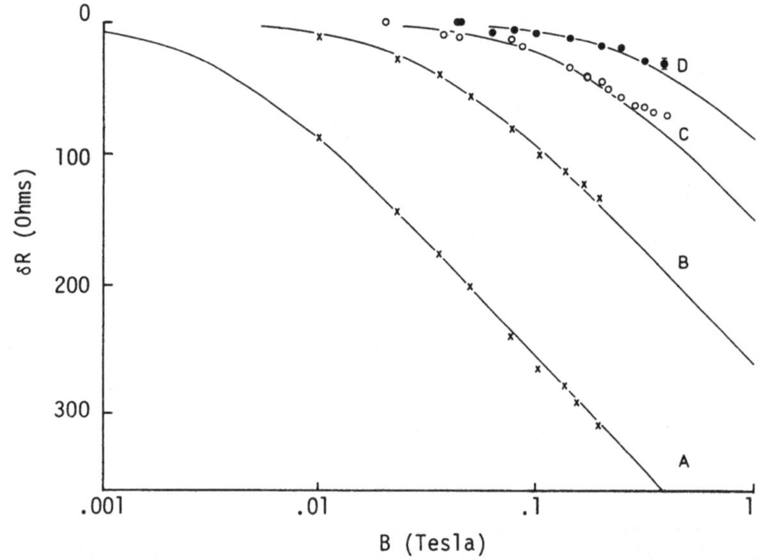


Figure 6. The change in resistance of an inversion layer (carrier concentration 3.8 10" $cm^{-2}$) plotted against log B. The curves are theoretical plots for $\alpha = 0.28$. A and B are for a lattice temperature of 50 mK and electron temperatures of 110 ± 10 mK and 550 ± 50 mK, respectively. C and D are in the ohmic regime and temperatures are 1.2 K and 2.1 K respectively, from Uren et al [38].

Finally, recent work by Poole, Pepper and Glew has investigated the situation where F is about 0.5, and both interactions and localization effects are present in the absence of a magnetic field. These experiments showed that the negative magneto-resistance, arising from the suppression of localization, does not give a correction to $R_H$, as suggested by Fukuyama [26]. However after suppression of the localization a logarithmic temperature dependence is still present, this is due to the interaction

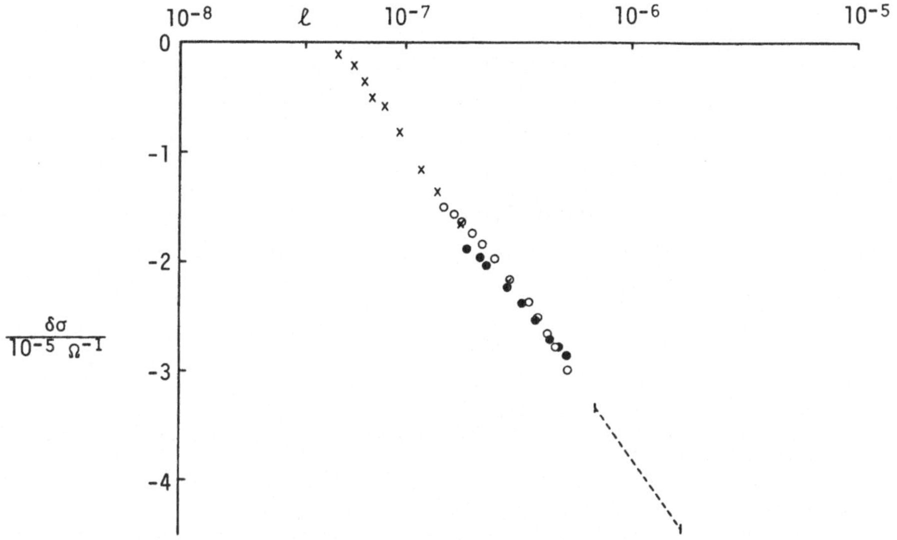


Figure 7. The change in conductance is plotted against the shortest length in the system. Points marked X are when the length is determined by the magnetic field, i.e. $L_c \ll L_{IN}$. The open and closed circles are when the length is the inelastic length determined by electron heating and lattice heating respectively. The dashed line indicates the results expected if the electric field determined the length scale, i.e. the energy change during an elastic mean free path determines the value of conductance correction. This is not found here, as the values of field are such that inelastic scattering dominates, from Kaveh et al [40].

mechanism and, now, the Hall ratio of 2 is found. The suggestion of Kaveh and Mott that both mechanisms add in producing the total conductance correction is consequently confirmed.

## CONCLUSION

It has proved possible to experimentally verify the predictions of both the localization and the interaction theories. These can now be incorporated into an understanding of two dimensional transport. The power law localization can be suppressed by a magnetic field, unlike exponential localization, and, under these circumstances, a discontinuity in conductance at zero Kelvin should be regained when the Fermi energy is at the mobility edge. This assumes that the interaction mechanism is a perturbation and does not lead to $\sigma = 0$ at $T = 0$.

Neither of these mechanisms affects the use of two dimensional systems as quantized Hall resistors in a strong magnetic field. Thus the localization is completely suppressed when $\omega\tau > 1$, and $e^2/\hbar$ is measured when $\sigma_{xx}$ is virtually zero and $I_x$ is determined by $\sigma_{xy}$. Interactions do not give rise to a correction in $\sigma_{xy}$, so resulting in the observation of the Hall ratio of 2 at low values of magnetic field.

ACKNOWLEDGEMENTS

The work reviewed here was performed in collaboration with R. A. Davies, M. Kaveh, D. A. Poole and M. J. Uren, and will be more fully described elsewhere and in the Ph.D. Theses of R. A. Davies, D. A. Poole and M. J. Uren. Throughout this work we have all enjoyed many discussions with Professor Sir Nevill Mott.

This research was supported by the Science Research Council, and experiments were performed at the S.R.C. Rutherford Laboratory with the advice and assistance of Dr S. F. J. Read and Mr G. Regan.

REFERENCES

1. G.J.Dolan and D.D.Osheroff, Phys. Rev. Lett. 43, 721 (1979)
2. F.Koch in "Narrow Gap Semiconductors, Physics and Applications", ed. W.Zawadzki, Springer, Berlin (1980) L511
3. D.C.Tsui and R.A.Logan, Appl. Phys. Lett 35, 99 (1979)
4. D.A.Poole, M.Pepper and R.J.Glew, to be published
5. N.F.Mott, M.Pepper, S.Pollitt, R.H.Wallis and C.J.Adkins, Proc. Roy. Soc. A345, 169 (1975)
6. F.F.Fang and A.B.Fowler, Phys. Rev. 169, 619 (1968)
7. F.Stern and W.E. Howard, Phys. Rev. 163, 816 (1967)
8. N.F.Mott, Electronics and Power 19, 321 (1973)
9. N.F.Mott, "Metal-Insulator Transitions", London: Taylor and Francis (1974)
10. M.Pepper, S.Pollitt and C.J.Adkins, J. Phys. C 7, L273 (1974)
11. M.Pepper, Proc. Roy. Soc. A353, 225 (1977)
12. M.Pepper, Phil. Mag. 38, 515 (1978)
13. M.Nakai and M.Pepper, to be published
14. C.J.Adkins, S.Pollitt and M.Pepper, J. Physique 37, C4 343 (1976)
15. D.J.Bishop, D.C.Tsui and R.C. Dynes, Phys. Rev. Lett. 44, 1153 (1980)
16. E.Abrahams, P.W. Anderson, D.C.Licciardello and T.V.Ramakrishnan, Phys. Rev. Lett. 42, 673 (1979)
17. B.L.Altshuler, A.G.Aronov and P.A.Lee, Phys. Rev. Lett. 44, 1288 (1980)
18. G.Polya, Math, Ann. 84, 149 (1921)
19. R.V.Haydock, Phil. Mag. B43, 203 (1981)
20. A.Houghton, A.Jericki, R.D.Kenway and A.M.M.Pruisken, Phys. Rev. Lett. 45, 394 (1980)
21. R.B.Allen, J. Phys. 13, L667 (1980)
22. C.H.Hodges, J. Phys. C 14, L247 (1980)
23. L.P.Gorkov, A.H.Larkin and D.Khmelnitzkii, J.E.T.P. Lett. 30, 299 (1979)
24. M.Kaveh and N.F.Mott, J. Phys. C 14, L177 (1981)
25. A slightly different form of power law localization is found by J.L. Pichard and G.Sarma, J. Phys. C, in the press
26. H.Fukuyama, J. Phys. Soc. Japan 48, 2169 (1980)
27. S.Hikami, A.I.Larkin and Y.Nagaoka, Prog. Theor. Phys. 63, 707 (1980)
28. B.L.Altshuler, D.Khmelnitzkii, A.I.Larkin and P.A.Lee, Phys. Rev. B22, 5142 (1980)
29. M.Pepper, J. Non Crystalline Solids 32, 161 (1979)
30. B.L.Altshuler and A.G.Aronov, Solid State Comm. 30, 115 (1979)
31. M.Kaveh and N.F.Mott, J. Phys. C 14, L183 (1981)
32. H.Fukuyama, J. Phys. Soc. Japan 49, 644 (1980)
33. P.A.Lee and T.V.Ramakrishnan, private communication, to be published
34. P.W.Anderson, E.Abrahams and T.V.Ramakrishnan, Phys. Rev. Lett. 44, 1288 (1979)
35. Y.Kawaguchi, Y.Kitahara and S.Kawaji, Solid State Comm. 26, 701 (1978)
36. Y.Kawaguchi and S.Kawaji, J. Phys. Soc. Japan 48, 699 (1980)
37. M.J.Uren, R.A.Davies and M.Pepper, J. Phys. C 13, L985 (1980)
38. M.J.Uren, R.A.Davies, M.Kaveh and M.Pepper, J. Phys. C, in the press
39. R.A.Davies, M.J.Uren and M.Pepper, J. Phys. C, in the press
40. M.Kaveh, M.J.Uren, R.A.Davies and M.Pepper, J. Phys. C, in the press
41. K.von Klitzing, G.Dorda and M.Pepper, Phys. Rev. Lett. 45, 494 (1980)

# CRITICAL BEHAVIOUR AT THE MOBILITY EDGE OF THE ANDERSON MODEL OF DISORDERED SYSTEMS

FRANZ WEGNER

Institut für Theoretische Physik, Ruprecht-Karls-Universität,
D-6900 Heidelberg, F. R. Germany

The talk covers the behaviour of a quantum mechanic particle moving in a random potential with special emphasis on the aspect of local gauge-invariance. Symmetry arguments are reviewed which allow the mapping of such a system onto a field theoretic model of interacting matrices. This model yields an expansion of the critical exponents at the mobility edge around the lower critical dimensionality two. Since most of this lecture has already been published as a contribution to the Les Houches institute 1980 "Common Trends in Particle and Condensed Matter Physics" I give here only some new results and refer the reader to reference [1].

## New Results

Hikami [2] has calculated the W-(ß-) function in three loop order for the orthogonal and symplectic case of the nonlinear $\sigma$-model. As a consequence the critical exponent s for the conductivity $\sigma \sim (E - E_c)^s$ of a system with spin-independent potential with time-reversal invariance reads

$$s = 1 + O(d - 2)^3.$$

Pruisken and Schäfer [3] have mapped the site-diagonally disordered electron system on a system obeying local gauge invariance. This supports strongly the assumption that the mobility edge behaviour is governed by local gauge invariant interactions.

The present author has proven [4] that for a class of tight-binding models with diagonal disorder and short-range one-particle interaction the density of states is positive (non-zero) and finite within the whole band. This class of systems includes models with rectangular, Lorentzian, and Gaussian distribution of the diagonal matrix elements. Systems in which the site-diagonal and the off-diagonal matrix elements are independent random variables with symmetric distribution (on a compact support) have a non-zero density of states everywhere inside the band. (In the case of off-diagonal disorder only, the proof does not work for E = 0). Thus, predictions that at the mobility edge the density of states for site-diagonal disorder would diverge ($4-\varepsilon$ expansions) or that it would vanish for off-diagonal disorder ($8-\varepsilon$ expansions) or predictions which allow for both options (inhomogeneous fixed point ensemble [5]) can be ruled out. The homogeneous fixed point ensemble [5], however, is in agreement with these findings.

In a region not to close to the mobility edge the measurements of ref. [6] on P doped Si are in good agreement with an exponent s = 0.55. However, in comparing with experiments [6,7] one has to keep in mind that the Anderson model does not incorporate interactions between electrons, which seems to be important for the mobility edge behaviour [8].

[1] F. Wegner, Physics Reports 67, 15 (1980)
[2] S. Hikami, Prog. Theor. Phys. 64, 1466 (1980)
[3] A. M. Pruisken, L. Schäfer, Phys. Rev. Lett. 46, 490 (1981)
[4] F. Wegner, in preparation
[5] F. Wegner, Z. Physik B 25, 327 (1976)
[6] T. F. Rosenbaum, K. Andres, G. A. Thomas, R. N. Bhatt, Phys. Rev. Lett. 45, 1723 (1980)
[7] B. W. Dodson, W. L. McMillan, J. M. Mochel, Phys. Rev. Lett. 46, 46 (1981)
[8] W. L. McMillan, preprint

# Numerical Results on the Anderson Localization Problem

U. Krey, W. Maaß, J. Stein

Naturw. Fakultät II - Physik - der Universität Regensburg, F.R.G.

Abstract:

The problem of the Anderson localization of electrons is studied numerically for large two-dimensional gauge-invariant disordered model systems, including systems with spin-flip scattering. The numerical treatment is based on recursive mappings of the original systems onto equivalent chains. This allows for accurate calculations of various Green's functions, and particularly for an exact real-space renormalization by decimation. Finally, also the conductivity is calculated directly from the Kubo formula.

1. Introduction: The Anderson localization problem, dating back already to 1958, see |1|, has regained an enormous interest since two or three years. Although the problem is far from being settled, |2,3|, there is a large amount of new and partly unexpected insights, which are mainly based on recent progress in analytical techniques |2,4|, but also partly on the development of powerful numerical algorithms |3,5-12|. Particularly, in a number of recent papers, two of the present authors have exploited recursive algorithms which are based on unitary mappings of the original Hamiltonian onto an equivalent semi-infinite chain with nearest-neighbour interactions |3,7,8,9-11|. This leads to continued-fraction expansions for various Green's functions appearing in certain localization criteria |3,10|, and also to an exact numerical real-space renormalization by decimation |3,7|. Furthermore, by an extensive application of the above-mentioned mappings also the Kubo formula for the electrical conductivity could be evaluated directly |8|.

In the following, after a short presentation of the models considered (Chap.2), and after a sketch of the numerical procedures (Chap.3), we present recent results for gauge-invariant models suggested recently by Wegner and coworkers, see |2|.

2. Models: We consider disordered tight-binding systems with a Hamiltonian

$$H = \sum_{l,m,\sigma,\sigma'} H_{l,m}^{\sigma,\sigma'} c_{l,\sigma}^{+} c_{m,\sigma'} \quad . \tag{1}$$

Here l and m denote the sites of a lattice, while the disorder is contained in the matrix elements $H_{l,m}^{\sigma,\sigma'}$. Square, triangular, diamond, and simple-cubic lattices have been considered hitherto |3,5-12|. $\sigma$ and $\sigma'$ (= +1 or = -1) are the spin indices, and $c_{l,\sigma}^{+}$ and $c_{m,\sigma'}$ denote the creation and destruction operators for electrons. Generally, in the following it is assumed that the matrix elements $H_{l,m}^{\sigma,\sigma'}$ vanish unless $l = m$ or $l = m+\Delta$, where $|\Delta|$ corresponds to a nearest-neighbour distance.

Numerically, hitherto almost exclusively the case of the original Anderson model has been considered, where the spin does not appear, and where the hopping matrix element V ($=H_{l,l+\Delta}$) between nearest neighbours is constant, the disorder being contained only in the single-site energies $\varepsilon_l (= H_{l,l})$ . However, it has already been pointed out in |3,7,8| that this model is not completely representative. Moreover, on the basis of recent field-theoretical results, Wegner and coworkers, see |2|, suggest the existence of three different universality classes which are represented by gauge-invariant ensembles: The simplest case is that of a real, spin-independent ensemble (RME), with real gauge-invariance, e.g. represented by a system with $\varepsilon_l = 0$, while the $H_{l,l+\Delta}$ are random, following e.g. a rectangular distribution of width $W_2$ , which should be centered around zero to yield the desired gauge invariance. This model, and the original Anderson model, should belong to the same universality class. The second class corresponds to real matries, too, but with spin (RMES model). There, too, we assume $\varepsilon_l = 0$, while the hopping matrix elements corresponding to potential scattering, i.e. without spin flip, $H^{\sigma,\sigma}_{l,m} (= H^{-\sigma,-\sigma}_{l,m} = H_{m,l} = H_{l,m})$, are rectangularly distributed with width $W_2$ around zero, and the spin-flip hopping matrix elements $H^{+1,-1}_{l,m}$ $(=-H^{-1,+1}_{l,m} = H^{-1,+1}_{m,l} = -H^{+1,-1}_{m,l})$ similarly, however with width $W_2'$ . For $W_2' = 0$, however also for $W_2 = 0$ in case of the square, simple cubic or body-centered cubic lattices, this RMES model reduces to the RME model (up to unitary equivalence in the case $W_2 = 0$). The matrix of the RMES model is not only hermitian, but also invariant against time reversal, which, in the general case, corresponds to the condition $H^{\sigma,\sigma'}_{l,m} = \sigma\cdot\sigma' \, (H^{-\sigma,-\sigma'}_{l,m})^*$.

Finally, the third universality class (CMES) can be represented by complex hermitian matrices fulfilling this last condition, with $\varepsilon_l = 0$, while the magnitudes $|H^{\sigma,\sigma}_{l,m}|$ and $|H^{\sigma,-\sigma}_{l,m}|$ are drawn e.g. from rectangular distributions with the limits 0 and $W_2/2$, and 0 and $W_2'/2$, respectively, whereas the phases of the complex matrix elements are randomly distributed between 0 and $360^0$.

In the field-theoretic approach, see |2,13|, these three models are represented by Schwinger functionals defined on noncompact spaces of certain quasi-orthogonal, -unitary, or -symplectic matrices, respectively.

3. Numerical Methods: Our methods are based on the fact that by introduction of a new orthonormal basis $|\phi_1\rangle, |\phi_2\rangle, \ldots$ a hermitian matrix can be tridiagonalized:

$$H^{\sigma,\sigma'}_{l,m} \Rightarrow \langle\phi_\mu | H \phi_\nu\rangle = \begin{pmatrix} a_1, b_1, 0, \ldots \\ b_1, a_2, b_2, 0, \ldots \\ 0, b_2, a_3, b_3, 0, \ldots \\ \ddots\ \ddots\ \ddots \end{pmatrix} \qquad (2)$$

Here the real quantities $a_\nu$ and $b_\nu$ represent single-site and hopping matrix elements for a semi-infinite chain, which is equivalent to the original system.

The $a_\nu$ und $b_\nu$ are generated <u>recursively</u>, see |7|, starting from an arbitrary $|\phi_1>$; usually, $|\phi_1>$ is chosen as a single-site state, i.e. $\langle 1,\sigma|\phi_1\rangle = \delta_{1,m}\,\delta_{\sigma,\sigma'}$, and in case of a system of linear diameter L (i.e. L = 128), one calculates up to $\nu \approx L/2$ (i.e. $\nu = 64$), while the rest of the coefficients is set constant (i.e. $a_\nu = a_{64}$ and $b_\nu = b_{64}$ for $\nu \geq 64$). In this way, see |7| one obtains very accurate continued-fraction representations (i.e. corresponding to a moment expansion with 128 exact moments) of the Green's functions

$$G_{1,1}^{\sigma,\sigma}(z) = \langle \phi_1|(z-H)^{-1}\,\phi_1\rangle = 1/(z-a_1-b_1^2/z-a_2-b_2^2/(z-a_3-b_3^2/(\ldots\quad , \qquad (3)$$

which allows for an accurate evaluation of the following quantity (with $z=E-i\varepsilon$):

$$f(E,\varepsilon) = \overline{|G_{1,1}^{\sigma,\sigma}(z)|^2/g_1^\sigma(z)} \quad . \qquad (4)$$

Here $g_1^\sigma(z) = (1/\pi)\,\mathrm{Jm}\,\{G_{1,1}^{\sigma,\sigma}(z)\}$, which in the limit $\varepsilon \to 0^+$ for $z=E-i\varepsilon$, represents the local density of states at site 1 with spin $\sigma$. The average in (4) is over a large number of sites and/or configurations.

According to |1,14|, the product $\varepsilon\cdot f(E,\varepsilon)$ should remain finite for $\varepsilon \to 0^+$, if the eigenstates at the energy E are localized, whereas for extended states it should converge to zero.

Another localization criterion based on the above-mentioned tridiagonalization is to perform a real space renormalization by decimation on the above-mentioned equivalent chain |3,7|: In this way, one obtains the effective potential fluctuations

$$\delta_n^2(E) = \overline{\left\{\frac{a_1^{(n)}}{b_1^{(n)}} - \left(\overline{\frac{a_1^{(n)}}{b_1^{(n)}}}\right)\right\}^2} \quad . \qquad (5)$$

Here the $a_1^{(n)}(E)$ and $b_1^{(n)}(E)$ are the renormalized $a_1$ and $b_1$ coefficients obtained after n iterations, where in each iteration every second site is eliminated |7|.

Finally, by systematic use of similar recursions, see |8|, one can also generate a number of different eigenstates $\psi_\alpha$ of (1) with given energy E, where the energy is sharp e.g. up to an inaccuracy of $10^{-3}$ times the total band width. With these eigenstates one can obtain the dc-conductivity at T = 0 K from the Eqn.

$$\sigma(E_F) \propto g(E_F)\cdot \overline{\mathrm{Jm}\{\langle v_x\,\psi_\alpha|(z-H)^{-1}\,v_x\,\psi_\alpha\rangle\}} \quad . \qquad (6)$$

The average (---) in (6) is performed over a large number, e.g. 25 to 50, of different eigenstates with the same energy $E = E_F$ (= Fermi energy); the proportionality constant in (6) is known; $g(E_F)$ is the density of states, and $(v_x)_{1,m}^{\sigma,\sigma'} = (i/\hbar)H_{1,m}^{\sigma,\sigma'}\cdot(x(1)-x(m))$ is the velocity operator. The matrix elements in (6) are calculated as described above, with $\phi_1 \propto v_x\,\psi_\alpha$

4. Results: In Fig. 1, for the RME model, and for the REMS model with $W_2 = W_2'$, the function $f(E,\varepsilon)$ of Equn. (4) is presented, calculated for 100x100-square lattices, with $\varepsilon = 10^{-10}$, from an average over 40 configurations.

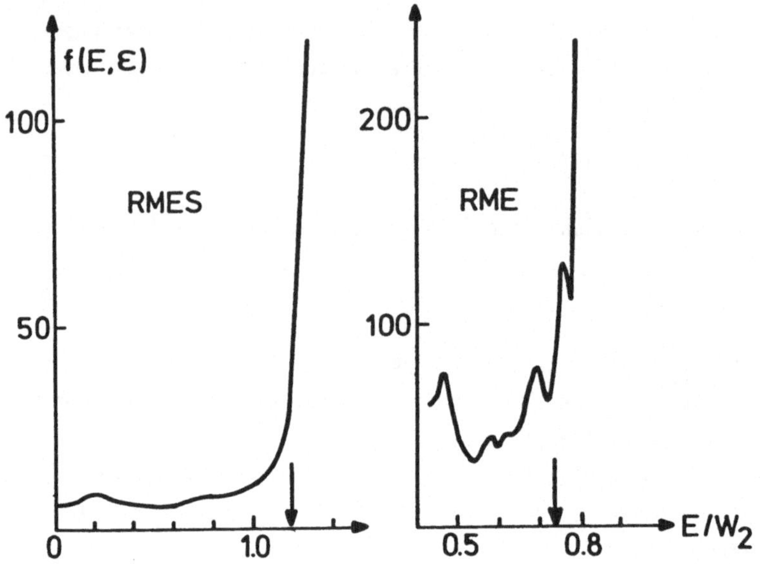


Fig. 1:The function $f(E,\varepsilon)$ is presented for the REMS and RME models

From the steep increase of $f(E,\varepsilon)$ for $E > 1.18\ W_2$ and $E > 0.73\ W_2$ one would conclude that these values denote critical energies $E_c$ separating localized states ($E>E_c$) from states which look extended, at least on the scales considered. A similar conclusion is obtained from Fig. 2, which presents the renormalized potential fluc-

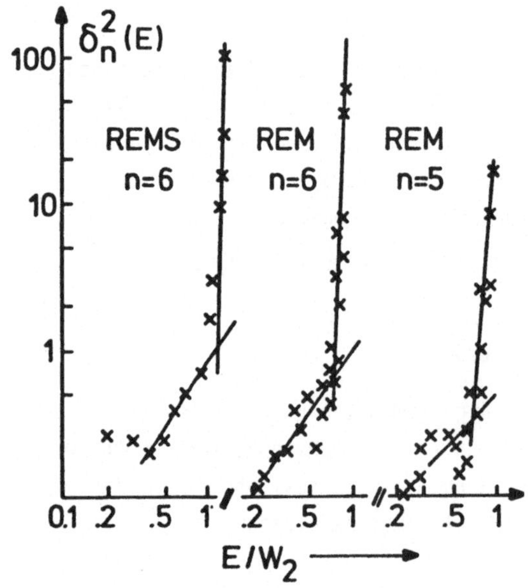


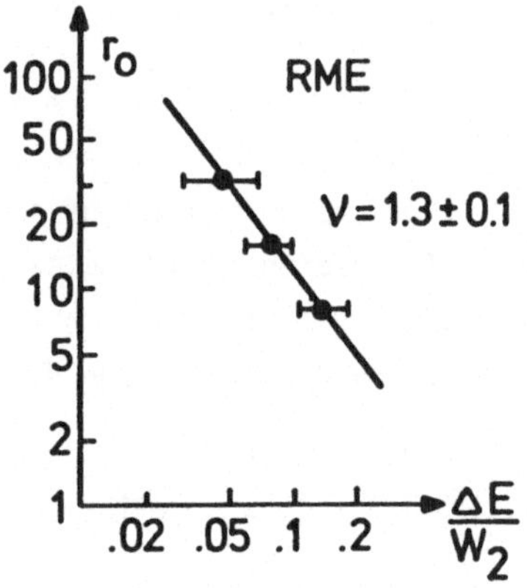


Fig.2 (left): Effective potential fluctuations (see text).

Fig.3 (above):Localization length $r_0$. $\Delta E = E - 0.73 W_2$.

tuations $\delta_n^{\,2}(E)$, see above, for n=6; for the RME model, also the case of n=5 has been plotted, where $E_c(n)$ is somewhat smaller, $E_c(5)=0.68\ W_2$, $E_c(6)=0.73\ W_2$. If one plots the characteristic length $r_o(n)=2^n$ over $(E_o-E_c(n))/W_2$ (=:$\Delta E/W_2$), then one obtains Fig. 3, there it has been assumed $E_o$ =0.73 $W_2$. From this double-logarithmic plot one obtains a critical exponent $\nu$=1.3±.1 ($r_o \propto \Delta E^{-\nu}$); this is the same value as the exponent $\nu_E$ which we obtained in |7| for Anderson's original model (where at the same time a second exponent $\nu_W$= 0.8 appeared, |7|). For the REMS model a value of $\nu$ around 2 was obtained, but still less well defined |15|. For the CMES model our calculations are not yet finished.

Finally, in Fig. 4, results for the dc-conductivity at T=OK are presented for the RMES and RME cases; both the reduced diffusivity $D(E_F) = \sigma(E_F)/g(E_F)$ and the density of states $g(E_F)$ are presented over the fermi energy, for 50 · 50 sites.

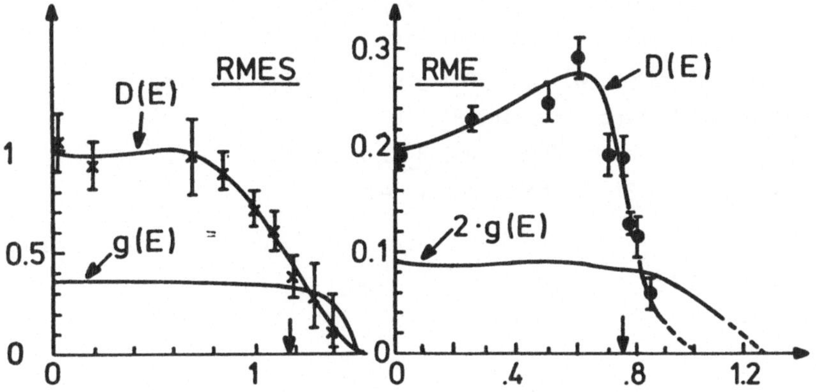


Fig. 4: Diffusivity D(E) and density of states g(E) over $E/W_2$.

At the respective critical values, $E_c$=1.18 $W_2$ (RMES) and $E_c$=0.73 (RME), see above, one finds $\sigma=(0.155\pm0.03)e^2/\hbar$ and $\sigma=(0.85\pm0.015)e^2/\hbar$. Thus for the RME case the conductivity is somewhat below, but still in the range, of the former values of the minimal metallic conductivity $\sigma=(0.11\pm0.02)e^2/\hbar$ obtained for the original Anderson model in |8|, whereas apparently the model with spin, RMES, may have a higher universal minimum metallic conductivity of its own.

5. Conclusions: To the accuracy of the present calculations there seems to be a localization transition both for the RMES and also for the RME models in two dimensions, with a finite, and universal minimum metallic conductivity, which seems to be different for the different universality classes. Our preliminary calculations for the CMES class (which we do not yet present) do also not show any striking qualitative difference with respect to other models. In any case, as explained in |3| and |10|, one should be well aware, however, that a definite conclusion cannot be drawn from numerical calculations as the present one, which are hampered both by finite-size effects and also by effective inelasticities |10|.

## References

|1| P.W. Anderson: Phys. Rev. 109, 1492 (1958)

|2| F. Wegner: Phys. Reports 67, 15 (1980), and lecture at this conference

|3| J. Stein, U. Krey: Physica 106A, 326 (1981)

|4| E. Abrahams, P. W. Anderson, D.C. Licciardello, T.V. Ramakrishnan: Phys. Rev. Letters 42, 673 (1979)

|5| S. Yoshino, M. Okazaki: J. Phys. Soc. Japan 43, 415 (1977)

|6| P. A. Lee: Phys. Rev. Letters 42, 1492 (1979)

|7| J. Stein, U. Krey: Z. Physik B 34, 287 (1979)

|8| J. Stein, U. Krey: Z. Physik B 37, 13 (1980)

|9| D. Weaire, B. Kramer: J. Noncryst. Solids 32, 131 (1979)

|10| J. Stein, U. Krey: Solid St. Comm. 36, 951 (1981)

|11| J. Stein, U. Krey: Solid St. Comm. 27, 797 (1978)

|12| J. L. Pichard, G. Sarma: J. Phys. C 14, L127 (1981), and preprint

|13| R. Oppermann, Heidelberg, preprint 96, 1980

|14| E.N. Economou, M.H. Cohen: Phys. Rev. B 5, 2931 (1972)

|15| U. Krey, W. Maaß, J. Stein: to be published

Acknowledgments:

The authors would like to thank the Deutsche Forschungsgemeinschaft for financial support, and the computer centre of the university of Regensburg for computing time on the TR 440 computer.

# CRITICAL PROPERTIES OF THE ANDERSON MOBILITY EDGE: RESULTS FROM FIELD THEORY

T.C. Lubensky
Dept. of Physics
University of Pennsylvania
Philadelphia, Pa. 19104/USA

## Abstract

The field theoretic formulation of the Anderson localization of an electron moving in periodic lattice with random on site potentials and/or hopping is reviewed. Mean field theory for Gaussian bond randomness is presented and found to predict a mobility edge with a vanishing density of extended states. Fluctuations about this mean field solution are studied. The upper critical dimension is eight, and to all orders in perturbation theory, this mobility transition is in the same universality class as that describing the statistics of lattice animals and the Yang-Lee singularity in a random imaginary field. Critical exponents are very accurately determined for all spatial dimension d, $2 < d < 8$, by a Flory approximation for the correlation length exponent: $\nu = 5/(2(d + 2))$. Localized states are studied via localized solutions of finite action of the field theory, which in the simplest treatment predict that the density of localized states vanish at the mobility edge. This singular behavior of the density of states violates general theorems for the case of pure Gaussian site randomness but may have a regime of validity when there is Gaussian bond randomness. A modified model with Lorentzian site randomness and Gaussian bond randomness is shown to have conducting states at all energies in mean field theory.

## I. Introduction

The Hamiltonian for a single particle moving in a periodic crystal in d dimensions with lattice sites $\vec{x}$ and lattice constant a can be expressed in terms of site local potentials $V(\vec{x})$ and hopping integrals $t(\vec{x},\vec{x}')$ between nearest neighbor sites $\vec{x}$ and $\vec{x}'$ as

$$\mathcal{H} = \sum_{\vec{x}} V(\vec{x}) + \sum_{\langle \vec{x},\vec{x}'\rangle} t(\vec{x},\vec{x}') \qquad 1.1$$

where $\langle\vec{x},\vec{x}'\rangle$ represents a bond between nearest neighbor sites. If $V(\vec{x})$ and $t(\vec{x},\vec{x}')$ are homogeneous, with values $V = 0$ and $-t_0$, there will be a band of electronic states with energy, E, between $-E_b$ and $+E_b \equiv zt_0$ where z is the co-ordination number of the lattice. The electron Green's function is

$$g_\pm(\vec{x},\vec{x}'; E_\pm) = \langle\vec{x}|\frac{1}{\mathcal{H}-E_\pm}|\vec{x}'\rangle \qquad 1.2$$

where $E\pm$ is in the upper (lower) half plane. From $g\pm$, one can obtain the density of states

$$\rho = \pm\frac{1}{\pi} \operatorname{Im} g_\pm(\vec{x},\vec{x}; E_\pm) \qquad 1.3$$
$$\sim |\Delta E|^{\frac{d}{2}-1} \quad \text{as} \quad |\Delta E| \to 0$$

where $\Delta E = E_b^{\,2} - E^2$. Similarly the Fourier transform of $g\pm$ has a singularity as $E_b$ is approcahed from outside the band of the form

$$g_\pm(\vec{q}, E_\pm) \sim |\Delta E|^{-1} f_\pm\left(q\,|\Delta E|^{-1/2}\right) \qquad 1.4$$

All the states have spatial wave functions that extend throughout the sample.

If V and/or t vary randomly throughout the sample, there will be states decaying exponentially away from specific lattice sites. These states are localized and cannot conduct electricity at zero temperature.[1.] If the randomness is not too strong, extended states that can conduct electricity also exist. Typically, extended states exist near the center of the band and localized states at larger values of $|E|$. The Energy $E_c$ separating localized from extended states is called the mobility edge and the transition from extended to localized states the localization or mobility transition. It is generally believed that the density of states is smooth at the mobility edge as shown in fig. 1a. The analysis of this paper leads to a mobility edge with $\rho = 0$ as shown in fig. 1b. The validity of this picture will be discussed briefly.

To study the electronic properties of random systems, it is convenient to introduce various Green's functions and their products averaged over the configurations of the random potentials:

$$G_{\pm}(\vec{x}, \vec{x}'; E) = [g_{\pm}(\vec{x}, \vec{x}'; E)] \qquad 1.5a$$

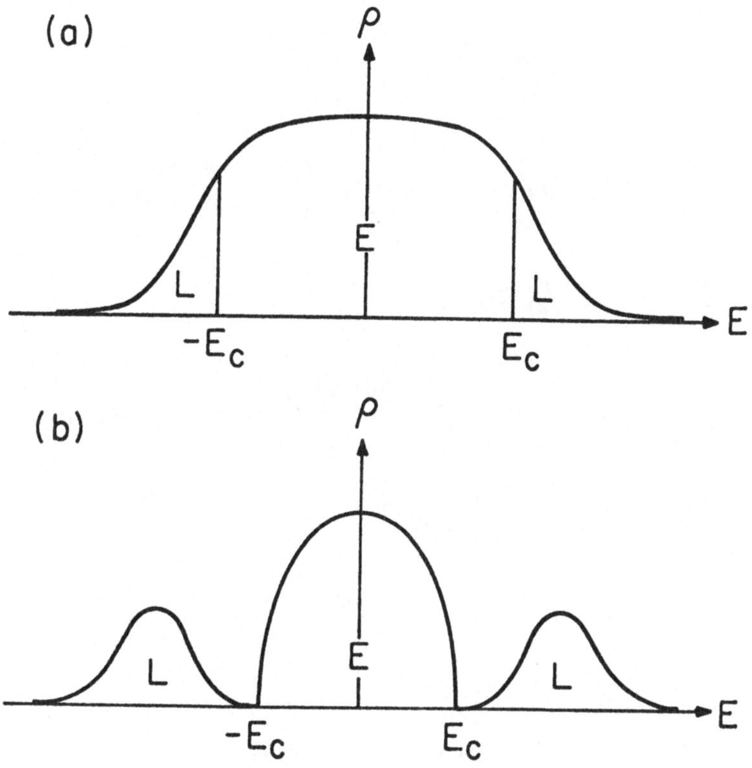


Fig. 1 Possible densities of states, $\rho$, for an electron moving on a lattice with random potentials. E stands for extended states and L for localized states. $E_c$ locates the mobility edge.

$$\mathcal{G}_{\sigma\sigma'}(\vec{x},\vec{x}';E_\sigma,E_{\sigma'}) = [g_\sigma(\vec{x},\vec{x}';E_\sigma)\,g_{\sigma'}(\vec{x},\vec{x}';E_{\sigma'})] \quad 1.5b$$

where the square brackets indicate an average over the random potentials. Extended states are distinguished from localized states by their conductivity. The conductivity at frequency $\omega$ of a state at energy can be obtained via the Kubo formula

$$\Sigma(\omega) \sim \omega^2 \lim_{q\to 0} \frac{\partial}{\partial q^2} \sum_{\sigma\sigma'} \sigma\sigma' \mathcal{G}_{\sigma\sigma'}(\vec{q};E+\omega,E) \quad 1.6$$

$\Sigma(E\ ) \equiv \Sigma(E\ ,\omega = 0)$ will be non-zero for $|E| < |E_c|$. Thus the transition from localized to extended states has the appearance of a phase transition, suggesting that we introduce scaling forms for $\Sigma$ and $\mathcal{G}$ in the vicinity of $E_c$:

$$\Sigma \sim |\Delta E|^{\mu} \quad 1.7a$$

$$\mathcal{G}_{\sigma\sigma'} \sim |\Delta E|^{-\gamma} f_{\sigma\sigma'}(q\xi, \xi^{\lambda_\omega}\omega) \quad 1.7b$$

$$\xi \sim |\Delta E|^{-\nu} \quad 1.7c$$

$$f_{\sigma\sigma'}(x) \sim |x|^{-(2-\eta)} \qquad x \to \infty \quad 1.7d$$

$$\rho_{ext} \sim |\Delta E|^{\beta} \quad 1.7e$$

where $\rho$ ext is the density of extended states and $\Delta E = E^2 - E_c^{\,2}$. If $\mu$ is zero, the conductivity jumps discontinuously at the mobility edge; similarly if $\beta = 0$, the density of extended states jumps discontinuously at $E_c$ and is finite at $|E| = |E_c|^-$.

In low spatial dimensions, it is generally believed[2.] that $\beta = 0$ but that $\mu \neq 0$. Wegner[3] has argued on general grounds that in this case, which he calls the homogeneous mobility edge, only $+\!\!-$ satisfies the Eq. 1.7b and $\mu = (d - 2)\nu$ , $\lambda_\omega = d$,

$\eta = 2 - d$ and $\gamma = d\nu$. This behavior has in fact been calculated in $2+\varepsilon$ dimensions using non-linear $\sigma$-models[4] and/or direct perturbation [5] theory with the result $\nu = (d-2)^{-1}$, $\mu = 1$. Wegner also considered the possibility of what he called an in homogeneous mobility edge with $\beta \neq 0$. We will return to a discussion of this critical point later in the paper.

The analogy between thermodynamic phase transitions and the localization transition is quite striking, and it is natural to apply to the localization problem the mathematical tools so successfully applied to phase transitions. In particular, one would like to carry out the following program for the localization problem: 1) Set up a field theory for the generating function, $\Xi$, for averaged Green's functions. This is the analog of establishing a field theory for the partition function of a thermodynamic system. 2) Study the properties of $\Xi$ and the correlation functions it generates in mean field theory. 3) Study fluctuations about mean field theory and determine the upper critical dimension, $d_c$, below which mean field theory breaks down. 4) Calculate critical properties for $d < d_c$ using renormalized perturbation theory and/or the renormalization group. 5) Locate the lower critical dimension $d^*$ below which the mobility edge disappears and study properties for $d > d^*$ again using some form of the renormalization group.

The last step in this procedure seems to have been successfully carried out for the homogeneous transition using non-linear $\sigma$-models and perturbation[4,5] theory and has been described at this meeting by F. Wegner. In this paper, we will outline how the first four steps[6,7] can be carried out for systems with independent Gaussian bond and/or site randomness with variances $\sigma_b^2$ and $\sigma_s^2$. This analysis can be done to all orders in perturbation theory[7] and leads only to conducting states, and a mobility edge which corresponds to the appearance of extended states. If $t_o$ is the average hopping, then for weak randomness, $\sigma^2 = z\sigma_b^2 + \frac{1}{2}\sigma_s^2 << t_o^2$, the mobility edge is in the same universality class as the band edge of a pure system. The weakly random behavior is inaccessible for $d < 4$. For $\sigma^2 >> t_o^2$, the mobility edge to all orders in perturbation theory is in the same universality class as the statistics of lattice animals[8] and the Yang-Lee edge singularity[9] for an Ising model in a random complex field.[10] Remarkably $\beta < 0$ for $d \lesssim 4\ 2/3$ indicating a divergent density of states from perturbation theory.

The renormalized perturbation expansion just described fails to produce localized states to any order in perturbation theory. To study localized states we investigate non-perturbative instantons or localized droplet solutions of finite action of the field theory. The most straightforward calculation of the instanton contributions to $\Xi$ leads to a density of states falling exponentially to zero at the mobility edge as

$$\rho_{LOC} \sim \exp\left(-\text{const.} \times |\Delta E|^{-5}\right) \qquad 1.8$$

Thus we predict a zero density of states at the mobility edge as shown in Fig. 1.

This surprising form for the density of states is surely wrong for the case of Gaussian site randomness since it violates a theorem due to Edwards and Thouless[11] that the density of states must be analytic for this case. The Edwards and Thouless theorem does not apply, if $\sigma_b^2 \neq 0$, and we believe that it may be possible for the density of states to be zero at the mobility edge for sufficiently large d and $\sigma_b^2 / t_0^2$. We will speculate how the present treatment may be extended to yield an analytic density of states.

## II. Formal Details

Our study begins with the now familiar generating function for Green's functions in terms of replicated 2n component random fields $\underset{\sim}{\Psi}$ with components $\Psi_i = \psi_\sigma^\alpha$, i = 1 ...2n, $\alpha$ = 1...n and $\sigma = \pm 1$: ($\alpha$ is the usual replica index and $\sigma$ specifies whether the field corresponds to E in the upper or lower half complex plane.)

$$\Xi = [Z^n] = \int \mathcal{D}\underset{\sim}{\Psi} \, [e^{-\mathcal{L}^\circ}] \qquad 2.1$$

where

$$\mathcal{L}^\circ = -\frac{1}{2} \sum_{\vec{x},\vec{x}'} \underset{\sim}{\Psi}^T(\vec{x}) \left[ \underset{\approx}{E}\, \delta_{\vec{x},\vec{x}'} - H(\vec{x},\vec{x}')\, \underset{\approx}{1} \right] \underset{\sim}{\Psi}(\vec{x}') \qquad 2.2$$

where $\underset{\approx}{E}$ is a 2n x 2n component matrix with components $E_{ij} = E_\sigma \delta_{\sigma\sigma'} \delta_{\alpha\alpha'}$, $\underset{\approx}{1}$ is the 2n x 2n component unit matrix with $H(\vec{x},\vec{x}') = V(x)\, \delta_{\vec{x},\vec{x}'} + t(\vec{x},\vec{x}')$. The functions $G_\sigma$ and $\mathcal{G}_{\sigma\sigma'}$ are related to correlation functions of $\psi_\sigma^\alpha$ :

$$G_\sigma(\vec{x},\vec{x}') = \langle \psi_\sigma^\alpha(\vec{x})\, \psi_\sigma^\alpha(\vec{x}') \rangle \qquad 2.3$$

$$\mathcal{G}_{\sigma\sigma'}(\vec{x},\vec{x}') = \langle \psi_\sigma^\alpha(\vec{x})\, \psi_{\sigma'}^\beta(\vec{x})\, \psi_\sigma^\alpha(\vec{x}')\, \psi_{\sigma'}^\beta(\vec{x}') \rangle \qquad 2.4$$

The average of $e^{-L^{o}}$ can be carried out explicitly for Gaussian randomness leading to

$$\Xi = \int \mathcal{D}\underset{\sim}{\Psi}\, e^{-L[\underset{\sim}{\Psi},\, \underset{\approx}{E}]} \tag{2.5a}$$

$$L = -\frac{1}{2}\sum_{\vec{x},\vec{x}'} \underset{\sim}{\Psi}^{T}(\vec{x})\left[\underset{\approx}{E}\,\delta_{\vec{x},\vec{x}'} + t_{0}(\vec{x},\vec{x}')\right]\underset{\sim}{\Psi}(\vec{x}')$$

$$-\frac{1}{2}\sigma_{s}^{2}\sum_{\vec{x}}\left(\underset{\sim}{\Psi}^{T}(\vec{x})\underset{\sim}{\Psi}(\vec{x})\right)^{2} - \frac{1}{4}\sigma_{b}^{2}\sum_{\vec{x},\vec{x}'}\gamma_{\vec{x},\vec{x}'}\left(\underset{\sim}{\Psi}^{T}(\vec{x})\underset{\sim}{\Psi}(\vec{x}')\right)^{2} \tag{2.5b}$$

where $\gamma_{\vec{x},\vec{x}'} = 1$ for $\vec{x}$ and $\vec{x}'$ nearest neighbors and zero otherwise, and where the contours for the $\underset{\sim}{\Psi}$ integrations are chosen to insure convergence of the integral. Note that for the Gaussian case, and the Gaussian case only, L has terms only up to order $\underset{\sim}{\Psi}^{4}$. Since we expect the critical properties of the mobility transition to be determined by fluctuations in the product order parameter $\Psi_i(\vec{x})\Psi_j(\vec{x})$, a further transformation of $\Xi$ is useful:

$$\Xi = \int \mathcal{D}\underset{\approx}{Q}\, e^{-L[\underset{\approx}{Q},\, \underset{\approx}{E}]} \tag{2.6a}$$

where

$$L[\underset{\approx}{Q},\underset{\approx}{E}] = \frac{1}{2}\mathrm{Tr}_{\vec{x},\vec{x}'}\left\{\ln\left[\left(\underset{\approx}{E} + \underset{\approx}{Q}(\vec{x})\right)\delta_{\vec{x},\vec{x}'} + t_{0}(\vec{x},\vec{x}')\underset{\approx}{1}\right]\right\}$$

$$+\frac{1}{4}\mathrm{Tr}_{\vec{x},\vec{x}'}\left\{\underset{\approx}{Q}(\vec{x})A^{-1}(\vec{x},\vec{x}')\underset{\approx}{Q}(\vec{x}')\right\} \tag{2.6b}$$

where $\underset{\approx}{Q}$ is a 2n x 2n component order parameter, $A(\vec{x},\vec{x}') = \frac{1}{2}\sigma_{s}^{2}\,\delta_{\vec{x},\vec{x}'} + \frac{1}{4}\sigma_{b}^{2}\,\gamma_{\vec{x},\vec{x}'}$ and $\mathrm{Tr}_{\vec{x},\vec{x}'}$ signifies a trace over both replica and space indices.

## III. Mean Field Theory for Gaussian Randomness

### A. Pure Random Hopping: $\sigma_b^2 \neq 0, t_o = 0, \sigma_s^2 = 0$.

This is by far the simplest case to study since the logarithmic term in Eq. (2.6b) is diagonal in $\vec{x}$ and $\vec{x}'$. To study mean field theory we minimize $L(\underset{\approx}{Q},\underset{\approx}{E})$ (Eq 2.6b) with respect to a spatially uniform $\underset{\approx}{Q}(\vec{x})$. $\underset{\approx}{Q}$ is a complicated tensor so we seek solutions of the form $Q_{ij} = Q_\sigma \delta_{\sigma\sigma'} \delta_{\alpha\beta}$ motivated by our knowledge that $\langle Q_{ij}\rangle \sim G_\sigma \delta_{\sigma\sigma'} \delta_{\alpha\beta}$. Minimizing L with respect to $\underset{\approx}{Q}$'s of this form, we find

$$\frac{\partial L}{\partial Q_\pm} = \frac{1}{2}\left(\frac{1}{z\sigma_b^2} Q_\pm + \frac{1}{E_\pm + Q_\pm}\right) = 0 \qquad 3.1$$

with solutions

$$Q_\pm = \begin{cases} \frac{1}{2} E\left[(1 - E_c^2/E^2)^{1/2} - 1\right] & E^2 > E_c^2 \\ \frac{1}{2} E\left[\pm i\,(E_c^2/E^2 - 1)^{1/2} - 1\right] & E^2 < E_c^2 \end{cases} \qquad 3.2$$

where $E_c = 2\sigma_b\sqrt{z}$ is the mobility edge. Eq. (3.2) yields a semi-circular density of states

$$\rho = \begin{cases} \frac{2}{\pi|E_c|}\sqrt{1 - \frac{E^2}{E_c^2}} & E^2 < E_c^2 \\ 0 & E^2 > E_c^2 \end{cases} \qquad 3.3$$

implying that $\beta = \frac{1}{2}$ in meanfield theory. For $E^2 > E_c^2$, we find that <u>all</u> $2n(2n+1)/2$ components of $Q_{ij}$ are simultaneously critical leading to

$$\mathcal{G}_{\sigma\sigma'}(q; E, E) \sim |\Delta E|^{-1/2}\left(1 + (q\xi)^2\right)^{-1} \qquad 3.4$$

where $\xi \sim |\Delta E|^{-1/4}$ implying $\gamma = \frac{1}{2}$ and $\nu = \frac{1}{4}$. For $E^2 < E_c^2$, $\mathcal{G}_{++}$ and $\mathcal{G}_{--}$ have the same form as Eq. (3.4) (they are pure complex as $E \to E_c$, $q \to 0$). $\mathcal{G}_{+-}$ on the other hand has a different form implied by a Ward[2,6] identity associated with the invariance of L under the transformation $\underset{\approx}{Q} \to \underset{\approx}{U}\underset{\approx}{Q}\underset{\approx}{U}^{-1}$, $\underset{\approx}{E} \to \underset{\approx}{U}^{-1}\underset{\approx}{E}\underset{\approx}{U}$ where $\underset{\approx}{U}$ is an orthogonal 2n x 2n matrix;

B. $t_0 \neq 0,\ \sigma^2 = z\sigma_b^{\,2} + \frac{1}{2}\sigma_s^{\,2} \neq 0$

When $t_0 \neq 0$ the analysis of mean field theory becomes considerably more complicated. When $\sigma^2 = 0$, we know that there is a band edge singularity of the type discussed in the introduction. When $\sigma^2 \ll t_0^{\,2}(d-4)$ for $d > 4$, the mobility edge is in fact determined by the band edge in mean field theory. We call this the weakly random mobility edge. The critical exponents for this mobility edge are $\gamma = 1$, $\beta = \frac{d}{2} - 1$, $\nu = \frac{1}{2}$, $\eta = 0$ and $\mu = 1$ where the exponents $\gamma$ and $\eta$ now refer to the behavior of G rather than $\mathcal{G}$. Notice that the weakly random regime becomes totally inaccessible for $d < 4$.

When $\sigma^2 \gg t_0^{\,2}$, there is a strongly random mobility edge with critical behavior identical to that of the simple case $t_0 = 0$, $\sigma_s^{\,2} = 0$ just discussed. At some intermediate value of $\sigma^2$ there is presumably a multi-critical point.

## IV. Renormalized Perturbation Theory

To keep the discussion in this section as simple as possible, we will restrict our attention to the ++ subspace of $Q_{ij}$. We first introduce the shifted field

$$\varphi^{\alpha\alpha'} = Q_{++}^{\alpha\alpha'} - Q_{+}\,\delta^{\alpha\alpha'} \qquad 4.1$$

in the usual way. It is convenient to break $\phi^{\alpha\alpha'}$ up into a diagonal and an off-diagonal part

$$\varphi^{\alpha\alpha'} = \varphi_\alpha\,\delta^{\alpha\alpha'} + \psi_{\alpha\alpha'}\,(1-\delta^{\alpha\alpha'}) \qquad 4.2$$

with $\psi_{\alpha\alpha'} = \psi_{\alpha'\alpha}$. In terms of these variables, to lowest order in the fields, we have

$$\begin{aligned} \mathcal{L} = &\int d^dx \left[ \tfrac{1}{2}\sum_{\alpha=1}^{n}\left(r\varphi_\alpha^2 + (\nabla\varphi_\alpha)^2\right) + \tfrac{1}{2}\left(r\,\mathrm{Tr}\,\underset{\approx}{\psi}^2 + \mathrm{Tr}(\nabla\underset{\approx}{\psi})^2\right)\right] \\ &+ \int d^dx \left[ \tfrac{1}{2}\tau\left(\sum_{\alpha=1}^{n}\varphi_\alpha\right)^2 + \tfrac{1}{3!}W\sum_\alpha \varphi_\alpha^3 \right] \\ &+ \int d^dx \left[ \tfrac{1}{2}W\sum_{\alpha\neq\beta}\varphi_\alpha\psi_{\alpha\beta}^2 + \tfrac{1}{3!}W\,\mathrm{Tr}\,\underset{\approx}{\psi}^3 \right] \end{aligned} \qquad 4.3$$

where $\tau$ and $w$ are simply related to E and $\sigma_b{}^2$ . The potential $\tau$ is zero in the initial theory. It is, however, generated at one loop order in perturbation theory and must be included to obtain the correct critical behavior. The $\psi'$ fields have the usual $(r + q^2)^{-1}$ propagator. The $\phi_\alpha$ fields on the other hand have the more complicated propagator

$$\langle \phi_\alpha \phi_\beta \rangle = \frac{1}{r+q^2}\,\delta_{\alpha\beta} - \frac{\tau}{(r+q^2)(r+q^2+n\tau)} \qquad 4.4$$

The term proportional to $\tau$ is of the same form as encountered in the study of an Ising model in a Gaussian random external magnetic field. It is more singular than the term proportional to $\delta_{\alpha\beta}$ in the limit $n \to 0$ and has the effect of enhancing the infrared divergences and raising the upper critical dimension, $d_c$ , by two. Since $d_c$ for a system with a cubic interaction is 6, we expect $d_c$ for this system to be 8. In fact, one can analyze the perturbation theory for Eq. (4.3) in exactly the same way as Parisi and Sourlas[10] did for the animals problem[8] and find that this localization and the animals problem are both in the same universality class as the Lee-Yang edge singularity[9] in a random imaginary field.

This model has only one independent critical exponent which we can take to be $\nu$ . Other exponents expressed in terms of $\nu$ are[7]

$$\gamma = 2 - (d-2)\nu \qquad \beta = (d-2)\nu - 1$$

$$\mu = (d-4)\nu \qquad \eta = 2 - d\nu^{-1} \qquad \lambda_\omega = \nu^{-1} \qquad 4.5$$

These scaling relations are identical to those proposed by Wegner[3] for the inhomogeneous fixed point provided d is replaced by d − 2 in his expressions.

A Flory-like approximation[14] gives values for $\nu$ that are remarkably accurate over the entire range of dimensions from d = 2 to 8; it is exact at d = 2, 3 and 8 and within a couple of percent of the best values obtained by series $\varepsilon$-expansion in all other dimensions. $\nu$ and other exponents in this approximation are

$$\nu = \frac{5}{2(d+2)} \qquad \beta = \frac{3d-14}{2(d+2)}$$

$$\gamma = \frac{18-d}{2(d+2)} \qquad \mu = \frac{5(d-4)}{2(d+2)} \qquad 4.6$$

The exponent $\beta$ describing the growth of $\rho_{ext}$ becomes negative for $d \lesssim 14/3$ indicating a divergent density of states at $E^2 = E_c^{2\,-}$. In addition $\mu$ becomes negative for $d < 4$, indicating a divergent conductivity. These divergences seem unphysical to us and possibly indicate that the type of mobility edge discussed here does not exist below 4 2/3 dimensions.

## V Localized States

Mean field theory and renormalized perturbation theory for Gaussian randomness produce only extended states. General variational arguments[6], however, rigorously establish that states exist at arbitrarily large $|E|$ for Gaussian randomness on a lattice. Since the generating function $\Xi$ formally reproduces the exact averages of $g\pm^{\kappa}$ for any $\kappa$, we expect $\Xi$ to contain these localized states as well as the extended ones. Cardy[15] showed that the Halperin-Lax-Langer-Zittartz[16,17] tails in the density of states for Gaussian white noise (independent Gaussian random potentials in a continuum rather than on a lattice) are reproduced by instanton solutions of finite action of the field theory of Eq. (2.5) with $\sigma_b^2 = 0$. An alternate but related mechanism for producing imaginary contributions to $\Xi$ occurs when the system is constrained to be in a metastable state as in the case of an Ising model constrained to have a positive magnetization in a negative magnetic field studied by Langer[18]. The first mechanism is appropriate to the $\psi$ formulation of $\Xi$ Eq. (2.5) and the second to the $\underset{\sim}{Q}$ formulation Eq. (2.6b).

We begin by seeking non-uniform extrema to L when $\sigma_b^2 = 0$. Since L is rotationally invariant in replica space, we seek solutions of the form $\psi_i(x) = \psi_\sigma(x)e^\alpha$ where $\Sigma e^\alpha e^\alpha = 1$:

$$\frac{\partial L}{\partial \psi_\sigma(\vec{x})} = \sum_{\vec{x}'} \left[E\,\delta_{\vec{x},\vec{x}'} + t_o(\vec{x},\vec{x}')\right]\psi_\sigma(\vec{x}') + \frac{1}{2}\sigma_s^2\,\psi_\sigma^3(\vec{x}) = 0 \tag{5.1}$$

If spatial variations of L are slow as they are far from the core of a localized solution, we can expand $t_o(\vec{x},\vec{x}')$ in powers of gradients and treat $\psi_\sigma(\vec{x})$ as a continuous field. If $E <- zt_o$, Eq. (5.1) reads

$$\left[\Delta E + C_o\nabla^2\right]\psi_\sigma(\vec{x}) + \frac{1}{2}\sigma_s^2\,\psi_\sigma^3(\vec{x}) = 0 \tag{5.2}$$

where $\Delta E = E + zt_o$ and $C_o = a^2 t_o$. For $E <- zt_o$, Eq. (5.2) has a scaled solution of the form

$$\psi_\sigma(\vec{x}) \sim \frac{1}{\sigma_s}(-\Delta E)^{1/2} f_\sigma\left(|\Delta E|(\vec{x}-\vec{x}_0)^2/c_0\right) \qquad 5.3$$

where $\vec{x}_0$ specifies the center of the localized solution. This solution is to be matched at short distances, $|\vec{x} - \vec{x}_0|$ , to a core-function that depends on the details of the lattice structure. The contributions to the instanton free energy, $L_I$, from the core and far field parts of $\psi_\sigma(x)$ are additive yielding

$$L_I = L_c + B[-\Delta E]^{2-d/2} \qquad 5.4$$

where $L_c$ is the core part. Finally, we have

$$\rho_{Loc} \sim e^{-L_I} \qquad 5.5$$

Note that the far field part leads to a divergent contribution to $L_I$ as $|\Delta E| \to 0$ for $d > 4$ forcing the density of states to zero at the weakly random mobility edge. For $d < 4$, the far field part yields the Halperin-Lax-Zittartz-Langer fails. We should note that for Gaussian white noise, there is no core, and solutions to Eq. (5.2) do not exist[19] for $d > 4$. Finally we note that we do not expect this picture to change substantially if $\sigma_b{}^2$ but non-zero small.

In random system, the replicated partition function $[z^n] = \Xi$ must satisfy the constraint: $\lim_{n \to 0} \Xi = 1$. All solutions we have considered so far satisfy this constraint. For E outside the band of extended states, however, there are solutions which we call bad solutions of the form

$$\varphi^{\alpha\beta}_{\sigma\sigma'} = \left(Q_\sigma \delta^{\alpha\beta} + R_\sigma e^\alpha e^\beta\right)\delta_{\sigma\sigma'} \qquad 5.6$$

which do not satisfy this constraint and which have lower free energy that the solutions which we call good solutions, that do. Thus localized droplets of the bad solution in the sea of the good solution will give rise to imaginary parts of $\Xi$ and contributions to $\rho_{LOC}$ . Analyzing these contributions, we find

$$\rho_{LOC} \sim \exp(-\text{const} \cdot |\Delta E|^{-\zeta}) \qquad 5.7$$

where

$$\zeta = \begin{cases} \frac{1}{3}(d-6) & d>8 \\ 2\nu & d<8 \end{cases} \qquad 5.8$$

Thus, we conclude that for both strong and weak Gaussian site and/or bond randomness, there is a solution with an external free energy associated with $\Xi$ with a density of states that is zero at the mobility edge. This picture is surely incorrect for the case of pure site randomness, $\sigma_b^2 = 0$, $\sigma_s^2 \neq 0$. Therefore, there must be another solution which produces an extremal free energy but with an analytic density of states. Preliminary calculations indicate that such a solution can be produced by extending the instanton contributions to $\Xi$ to complex values of the energy and seeking solutions for $\underset{\sim}{Q}$ which minimize the total free energy including the nonperturbative as well as the perferbative parts. In high dimensions, it appears that the solution with the zero density of states at the mobility edge may be used as a starting point to seek the other solution of perturbation theory in the density of states at the mobility edge. This will be discussed in a future publication.

## VI. Modified Lloyd Model

In the previous sections we indicated how the density of states and other properties in random systems can be calculated approximately using renormalized perturbation theory and instantons of finite action. The $\rho(E)$ we found in this way is non-analytic at the mobility edge and is, therefore, incorrect for pure Gaussian site randomness but is possibly correct in sufficiently high dimension when $\sigma_b^2$ is non-zero. There is one probability distribution, P(V), for site energies for which the density of states can be obtained exactly: the Lorentzian distribution studied by Lloyd[20]

$$P(V) = \frac{1}{\pi} \frac{\Gamma}{V^2 + \Gamma^2} \qquad 6.1$$

The density of states in this model is analytic for all 0 - ∞ < E < ∞ . Unfortunately, the two particle propagator $\mathcal{G}_{+-}$ which contains information about Σ(E) cannot be calculated easily. Thus there is no precise information about the mobility edge in this model.

In this section, we will briefly show that a mean field theory for Gaussian bond randomness and Lorentzian site randomness can be solved exactly. The density of states is analytic and non-zero for all - ∞ < E < ∞ , and <u>all</u> states are extended.

We begin with Eq. (2.1), perform the Gaussian average over bonds and transform to Q variables to obtain

$$\Xi = \int \mathcal{D}\underset{\sim}{Q}\, e^{-\mathcal{L}[\underset{\sim}{Q},\, \underset{\sim}{E}]} \tag{6.2}$$

$$\mathcal{L}[\underset{\sim}{Q}, \underset{\sim}{E}] = \frac{1}{4}\, \mathrm{Tr}_{\vec{x},\vec{x}'}\, \underset{\sim}{Q}(\vec{x}) A^{-1}(\vec{x},\vec{x}') \underset{\sim}{Q}(\vec{x}') - \ln\, [W^n]_V \tag{6.3}$$

where $[\;]_V$ signifies a Lorentzian average over site potentials and

$$[W^n]_V = \left[ \int \mathcal{D}\underset{\sim}{\Psi}(\vec{x})\, e^{\frac{1}{2}\sum_{\vec{x}} \underset{\sim}{\Psi}^T(\vec{x})[\underset{\sim}{E} + \underset{\sim}{Q}(\vec{x}) - V(\vec{x})\underset{\sim}{1}]\underset{\sim}{\Psi}(\vec{x})} \right]_V . \tag{6.4}$$

The mean field equations for Q± are

$$\frac{\partial \mathcal{L}}{\partial Q_\pm(\vec{x})} = \frac{1}{2} A^{-1}(\vec{x},\vec{x}') Q_\pm(\vec{x}) - \left[ \frac{1}{E_\pm + Q_\pm - V} \right]_V$$

$$\frac{1}{z\sigma_b^2} Q_\pm + \frac{1}{E_\pm + Q_\pm \pm i\Gamma} = 0 \tag{6.5}$$

Im Q± from this equation and thus ρ in non-zero for all values of E.

The two particle propagator $\mathcal{G}_{+-}$ is related to the second derivative $\Gamma^{(2)}$ of L with respect to $\underset{\approx}{Q}$

$$\Gamma^{(2)}_{+-} = \frac{1}{2\sigma_b^2 \delta(\vec{q})} - \frac{1}{2}\left[\frac{1}{E_+ + Q_+ - V}\;\frac{1}{E_- + Q_- - V}\right]_V \qquad 6.6$$

$$\Gamma^{(2)}_{+-}(q; E+\omega, E) = \frac{1}{2\sigma_b^2 \delta(q)} - \frac{1}{2}\,\frac{1}{\omega + Q_+ - Q_-}\left(\frac{1}{E + Q_- - i\Gamma} - \frac{1}{E + \omega + Q_+ + i\Gamma}\right)$$

$$\sim \frac{q^2}{2z^2\sigma_b^2} - \frac{1}{2z^2\sigma_b^2}\,\frac{i\omega}{\pi\rho} \qquad 6.7$$

where we used $Q_+ = Q_-^*$ and $\mathrm{Im}\, Q_\pm = \pm\pi\sigma_b^2 z\rho$. Thus we see that all states are extended in this model.

## Acknowledgements

Most of the work reported here was a collaborative effort with A.B. Harris and would not have been completed without his aid. The work on connection between the critical properties of the mobility edge studied here and those of lattice animals was done in collaboration with A.J. McKane. We are grateful for financial support to N.S.F. under contract No. DMR79-10153 and O.N.R. under contract No. N00014-0106.

References


1. P.W. Anderson, Phys. Rev. 109, 1492 (1958).

2. I.M. Lifshitz, Adv. Phys. 13, 483 (1964);D.J. Thouless, Phys. Reports C13 94 (1974); N.F. Mott and A.E. Davis, Electronic processes in Non-Crystalline Materials 2nd. Ed. (Oxford: Pergamon, 1979).

3. F.J. Wegner, z. Physik B25 327 (1976).

4. F. Wegner, z. Physik B35, 207 (1979); A.J. McKane and M. Stone, Annals of Physics (N.Y.) 131, 36 (1981); A. Houghton, A. Jevicki, R. Kenway and A.M.M. Pruisken, Phys. Rev. Lett. 45, 394 (1980).

5. A. Abrahams, P.W. Anderson, D.C. Licciardello and T.V. Ramakrishnan, Phys. Rev. Lett. 42, 673 (1979); L.P. Gorkov, A.I. Larkin and D. Khmel'nitzkii JETP Lett. 30, 229 (1979); W. Gotze, Solid State Commun. 27, 1391 (1978); D. Vollhardt and P. Wolfe, Phys. Rev. Lett. 45, 842 (1980).

6. A.B. Harris and T.C. Lubensky, Solid State Commun, 34, 343 (1980), Phys. Rev. B23, 2640 (1981).

7. T.C. Lubensky and A.J. McKane, submitted to J. Physique Lettre.

8. T.C. Lubensky and J. Isaacson, Phys. Rev. A20, 2130 (1979).

9. M.E. Fisher, Phys. Rev. Lett. 40, 1610 (1978); D.A. Kurtz and M.E. Fisher, Phys. Rev. B20, 2785 (1979).

10. G. Parisi and N. Sourlas, Phys. Rev. Lett. 46, 871 (1981).

11. J.T. Edwards and D.J. Thouless, J. Phys. C4, 453 (1971).

12. B. Velicky, Phys. Rev. 184, 614 (1969).

13. Y. Imry and S. -K. Ma, Phys. Rev. Lett. 35, 1399 (1975); A.P. Young, J. Phys. C10, L257 (1977); G. Parisi and N. Sourlas, Phys. Rev. Lett. 43, 744 (1979).

14. J. Isaacson and T.C. Lubensky, J. de Physique 41, L469 (1980).

15. J.L. Cardy, J. Phys. C11, 1321 (1978); M.V. Sadovskii, Sov. Phys. Solid State, 21, 435 (1979).

16. B.I. Halperin and M. Lax, Phys. Rev. 148, 722 (1966).

17. J. Zittartz and J.S. Langer, Phys. Rev. 148, 741 (1966).

18. J.S. Langer, Annals of Physics 41, 108 (1967).

19. S. Coleman, V. Glaser and A. Martin, Commun. Math Phys. 58, 211 (1978).

20. P. Lloyd, J. Phys. C2, 1717 (1969).

# LOCALIZATION THEORY : SOME RECENT RESULTS

Hervé Kunz
Lab. de Physique Théorique
EPF - Lausanne
Suisse

Bernard Souillard
Centre de Physique Théorique
Ecole Polytechnique
F-91128 Palaiseau Cedex
France

We present here results concerning the localization theory for disordered systems. We will study Hamiltonians describing the motion of independant electrons at zero temperature in the tight binding approximation :

$$H = \sum_{x} \sum_{y \in \mathcal{N}(x)} |x\rangle\langle y| + \sum_{x} V(x) \; |x\rangle\langle x|$$

where $x$ is a site of a lattice $L$, e.g. the d-dimensional cubic lattice $\mathbb{Z}^d$, or a Bethe tree, and the sum on $y$ runs over the set $\mathcal{N}(x)$ of sites of the lattice which are nearest neighbour of $x$. The potential variables $V(x)$ will be independent random variables with identical distribution of density $\rho$, e.g. Gaussian, Lorentzian, or rectangular ...

We have been interested in obtaining mathematically exact results concerning the nature of the states of such an Hamiltonian, i.e. are these states localised or extended ? We first describe these results below.

Our first results[1] are the following : consider the quantity

$$\overline{\rho}(0,y;\, A) = \lim_{\Lambda \uparrow L} \overline{\rho}_{\Lambda}(0,y\,;\, A) = \lim_{\Lambda \uparrow L} \left\langle \sum_{e_\alpha \in A} \frac{|\psi_\alpha(o)\;\psi_\alpha(y)|}{\|\psi_\alpha\|^2} \right\rangle_{\Lambda}$$

where the limit is taken over a sequence of boxes increasing in order to cover the whole lattice $L$, where the brakets $\langle . \rangle_{\Lambda}$ denote averaging with respect to the potential variables $\{V_x\}$, and where for a given interval $A$, $(e_\alpha, \psi_\alpha)$ denote respectively the eigenvalues lying in $A$ and the corresponding eigenfunctions, for the Hamiltonian $H$ restricted to the finite box $\Lambda$.

Then we can prove (and this is intuitively natural) that :

- if $\sum_{y \in L} \overline{\rho}(0,y\,;\, A) < \infty$, then with probability one, all states with energy lying in $A$ are localised.
- <u>Remarks</u> : . in contrast, for the ordered case $(V_x \equiv 0)$, one checks easily that

$$\lim_{y\to\infty} \bar{\rho}(0,y\ ;\ A) = Cte > 0.$$

. The transition from localised to delocalised states appears then as analogue to a phase transition in statistical mechanics where the integrability of the two-points correlation function corresponds to the absence of ordered phase, and on contrary if the two-points function does not tend to zero we have long range order and some ordered phase.

. Note that an intermediate situation is possible, namely that $\bar{\rho}(0,y\ ;\ A)$ can tend to zero when $|y| \to \infty$ but too slowly for being summable with respect to $y$. This situation appears in statistical mechanics in the presence of a continuous symmetry (e.g. for the x-y model) and as we will see later it seems also to appear for localisation transition in two dimensional systems.

- If $\lim_{\Lambda\uparrow L} \sum_{y\in A} |y|^2\, \bar{\rho}_\Lambda(0,y\ ;\ A) < \infty$ then with probability one, the static conductivity vanishes when the Fermi level lies in $A$.

The next step in our approach[1] is to compute "one-energy quantities", i.e. quantities of the form :

$$\left\langle \sum_{e_\alpha\in A} f\left(e_\alpha\, ,\, \frac{\Psi_\alpha}{\|\Psi_\alpha\|}\right)\right\rangle_\Lambda \equiv \int\left\{\sum_{e_\alpha} f\left(e_\alpha, \frac{\Psi_\alpha}{\|\Psi_\alpha\|}\right)\right\} \rho(V_1)\, \rho(V_2)\, \ldots\, \rho(V_\Lambda)\, dV_1\, \ldots\, dV_\Lambda \qquad (2)$$

where $\rho$ denotes the common density distribution of the random variables $V_x$. Examples of such one-energy quantities are the density of states, the correlation function $\bar{\rho}_\Lambda(0,y\ ;\ A)$, the inverse participation ratio. We want to compute them without solving explicitly the eigenvalue, eigenvector problem. For this we remark that the converse problem, namely given $e,\Psi$ find $V$ such that $H\Psi = e\Psi$, is an easy problem , solved by

$$V_{(x)} = e - \frac{\sum_{y\in\mathcal{N}(x)} \Psi(y)}{\Psi(x)} \qquad (3)$$

This allows us to make a change of variables in the integral (2), which leads to

$$\left\langle \sum_{e_\alpha\in A} f(e_\alpha,\Psi_\alpha)\right\rangle_\Lambda = \int de \int d\Psi_1\, \ldots\, d\Psi_\Lambda\;\; e^{-\sum_x \Psi(x)^2}\;\; f\left(e,\frac{\Psi}{\|\Psi\|}\right) \prod_{x\in\Lambda} \rho\left(e - \frac{\sum_{y\in\mathcal{N}(x)}\Psi(y)}{\Psi(x)}\right) J(e,\Psi)$$

where $J(e,\Psi)$ is the Jacobian of the transformation (3); although this change of variables is rather complicated, it turns out that an exact explicit expression can be found for it. Hence we are driven to the study of the observables of system of statistical mechanics of continuous spin systems with complicate interactions, and in particular to study the two points correlation function of such a system.

These are general exact results. We can go further however with mathematically exact results if we consider more specific models : the simplest is the one dimensional model. In that case we can prove that :

Theorem : In dimension $d=1$, i) $\overline{\rho}(0,y\,;\,A) \leq e^{-\gamma(A)|y|}$ for all $A$, ii) with probability 1, all states are exponentially localized iii) with probability 1, the static conductivity is zero.

The result of ii) had been proved previously by another method for a class of Schrödinger equation with a random potential by Gold'sheid, Molchanov et Pastur[2]. The result iii) although widely expected among physicists had been challenged by numerical computations in the last years.

The next simpler model is the case of a Bethe tree. In that case, there are partial results for a transition : first the quantity $\sum_y \overline{\rho}(0,y\,;\,e)$ is finite or infinite if the maximal eigenvalue $\lambda(e)$ of some integral equation is smaller or larger than 1. The equation $\lambda(e) = 1$ is then the equation for the mobility edge in our approach. It appears to be exactly the same than the one found by other methods by Abou-Chacra, Anderson et Thouless[3]. If $\rho$ is a Lorentzian, then $\lambda(e)$ is larger or smaller than 1 according to the disorder or to the value of $e$, hence there is a transition in that case. Moreover, if one defines the localisation length $\xi(e)$ by $\xi^2(e) = \sum_y y^2 \,\overline{\rho}(0,y\,;\,e)$ then it diverges at the mobility edge like $(e-e_M)^{-1}$, giving the value $\nu = 1$ for the correlation length exponent. On the other hand if $\rho$ has enough moments, one can prove for low disorder the breaking of the symmetry $O(n,n)$ in the replica representation of the two-points Green's function, which implies that the spectrum of $H$ has an absolutely continuous part. Such a mechanism was first proposed by Parisi[4].

Concerning the situation for "real" d-dimensional systems, we have so far no rigorous results when $d > 1$. However we have some indications on what happens. We found useful to discuss the problem with the two following quantities :

- the average probability of return of the particle in a ball B,

$$R = \lim_{T\to\infty} \frac{1}{T} \int_0^T \|P_B \,\psi_t\|^2 \,dt$$

- the time spent by the particle in the ball B,

$$T = \int_0^\infty \|P_B \,\psi_t\|^2 \,dt \quad .$$

Computing these quantities to second order in the potential, for energies in the middle of the band we obtained in $d = 1$ : $R > 0$ and $T = \infty$, in $d = 2$ : $R = 0$

and $T = \infty$, in $d = 3$ : $R = 0$ and $T < \infty$. These results correspond to complete localisation for $d = 1$, and to existence of extended states for $d = 3$. For $d = 2$ we find existence of non localised states ($R = 0$), which are even not square integrable, however these states are not extended ($T = \infty$). This transition when $d = 2$ is compatible with recent results of Pichard and Sarma[5] and of Kaveh and Mott[6].

Finally we mention that in the same approximation we obtain that :

- tight-binding Hamiltonians with pure off-diagonal disorder exhibits the same dependance with respect to the dimension, than those with diagonal disorder.

- continuous three-dimensional systems in a constant magnetic field present $R > 0$ $T = \infty$, which suggests that all states could be localised for them. A physical picture for this possibility can be easily seen.


[1] H. Kunz and B. Souillard : Commun. Math. Phys. 78, 201 (1980).

[2] I. Gold'sheid, S. Molchanov and L. Pastur : Funkts. Anal. Prilozhen 11, 1 (1977), and S. Molchanov : Math. USSR Izvestiga 42 (1978).

[3] R. Abou-Chacra, P.W. Anderson, and D.J. Thouless : J. Phys. C, 6, 1734 (1973).

[4] G. Parisi : J. Phys. A 14, 735 (1981).

[5] J.L. Pichard and G. Sarma : J. Phys. C Lett 14, L127 (1981) and to be published.

[6] M. Kaveh and N.F. Mott : J. Phys. C Lett. 14, L 177 (1981).

# CORRELATION EFFECTS IN METAL-INSULATOR TRANSITIONS

T. M. Rice
Bell Laboratories
Murray Hill, NJ

Recent progress on the understanding of correlation effects in the metal-insulator transition is reviewed. The theories of correlation effects in an ordered system and of the effects of disorder in a one-electron approximation have been well studied for some time, but there has not been much progress on the combination of both effects until recently. Bhatt and Rice have shown that short range correlations do not distinguish, in principle, between compensated and uncompensated samples. At low densities Efros and Shklovskii have demonstrated that correlations, imposed by the long range nature of the Coulomb interaction, lead to important modifications of the single particle density of states. At high densities, in the metallic state, Altshuler and Aronov have shown that even weak disorder leads to departures from Landau theory of Fermi liquids. A unified scaling theory which connects these two limits has been proposed very recently by McMillan.

## 1. Introduction

Over the years a fairly complete understanding of the metal-insulator transition in an ordered system, including the electron correlations caused by the interaction among the electrons, has been achieved. The transition in a random potential in a one electron approximation, ignoring the interactions and the ensuing correlations, is also quite well understood. The difficult problem of putting these two effects together has recently been the focus of activity and progress. This review will cover briefly the first two topics and will concentrate on the description of the transition in a disordered system including the role of electron correlations. It will further concentrate attention on a single material system namely a random array of donors doped into a semiconductor. This system has been the classic system in which to study the Mott, or metal-insulator, transition and has recently been the object of more experimental activity with better optical measurements[1] and extended studies at very low temperatures.[2,3] These studies are just now appearing in print but only theoretical aspects of the problem are covered here and the interested reader is referred elsewhere for accounts of the experiments.

In Section 2 the theory of the metal-insulator transition in an ordered lattice of hydrogenic donors is reviewed. The Coulomb interaction forces the transition to be first-order at a temperature below a critical temperature and the size of the first order transition can be estimated using the theory of the electron-hole liquid.[4] The results are contrasted with the behavior expected at an Anderson transition in the one-electron approximation.

The first part of Section 3 is devoted to the effect of short range correlation among the electronic states of donors in a semiconductor at low densities and later the effect of the long range Coulomb interactions is discussed. The theory of Efros and Shklovskii[5] for the effect of long range correlations on the single-particle density of states is presented. At high densities the overlap of the electron wavefunctions is strong and electrons form a degenerate Fermi liquid. Scattering processes off the random array of donors have recently been shown by Altshuler and Aronov[6] to have important effect at low temperatures and to lead to severe departures from Landau Fermi liquid theory. Finally in the last part of the section the scaling theory of McMillan[7] which interpolates between the two limits is very briefly discussed.

## 2. Metal-Insulator Transition

### a) Correlation without Disorder.

Suppose that one could grow a sample with the donor atoms substituted on an ordered lattice of sites in the host crystal. Within the effective mass approximation for the donors the problem is the same as a lattice of H atoms with different values of the lattice constant. At high densities the H atoms overlap strongly and the system is metallic -- the electrons forming a degenerate Fermi liquid. The properties of the electron liquid can then be described by the Landau theory of Fermi liquids similar to the normal state of any crystalline metal.

At low densities the system reduces to a lattice of isolated H-atoms with only a weak overlap of the electron wavefunctions. Short range correlations insure that this system is an insulator. These short range correlations are described by the Hubbard model[8]

$$H = \sum_{ij} t_{ij} a^{+}_{i\sigma} a_{j\sigma} + U \sum_{i} n_{i\uparrow} n_{i\downarrow} \qquad (1)$$

where $t_{ij}$ is the hopping integral and U is intra-site Coulomb repulsion. In the low density limit U >> t the ground state has one electron per site and has an antiferromagnetic spin structure (exchange energy $\sim t^2/U$). The single-particle density of states describes the density of states to add or subtract an electron. Adding an electron causes a site to be doubly occupied i.e. to have the configuration ($H^-$ or $D^-$) and thereby requires an energy U. Removing an electron leaves behind an empty site ($D^+$). A doubly occupied site or an empty site can propagate through the lattice acquiring a bandwidth $\sim$ zt (z: coordination number). Therefore at low densities there is a band gap between the upper Hubbard band ($D^-$) and the lower Hubbard band ($D^+$) whose magnitude is $\gtrsim U - 2zt$. This vanishes at $U \gtrsim 2zt$ -- a value which is close to Hubbard's criterion [8] for the metal-insulator transition.

The Hubbard Hamiltonian describes the essential physics of the low density state but if we wish to have a full description, the Coulomb interaction must be included. Recently detailed calculations of the band gap for a low density lattice of H-atoms were reported by Bhatt and Rice[4] and by Mott and Davies.[9] The band edges for the $D^-$ and $D^+$-bands were calculated using a Wigner-Seitz approximation. In the $D^-$ band an electron is propagating through a lattice of neutral hydrogenic donors and the potential it sees is short range $\sim R^{-4}$ at large distances R from the donor. At short distances the electron sees the full Coulomb potential. Using the method of polarized orbitals, a potential for the neutral hydrogenic donor which is correct in the two limits, large R and small R, was derived.[4] In the $D^+$-band the electron sees the full Coulomb potential of the donor in the central cell. The band gap is narrowed by polarization effect of the neighboring cells. The result of their calculation which includes these effects is shown in Fig. 1. The two curves, marked M-V and S-V, show the difference expected between many valley and single valley semiconductors -- the $D^-$ band of the latter being narrowed by strong spin scattering. The calculations show that the narrowing of the energy gap between $D^-$ and $D^+$ states occurs quite rapidly for $R \lesssim 7a_B$ with the gap collapsing at $R \approx 4a_B$ a value incidentally which agrees very well with Mott's criterion ($n_c^{1/3} a_B = 0.25$).

The energy density functional approach has also been applied to this system.[10,11] In this first principles method, the ground state energies of paramagnetic, ferromagnetic, and antiferromagnetic states are calculated and compared. The latter states undergo metal-insulator transition as the density is reduced. The antiferromagnetic state is lower in energy then the ferromagnetic state at all densities and the value of the density at which the metal-insulator occurs in the anti-ferromagnet is close to the value obtained by the Wigner-Seitz calculation. However the energy bands obtained in the density functional method do not agree with the $D^-$ and $D^+$ bands at low density.[11]

At the metal-insulator transition the long-range of the Coulomb effect has an important effect. There is an old argument, due to Peierls and Mott, that a metal cannot exist with an arbitrarily small number of carriers because the long-range Coulomb interaction among the electrons and holes would cause them bind. The whole situation was greatly clarified with the understanding of the electron-hole liquid. The theory of the electron-hole liquid was developed to describe the behavior of electrons and holes introduced into a semiconductor, but the same formalism may be applied to the case at hand, where the holes are introduced into the $D^+$-band rather than the valence band.
At the metal-insulator transition we need to consider the energy of

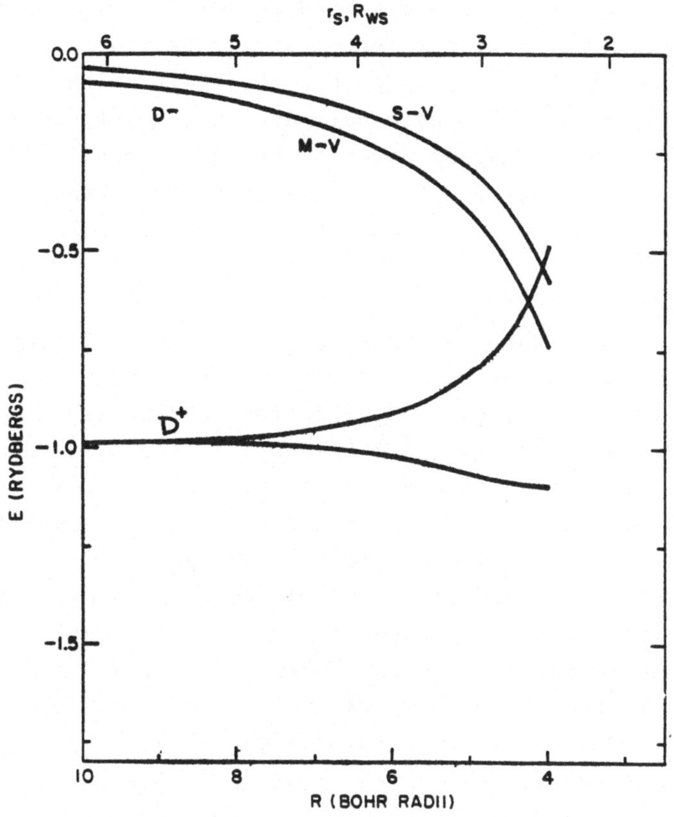


Figure 1. The maximum and minimum energies of the $D^+$-band and the minimum energy of the $D^-$-band in single-valley (S-V) and many-valley (M-V) semiconductors for an SC lattice of hydrogenic donors (lattice constant R). $R_{WS}$ is the radius of the Wigner-Seitz sphere and is equal to the usual electron gas parameter $r_s$. The energies were calculated[4] using a Wigner-Seitz method and potentials as described in the text.

promoting a small number of electrons to the $D^-$-band leaving holes in the $D^+$-band. The energy of the Fermi liquid of electrons and holes, per electron-hole pair, is composed of kinetic, exchange and correlation contributions and is minimized at a finite density of electron-hole pairs. The energy and length scale are set by the exciton parameters of the electron and hole, which in turn are determined by the electron and hole effective masses and the dielectric constant including interband polarization effects. A first-order transition occurs at the value of the energy gap when the shift of the chemical potential to create an electron-hole pair due to the interactions among electrons and holes exactly cancels the energy gap.[12] For smaller values of the energy gap the energy gain due to many body effects is greater than the energy cost to excite electrons across the energy gap.

Numerical estimates show that the critical value of the energy gap is approximately equal to the exciton Rydberg. At that value of the energy gap a finite density, $n_{e,h}$, of electrons and holes is spontaneously created and its value is $n_{e,h}^{1/3} \approx 1/16a_B$.[4] The phase diagram then has a first order transition at low temperatures but at temperatures above a critical temperature (estimated from electron-hole liquid theory as roughly a fifth of the exciton Rydberg) there is no phase transition and all properties are continuous. The key feature of the theoretical results for an ordered system, when the long range Coulomb force is included, is the first order transition which occurs as the density is varied at low temperatures. The first order transition is also observed as the temperature is varied at fixed density, in a narrow range of densities near the critical value.

b) Disorder without Correlation.

The localization of a single electron moving in a random potential is the subject of the lectures by F. Wegner and T. Lubensky and will not be discussed here. For the present purpose it is sufficient to note that the effect of randomness will be to give localized states at the band edges and eventually, as the band edges overlap sufficiently, an Anderson transition from localized to extended states at the Fermi level. The Anderson transition is not believed to have a direct effect on the density of states. In the localized regime there is a finite density of states at the Fermi level and a gap to the mobility edges for electrons and holes. The Anderson transition occurs when a mobility edge passes through the Fermi energy.

## 3. Correlation and Disorder

### a) Low Density-Short Range Correlation.

We begin the discussion of the combined effects of correlation and disorder by examining a low density of donors. The disorder arises because of the random positions of the donors. This in turn causes some donors to be much closer together than the average. Starting from a very dilute concentration of donors the principal effect of the increasing concentration will be to increase the absolute and relative numbers of close donor pairs, triples etc. This suggests that if we wish to examine a local property such as the density of states we should examine the effect of small clusters on the density of states.

This approach has been taken by two groups. Golka and Stoll[13] have looked at the energies to add and subtract an electron from clusters of two and three H atoms close together which will characterize donor clusters in a single-valley semiconductor,while Bhatt and Rice[4] have examined small clusters of donors in a many-valley semiconductor. The results in the two cases are quite different. For a single valley semiconductor the electron states at the minimum in the conduction band have only a twofold spin degeneracy and the energy levels of small clusters are similar to that of small clusters of atoms. The results are that for any value of the inter-donor separation the electron affinity of small clusters of donors never gets large. The Pauli exclusion principle restricts the number of electrons in the lowest s-state to two and forces occupancy of higher states. The result of the small affinity means that the tail of the $D^-$-band to lower energies will not extend far down e.g. both 2 and 3 atom clusters are restricted to affinities $< 0.1$Ry.[13] The situation for the lower Hubbard or $D^+$-band is quite different. Here the effect of forcing an electron into a 2s or 2p state in a 3-atom cluster causes it to be much less bound and as a result the ionization energy of such clusters is driven down to a value as low as $\sim 0.4$Ry. The result is the tail of the $D^+$-band extends to high energies. Therefore the randomness of the donor array leads to substantial broadening of the Hubbard bands but essentially all the broadening occurs on the lower Hubbard band for a single valley semiconductor. This marked asymmetry of the $D^-$ and $D^+$ bands is outside the scope of the simplified Hubbard model, Eq. (1).

The band broadening effects are quite different in a many-valley semiconductor. The degeneracy of the electron states at the conduction band edge is now twice the number of equivalent valleys and will be high (8 for Ge and 12 for Si). As a result the Pauli principle does not restrict the occupancy of electron states in small clusters. The result is that all the electrons can be placed in the lowest s-state and their binding energy is correspondingly increased. This in turn results in an increased ionization energy and electron affinity for a cluster. This leads to downward broadening of both the $D^-$ and $D^+$-bands. This downward broadening of the upper or $D^-$-band is so large that it can actually lead to an overlap of the tail of the $D^-$-band with the top of the $D^+$-band. Bhatt and Rice[4] have estimated using a density functional theory that four donors, in a many-valley semiconductor, within a radius $R \lesssim 2a_B$ have an electron affinity > 1Ry. Such clusters are sufficiently electronegative that they can attract an electron off an isolated donor. In effect these clusters act as compensation centers which occur in a sample which is nominally uncompensated. At low densities the number of these small dense clusters is very small, so their effect will be hard to pin down experimentally. However in principal, they lead to a finite overlap of $D^-$ and $D^+$ bands and remove the distinction between compensated and uncompensated samples. A word of caution concerns the effect of central cell corrections, which for many donors in many-valley semiconductors reduces the degeneracy of the donor ground state. Such donors will be intermediate between the two limits discussed here.

The importance of small clusters, which arise purely from statistics, is seen dramatically in the optical and magnetic properties. These properties probe different particle-hole densities of states, not the single particle density of states discussed above. Recent studies of the optical properties of a series of Si:P samples by Thomas, Capizzi and coworkers[1] show that at low densities specific structure associated with charge transfer excitations of donor pairs can be identified. At higher density there is an absorption edge, which varies rapidly with photon energy and density, whose behavior is in excellent agreement with cluster models. Similarly the magnetic properties are also well described by cluster calculations.[14]

It is possible to change the number of electrons per donor by compensating the semiconductor i.e. introducing acceptors which attract electrons off the donors. By varying the degree of compensation, or relative concentration of acceptors, the Fermi level is shifted across the $D^+$-band. Short-range correlations cannot cause insulating behavior for a partially filled band. Therefore the insulating behavior of compensated samples at low density must be due to Anderson localization of the electrons at the Fermi level due to the random positions of the donors and to the random electric fields at the donor sites from the acceptors. Now we have argued above that in most cases even uncompensated samples have the Fermi level in the $D^+$-band due to overlap of the $D^-$ and $D^+$ bands at any density. Certainly we expect such an overlap for all samples at densities higher than some critical value. This is the point of view that Mott has put forward for some time.[9] The randomness is in a sense more important than the short range correlations and there is no essential distinction between compensated and uncompensated samples. In either case the short range correlations cannot force a gap in the single-particle density of states at densities near the metal-insulator transition. The transition take place as an Anderson transition when the character of the states at the Fermi level changes from localized to extended.

This description is similar to the single-particle theory of the Anderson transition. One important difference between the single-particle and the many-body theories concerns a mobility edge away from the Fermi energy. When one allows for electron-electron scattering then all the quasiparticle states away from Fermi level develop a finite lifetime. Therefore one cannot classify them as extended or localized. Since an electron will only spend a finite amount of time in such states no rigorous definition can be made of localization and one can only use criteria based on the relative sizes of elastic versus inelastic scattering times. At the Fermi level the states have infinite lifetime and the distinction between localized and extended can only be made at that energy.

b) Low Density-Long Range Correlation.

So far I have discussed only short range correlations. What of the long range nature of Coulomb interaction? We saw that this has an important effect on the ordered array. It forced a first order transition from a state with zero density of states at the Fermi level to a state with a finite value. In other words it caused a discontinuous jump in the density of states as the concentration was varied at zero temperature. Some years ago Efros and Shklovskii[5] introduced an ingenious argument to show that the long range nature of the Coulomb interaction modified the single-particle density of states at the Fermi level in a disordered system. They considered a low density random array of donors in a compensated sample. The Coulomb interaction is unscreened because of localization of the electrons at the Fermi level. The relevant Hamiltonian (ignoring spin) is

$$H = \sum_i w_i n_i + \frac{1}{2} \sum_{i \neq j} \frac{e^2}{\kappa_0 r_{ij}} n_i n_j \qquad (2)$$

where $w_i$ are the energy of the localized states in the absence of the Coulomb interaction. There is on the average less than one electron per site and these electrons will be distributed so as to minimize the total energy. Clearly if the energy is a minimum then the ground state is stable against rearrangement. The simplest rearrangement is to move an electron from a filled site to an empty one. Let $E_i$ be the Hartree energy of the site i, i.e.

$$E_i = w_i + \sum_{j \neq i} \frac{e^2}{\kappa_0 r_{ij}} \langle n_j \rangle_G \qquad (3)$$

where $\langle n_j \rangle_G$ is the occupation number in the ground state. Then if one moves the electron from a filled site i into an empty site k then $E_i < E_F$ and $E_k > E_F$ and the excitation energy is

$$\Delta E_{ik} = E_k - E_i - \frac{e^2}{\kappa_0 r_{ik}} \qquad (4)$$

But the stability of the ground state demands that $\Delta E_{ik} > 0$ for all pairs i and k. Therefore if $E_i$ is arbitrarily close to $E_F$ there is a severe restriction on the available empty states $E_k$. In fact

$$r_{ik} > e^2/\kappa_0 (E_k - E_i) \qquad (5)$$

This large excluded volume around each site i with $E_i$ close to $E_F$ implies that in a three dimensional sample

$$\int_{E_F}^{E_F+\varepsilon} N(\varepsilon')\, d\varepsilon' \leq (a/R)^3 \qquad (6)$$

where $e^2/\kappa_0 R = \varepsilon$ and a is average donor separation. Solving for $N(\varepsilon)$ this gives a bound on the density of states near the Fermi level

$$N(E_F+\varepsilon) \lesssim \varepsilon^2 \kappa_0^3/e^6 \qquad (7)$$

The long range correlations in the ground state caused by the Coulomb interaction forces the single-particle density to go to zero at the Fermi level.

The density of states under discussion here is the single-particle density of states defined by adding or removing an electron from the system. It is related to the imaginary part of the single-particle Green's function

$$N(E) = N^{-1} \sum_{i,n} |\langle\psi_n|a_i^+|\psi_G\rangle|^2 \delta(E_n - E_G - E) \tag{8}$$

where $\psi_G$ is the ground state of the N-particle system and $\psi_n$ a state of the N+1-particle system. The Coulomb gap found by Efros and Skhlovskii[5] is tied to $E_F$ and moves with $E_F$ similar to a many body energy gap introduced by a broken symmetry such as a charge or spin density wave distortion. In both cases there are long range correlations introduced in the electron occupation numbers which cause the energy gap.

A related density of states may be defined by considering the variation of the Fermi energy with a change in electron density n. This is

$$\tilde{N}(E_F) = dn/dE_F \tag{9}$$

Clearly $\tilde{N}(E_F)$ varies smoothly as compensation sweeps $E_F$ across the $D^+$-band and has no gap. The finite density of states $\tilde{N}$ involves large scale rearrangements of the electron occupation with minute changes in the electron density whereas the single-particle density involves small rearrangements. Note it is the density of states $\tilde{N}$ which enters in the standard Thomas-Fermi theory of screening. Indeed the Thomas-Fermi wave vector $\lambda$ is defined as

$$\lambda^2 = 4\pi e^2 \tilde{N}(E_F)\kappa_0^{-1} \tag{10}$$

However Thomas-Fermi theory assumes that the electron distribution can flow to rearrange itself and keep $E_F$ constant. For a system with <u>zero</u> conductivity no rearrangement is possible. However for systems with a finite, but small, conductivity, $\sigma$, (say at $T > 0°K$) then rearrangement can take place but on a very slow time scale.[15] The dielectric constant $\kappa(\vec{k},\omega)$ in the <u>static</u> limit will have the Thomas-

Fermi form in this case

$$\kappa(\vec{k},\omega=0) = \kappa_0(1+\lambda^2/k^2) \tag{11}$$

For $\omega>\omega_c(=\sigma k^2/\lambda^2\kappa_0)$ the necessary rearrangement will no longer follow the driving field and the Thomas-Fermi form breaks down and $\kappa \approx \kappa_0$. $\kappa_0$ is a finite dielectric constant due to transitions between localized states and transitions to higher bands.

The exact form of the d.c. conductivity at low temperatures when the long range correlations are included has been the subject of some controversy. Mott's theory of variable range hopping which predicts the famous $T^{1/4}$ law is based on the premise that there is a finite density of empty states, into which an electron can hop, arbitrarily close to Fermi energy. The Efros-Shklovskii theory questions this premise and indeed by applying their theory Efros, Lien and Shklovskii[16] came up with a different form; to Mott's form in three dimensions $\sigma(T) \sim \sigma_0 \exp(-(T_0/T)^{\frac{1}{2}})$. Recently Pollak[17] and Mott and Davies[9] have argued that the conductivity at low temperatures will be dominated by correlated many-electron hops and the final form for $\sigma(T)$ is not clear.

It is interesting to look at the Hamiltonian (2) in terms of a spin glass model. Each site may be expressed as a two-state Ising model with up(down) corresponding to filled(empty). In this language the first term in Eqn. (2) represent a random magnetic field while the second term is a long range exchange interaction. Most of the work on spin glasses have concentrated on shorter range interactions than the $r^{-1}$ form in Eq. (2). The density of states to add (or subtract) a particle is the same as the density of states to turn over a spin. The Efros-Shklovskii argument, which gives a bound on this density of states, is equivalent to giving a bound on the distribution of local magnetic fields.

The Hamiltonian (2) neglects hopping between the localized states. This will add to Eq. (2) a term $\sum_{ij\sigma} t_{ij} a^+_{i\sigma} a_{j\sigma}$. Such a term is not expressible in general in terms of the Ising spin operators and takes the problem outside the spin-glass class. In one dimension however by using the Jordan-Wigner transformation one can represent it as a transverse coupling and the combined Hamiltonian can be represented as an anisotropic Heisenberg model. In higher dimensions the Jordan-Wigner transformation does not describe the hopping term.

c) High Density.

In the high density limit the donor wavefunctions overlap strongly and the system is metallic. A good description is obtained over most of the temperature range by using conventional transport theory for metals. Until recently it was generally accepted that the electronic properties of a disordered metallic state were described by Landau Fermi liquid theory. The resistivity at low temperatures would be governed by Mattheissen's rule in which the elastic, or residual impurity, scattering rate is added to an electron-electron scattering rate which varied as $T^2$. No anomalous behavior was expected in the behavior of the single-particle Green's function. Detailed experimental studies of the low temperature behavior of metals and metallic alloys did not show anomalous behavior which deviated from the conventional theory.

Recently, important corrections have been found to this anomalous behavior. In this brief review only a summary of the new results will be given. Altshuler and Aronov[6] found that in perturbation theory, the set of terms associated with the modification of the screening of the Coulomb interaction by the diffusive processes, when summed to all orders, led to surprising results. Specifically for a three-dimensional sample, with elastic scattering time $\tau$, they found that the single-particle density of states

$$N(\varepsilon+E_F) = N(E_F)(1 + \alpha|\varepsilon\tau|^{\frac{1}{2}}(E_F\tau)^{-2}) \qquad |\varepsilon\tau| << 1 \tag{12}$$

where $\alpha$ is a numerical constant $\sim 1$. This result gives a dip in the density of states at the Fermi level and a non-analytic behavior for N(E) which is outside Landau Fermi liquid theory. The Altshuler-Aronov theory[6] is an expansion about the weak scattering limit, $E_F\tau>>1$, and Eq. (12) is only the lowest order correction. There is a corresponding anomaly predicted in the d.c. conductivity, $\sigma(T)$

$$\sigma(T) = \sigma(0)(1+\alpha'(T\tau)^{\frac{1}{2}}(E_F\tau)^{-2}) \tag{13}$$

where $\alpha' \sim 1$. This square root behavior of the leading temperature correction is in contrast to the standard theories which give $T^2$ (electron-electron) or higher powers due to electron-phonon scattering. The coefficient of the $T^{\frac{1}{2}}$ term involves a characteristic energy scale $E_F(\sim\tau^{-1})$ which will make it a small effect and presumably accounts for the absence of such anomalies in the earlier experimental literature.

These results have been extended to lower dimensions where the effects are even more striking.[18]

The theoretical predictions have led to several experimental investigations. Careful measurement of the conductivity of Si:P samples down to milli-kelvin temperatures have shown corrections to the conductivity which vary as $T^{\frac{1}{2}}$.[3] The sign of these corrections depends on the density but it is in agreement with detailed calculations using the above methods. The single-particle density of states is not experimentally accessible in the Si:P samples. It is measured in tunneling experiments. Recently a series of tunneling measurements on granular $A\ell$[19] and on $Ge_{1-x}Au_x$ films[20] have shown large anomalies at zero bias which is interpreted as evidence of the dip in the single-particle density of states. As the composition is varied, and the conductivity decreases, the dip deepens until finally, in insulating samples, the density of states vanishes at the Fermi surface.

d) A Unified Scaling Theory.

McMillan[7] has proposed a scaling theory to interpolate between the high and low density limits. It is a generalization of the original scaling hypothesis introduced by Abrahams et al[21] to include also the electron-electron interaction effects. A coupled series of scaling equations are obtained to describe the behavior of the conductance and a dimensionless interaction strength as a function of the length scale. Solving these equations McMillan[7] obtains a solution in which the single-particle density of states at the Fermi level is coupled to the d.c. conductivity, so that for finite conductivity the density of states is finite, but has a dip of the form Eq. (12), while for zero conductivity the density of states is zero at $E_F$.

## 4. Conclusions

Both the Coulomb interaction and the disorder act to localize electrons and in general one expects the two effects to reinforce one another. Considerable progress has been made in understanding each effect separately but until recently not much progress had been made on their combined effects. Recently some unexpected results have been found in both the low density and high density limits. McMillan's theory is the first to attempt to connect these two and to propose a universal theory incorporating both effects and applicable to the metal-insulator transition. There is now reason to believe that a solution to this long standing problem is near.

## 5. Acknowledgements

It is a pleasure to acknowledge many useful conversations with my colleagues R. N. Bhatt, M. Capizzi, J. H. Davies, P. A. Lee, and G. A. Thomas on these topics. The author also is grateful to P. W. Anderson for remarks concerning the spin glass analogy.

References


[1] G. A. Thomas, M. Capizzi and F. DeRosa, Phil. Mag. B42, 913 (1980) also G. A. Thomas, M. Capizzi, F. DeRosa, R. N. Bhatt and T. M. Rice, Phys. Rev. B (in press).

[2] T. F. Rosenbaum, K. Andres, G. A. Thomas and R. N. Bhatt, Phys. Rev. Lett. 45, 1723 (1980).

[3] T. F. Rosenbaum, K. Andres, G. A. Thomas and P. A. Lee, Phys. Rev. Lett. 46, 568 (1981).

[4] R. N. Bhatt and T. M. Rice, Phil. Mag. B42, 859 (1980); Phys. Rev. B23, 1920 (1981).

[5] A. L. Efros and B. I. Shklovskii, J. Phys. C 8, L49 (1975); A. L. Efros, J. Phys. C 9, 2021 (1976).

[6] B. L. Altshuler and A. G. Aronov, Sol. State Comm. 30, 115 (1979) ZhETF 77, 2028 (1979) [Sov. Phys. JETP 50, 968 (1979)].

[7] W. L. McMillan, preprint.

[8] J. Hubbard, Proc. Roy. Soc. A277, 237 (1964).

[9] N. F. Mott and J. H. Davies, Phil. Mag. B 42, 845 (1980).

[10] J. H. Rose, H. B. Shore and L. M. Sander, Phys. Rev. B21, 3037 (1980) and to be published.

[11] P. Kelly, O. K. Andersen and T. M. Rice (to be published).

[12] W. F. Brinkman and T. M. Rice, Phys. Rev. B7, 1508 (1973).

[13] J. Golka and H. Stoll, Sol. State Comm. 33, 1183 (1980).

[14] K. Andres, R. N. Bhatt, P. Goalwin, T. M. Rice and R. E. Walstedt, Phys. Rev. B (in press).

[15] J. H. Davies, T. M. Rice and P. A. Lee, Bull. Am. Phys. Soc. 26, 389 (1981).

[16] A. L. Efros, N. V. Lien and B. I. Shklovskii, Sol. State Comm. 32, 851 (1979).

[17] M. Pollak, Phil. Mag. B 42, 781 (1980).

[18] B. L. Altshuler, A. G. Aronov and P. A. Lee, Phys. Rev. Lett. 44, 1288 (1980).

[19] R. C. Dynes and J. P. Garno, Phys. Rev. Lett. 46, 137 (1981).

[20] W. L. McMillan and J. Mochel, Phys. Rev. Lett. 46, 556 (1981).

[21] E. Abrahams, P. W. Anderson, D. C. Licciardello and T. V. Ramakrishnan, Phys. Rev. Lett. 42, 673 (1979).

# OPTICAL AND PRECURSIVE PROPERTIES APPROACHING THE METAL TO INSULATOR TRANSITION IN HIGHLY DOPED Si

M. Capizzi,[*] T.F. Rosenbaum,[§] K.A. Andres,[†] G.A. Thomas, R.N. Bhatt and T.M. Rice
Bell Laboratories, Murray Hill, N.J. 07974

Detailed measurements are discussed of the far-infrared absorption coefficient and dc conductivity of phosphorous donors in uncompensated silicon at very low temperature throughout the metal-insulator transition. The results obtained support an Anderson like model and formally agree with a general scaling description of the transition.

We present the main results of a detailed experimental and theoretical investigation of the far-infrared absorption of the nearly ideal, random, three-dimensional system formed by phosphorous donors in uncompensated Si. We argue that large scale potential fluctuations, rather than a uniform shrinking of the Hubbard gap,[1] dominate the optical properties of Si:P approaching the metal-insulator transition (MIT).[2] We also show that the static dielectric constant, evaluated from the optical absorption data, critically diverges near the MIT. This result, together with recent measurements of the zero-temperature dc conductivity at donor concentrations $n_D \geqslant n_{MI}$, supports a general scaling description of the precursive behavior for $[n_D/n_{MI}-1] \leqslant 1$.[3,4] Two regimes of density are illustrated in Fig. 1. In the lower density spectrum six prominent absorption lines are seen. These arise from well known transitions[5] of the outermost electron on the P atoms between the ground state and a series of hydrogen-like states. At this low density the broadening is almost completely accounted by the experimental resolution. However the short range nature of the donor-donor interaction together with the randomness of donor position naturally leads to the conclusion that an effect on the isolated donor lines, due to pairs which are closer than the average, has to be

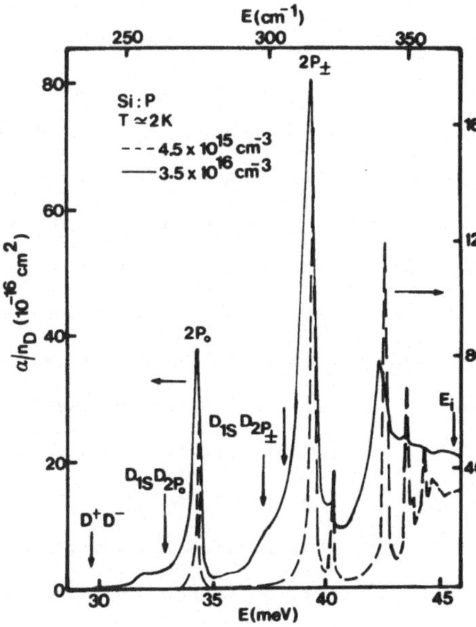


Fig. 1. Absorption cross-section $\alpha/n_D$ vs. photon energy E. The vertical arrows show the theoretical energies for the pair features discussed in the text. $E_i$ is the low density ionization energy.

observed as the impurity concentration is increased. A very asymmetric broadening is indeed observed for $2\times10^{16} \leqslant n_D \leqslant 2\times10^{17}$ $cm^{-3}$. Several spectral features due to pair transitions have also been detected in this density range (see spectrum at higher density in Fig.1). We identify the origin of these absorption bands evaluating and fitting their energy positions, their lineshapes and their intensities.[2] Two types of pair transitions are involved. The first is due to the excitation of one electron from the ground 1S state to an excited state on the same atom, but with another atom nearby in the ground state so that the energies of both the ground and the excited states are reduced ($D_{1S}D_{2P}$ transitions in Fig.1). The second band, called $D^+D^-$, is due to transitions with transfer of one electron between two nearby donors, both in their ground state. These transitions are quite important because they involve a large energy shift[6] and prove that simplified Hubbard models,[1] with only off-diagonal elements and on-site Coulomb interactions, are inadequate for describing the optical absorption spectra of doped semiconductors. $D^+D^-$ pairs are therefore donor excitons in the Hubbard gap, defined as the energy difference between the ground state and the band formed by $D^-$ states neglecting the interaction with $D^+$ states. However donor excitons, being fixed spatially at a random separation, cannot diffuse to the separation of minimum energy. At density higher than $2\times10^{17}$ $cm^{-3}$, but still below the MIT, the isolated donor lines get washed out and the absorption spectrum is essentially featureless, but for a bump which shows up at 12 meV (see arrow in Fig.2), an energy roughly equal to the valley-orbit splitting of the ground state of phosphorus [7]. At these densities a description of the excited states begins to involve clusters formed by N>2 donors, which do not show visible density of states features.[2] In order to find evidence of such clusters, we make a log-log plot of the absorption coefficient $\alpha$ at different fixed photon energies $\omega_N$ vs. $n_D$, as shown in Fig.3. The linearity in this plot is a strong argument that clusters of increasing size (N= 4,5,8,10 in the figure) play a central role for decreasing photon energies. Large clusters have indeed extremely high electron affinity because, *due to the many*

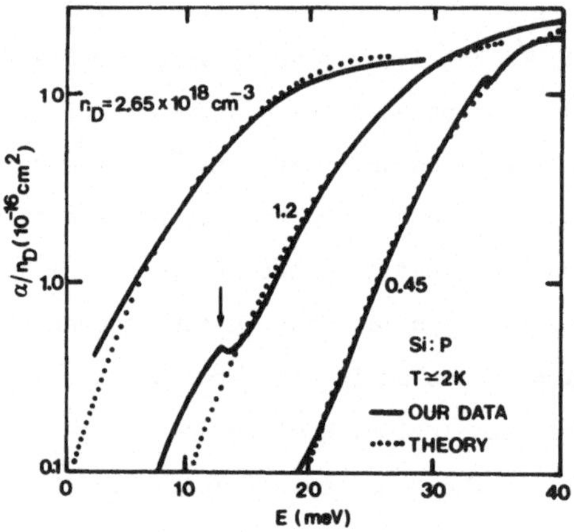


Fig. 2. Absorption cross section $\alpha/n_D$ vs. photon energy E. The solid lines are data and the dotted lines are theory.

*valley structure of the conduction band in Si*, it is possible to put up 12 electrons in the 1S state of a dense cluster without violating the Pauli principle.[8] At donor densities greater than $2\times10^{17}$ $cm^{-3}$ we apply a statistical theory of optical absorption by clusters, as developed in the case of expanded fluid mercury.[10] The model involves three parameters which are determined by fitting the data at $n_D = 4.5\times10^{17}$ $cm^{-3}$. Then, with the same values of the parameters, the model fits the data just scaling the density, as shown in Fig.2. The excellent agreement strongly supports the clusters rather than a uniform Hubbard model: instead of a simple reduction of the Hubbard gap, we are confronted with a growth in probability of random clusters that have absorption at lower energies. Calculations for the same model also show that clusters of four donors or more can attract an electron from an isolated donor. As a result *there is no gap due to short-range correlation in many valley semiconductors* and their insulating property is due to Anderson localization. The very large fluctuations in the one-electron potential make possible an Anderson transition to the metallic state. In order to get a closer insight on the MIT we transform $\alpha(\omega)$ to obtain values of the donor electric susceptibility $\chi$ of phosphorous using the Kramers-Krönig relation and some algebra.[3] These values are drawn in a suitable log-log plot in Fig.4, together with values of the dc zero-temperature conductivity $\sigma(0)$, extrapolated from very careful measurements performed down to 1 mK,[4] and normalized to Mott's $\sigma_{min}$.[1] Scaling theory of localization[11,12] propose that a variable length $\xi$ is the only pertinent scale near the MIT. Following these theories,

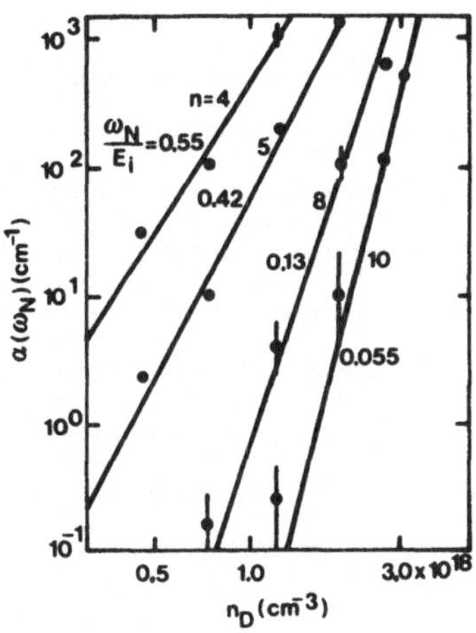


Fig. 3. The absorption coefficient $\alpha$ at four fixed frequencies $\omega_N$ vs. $n_D$.

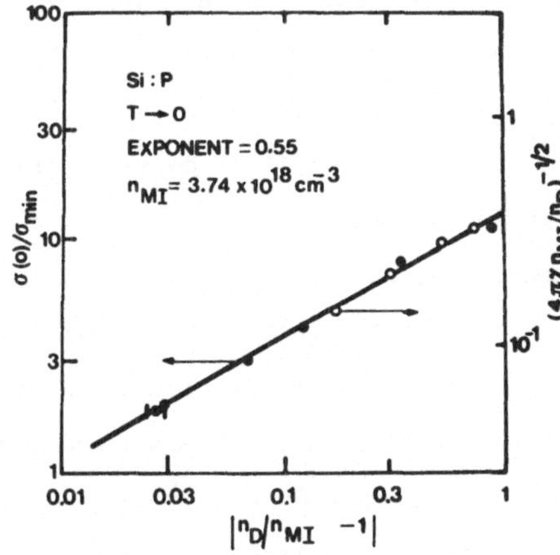


Fig. 4. Log-log plot of $\sigma(0)/\sigma_{min}$ and the dielectric susceptibility $4\pi\chi$ in normalized form vs. $|n_D/n_{MI}-1|$. The solid line determines the exponent.

$$\sigma(0) = C/\xi = \sigma_o[n_D/n_{MI}-1]^{\nu} \qquad (1)$$

$$\chi = K\, n_D \xi_L^2 = \chi_o\, n_D/n_{MI}[1-n_D/n_{MI}]^{-2\zeta} \qquad (2)$$

above and below the transition respectively, where $\xi_L$ is the localization length in the insulator.[4,12] The linearity found in Fig. 4 and the observed symmetry of the divergent lengths support a true critical point and a general scaling description of the precursive behavior for $[n_D/n_{MI} - 1] \leqslant 1$ in this random 3-D system. However these scaling theories give $\nu \leqslant 1$, and $\sigma_0 \sim \sigma_{min}$, while we find $\nu = \zeta = 0.55 \pm 0.1$ and $\sigma_0 = 13\ \sigma_{min}$. Classical bond and site percolation theories, which arrive to the same form of Eqs. (1) and (2), give instead $\nu = 1.6$[13] and $\zeta = 0.3$,[14] even inconsistent with our results. It must be remarked that the power law is obeyed over a surprisingly large range in $|n_D/n_{MI} - 1|$. Similar behavior was observed in the system KCl:Ag[15] where a more classical value $\zeta = 0.36$ was determined. A different approach is shown in Fig. 5, where the same values of $\sigma(0)/\sigma_{min}$ are plotted on a semilog scale. Three samples with values of $\sigma(0)$ below $\sigma_{min}$ have been added. These samples are still metallic, as shown by the temperature dependence of their $\sigma(0)$ for $T \to 0$ (see Fig. 2 in Ref. 4), but have not been considered in Fig. 4 because they are affected by a too large uncertainty in $[n_D/n_{MI} - 1]$ (the error in determining $n_D$ is $\simeq 1\%$). The existence itself of metallic samples with $\sigma(0)/\sigma_{min} \sim 10^{-3}$ further supports a continuous decrease to zero of $\sigma(0)$, accordingly to scaling theories. Unfortunately density inhomogeneities of the order of 1%, which cannot be ruled out in these samples, might broaden the discontinuity at $\sigma(0)/\sigma_{min} = 1$ expected in the Mott model.[1] We report also in Fig. 5 similar data for the system $Ge_{1-x}Au_x$ (dots). These data have been claimed to be an evidence of the failure of the Mott model.[16] Their scattering in our plot seems instead to indicate

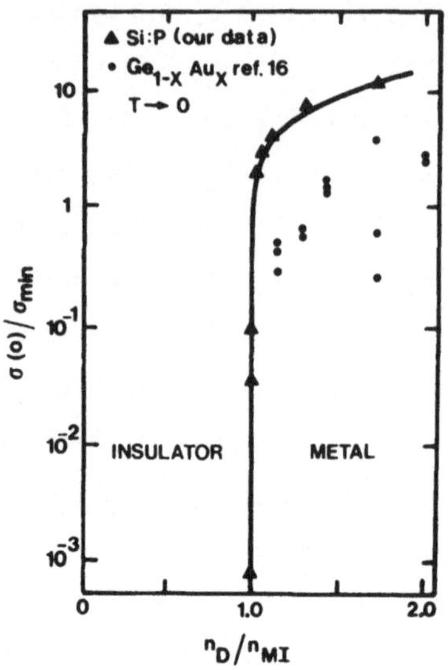


Fig. 5. Semilog plot of the zero temperature conductivity $\sigma(0)$ vs. $n_D$, normalized to $\sigma_{min}$ and $n_{MI}$ respectively. The solid line is a fit by Eq. (1) above $\sigma_{min}$ of our data. Data for $Ge_{1-x}Au_x$ from Ref. 16 are also shown (dots).

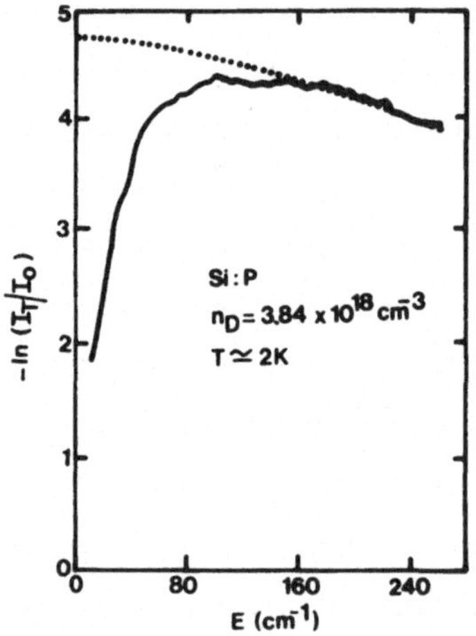


Fig. 6. Log of the transmitted ligth vs. photon energy for a just metallic sample. Drude behavior (dotted line) is shown for comparison.

# LOCALIZATION IN ORDERED SYSTEMS

C.Castellani

Istituto di Fisica "G.Marconi", Università di Roma, Roma, Italy, and
Istituto di Fisica, Università dell'Aquila, L'Aquila, Italy, and
G.N.S.M. del CNR, Sezione dell'Aquila, Italy.

C.Di Castro and L.Peliti

Istituto di Fisica "G.Marconi", Università di Roma, Roma, Italy, and
G.N.S.M. del CNR, Sezione di Roma, Italy

- ABSTRACT -

We give a short review of the attempts to understand the correlation-induced metal insulator transition as described by the Hubbard model within the general frame of critical phenomena. We describe two recent approaches: the first one introduces an effective Hamiltonian to make explicit the mechanisms of various transitions expected for the model, the second one allows for a Monte Carlo simulation of condensed matter systems involving fermionic degrees of freedom, useful for an exploration of the phase diagram of the system.

## 1. - INTRODUCTION -

There is now much hope of understanding the disorder-induced metal-insulator transition within the general framework of the theory of phase transitions(1,2). Some attempts have also been made to include electron-electron interactions in this scheme(3). On the contrary we know of no successful attempt to understand within the same framework the metal-insulator transition induced by correlation in the opposite limit, i.e. in absence of disorder. This was the phenomenon originally considered by Mott(4) and to which Hubbard(5) gave major contributions. It appears in transition-metal compounds with unfilled d-band, where the metal-insulator transition may be obtained upon varying the temperature or the pressure or doping. In these compounds the independent electron scheme breaks down.

It is believed that the phenomenon essentially reduces to the localization of 3d electrons around the nuclei due to the Coulomb interaction. The simplest model in which both the delocalizing effect of the interaction-free Hamiltonian and the localizing effect of electron-electron interactions appear is the well known Hubbard Hamiltonian(5):

small macroscopic density inhomogeneities, not discernable in electron microscopy. Thus it may be dangerous to draw conclusions on the nature of the MIT near $\sigma_{min}$ from these data.[17] A last intriguing result is reported in Fig. 6,[18] where is shown the first direct observation, at least at our knowledge,of a pseudogap in $-\ln(I_T/I_0)$, i.e. in Re $\sigma(\omega)$. This behavior is explained by a recent analysis of the mobility of a quantum particle in a 3-D random system.[19] However, a strong dip in the density of states is also predicted in the same region for an half filled Hubbard band,[1] namely for uncompensated samples such as ours.

ACKNOWLEDGEMENTS - The authors would like to acknowledge helpful discussions with P.A. Lee.

REFERENCES

* Permanent address: Istituto di Fisica G. Marconi, P.le A. Moro, 5, Rome, Italy.
§ Also at Joseph Henry Laboratories, Princeton Univ., Princeton, N.J. 08544.
† Permanent address: ZTTF, Munich, F.R.G.
1) N.F. Mott, Metal-Insulator Transitions (London, Taylor & Francis Ltd, 1974).
2) G.A. Thomas, M. Capizzi, F. De Rosa, R.N. Bhatt, and T.M. Rice, to be published in Phys. Rev. B (1981); and references therein.
3) M. Capizzi, G.A. Thomas, F. De Rosa, R.N. Bhatt, and T.M. Rice, Phys. Rev. Letters 44, 1019 (1980).
4) T.F. Rosenbaum, K. Andres, G.A. Thomas, and R.N. Bhatt, Phys. Rev. Letters 45, 1723 (1980).
5) P. Fisher and K.A. Ramdas, Physics of the Solid State, Edited by S. Balakrishna, M. Krishnamurti, and B. Ramachandra (London, Academic Press, 1969), p. 149.
6) The binding energy of the $D^-$ state in Si:P is $E_{D^-} = 1.7$ meV, while the $D^+D^-$ binding energy is $E_{D^+D^-} - E_{D^-} = 14.0$ meV.
7) Transitions within the split states, dipole forbidden in isolated atoms, are allowed in pairs, with an absorption integrated intensity that increases quadratically with $n_D$, as experimentally verified.
8) The $D_N^-$ state is estimated to have, for N=2, a binding energy roughly equal to 0.4 Rydberg, value that increases rapidly with N, as discussed in Ref. 9.
9) R.N. Bhatt and T.M. Rice, Phys. Rev. B 23, 1920 (1981).
10) R.N. Bhatt and T.M. Rice, Phys. Rev. B 20, 466 (1979).
11) E. Abrahams, P.W. Anderson, D.C. Licciardello, and T.V. Ramakrishnan, Phys. Rev. Letters 42, 673 (1979); F.I. Wegner, Phys. Rep. 67, 15 (1980); and references therein.
12) W. Götze, J. Phys. C 12, 1279 (1979).
13) S. Kirkpatrick, Rev. Mod. Phys. 45, 574 (1973).
14) D.J. Bergman and Y. Imry, Phys. Rev. Letters 39, 1222 (1977).
15) D.M. Grannan, J.C. Garland, and D.B. Tanner, Phys. Rev. Letters 46, 375 (1981).
16) B.W. Dodson, W.L. Mc Millan, J.M. Mochel, and R.C. Dynes, Phys. Rev. Letters 46, 46 (1981).
17) G.A. Thomas, T.F. Rosenbaum, and R.N. Bhatt, to be published.
18) T.F. Rosenbaum, K.A. Andres, G.A. Thomas, and P.A. Lee, Phys. Rev. Letters 46, 568 ( 1981).
19) W. Götze, Phil. Mag. B 43, 219 (1981).

$$H = \sum_{<ij>} \sum_{\sigma} t\, c^{+}_{i\sigma}\, c_{j\sigma} - \sum_{i\sigma} \mu\, n_{i\sigma} + \sum_{i} U\, n_{i\uparrow}\, n_{i\downarrow} \quad ; \tag{1.1}$$

$$n_{i\sigma} = c^{+}_{i\sigma}\, c_{i\sigma} \; ; \qquad \sigma = \uparrow,\downarrow \; ; \tag{1.2}$$

where <ij> indicates a sum over nearest neighbour site pairs. We shall consider the case of one electron per site (which sets the chemical potential $\mu$ equal to $U/2$). In order to reach Eq. (1.1) one assumed that the degeneracy of d-band levels and the long range nature of the Coulomb interaction need not be taken into account for a qualitative understanding of the phenomenon: these points are however still controversial. The model so introduced plays for the metal-insulator transition in ordered systems almost the same role as the Ising model in magnetic transition. Its understanding is a necessary step before more complex phenomena can be considered. The difficulty is that the order parameter of this phase transition, if it exists at all, is not yet identified. In fact no well established mean-field theory (or the analogue of a Bragg-Williams approximation for the Ising model) is yet available. Neither are available nontrivial exactly soluble limits such as the d or n equal to infinity limits for ordinary critical phenomena. As long as such an understanding is not reached it is hopeless to apply more sophisticated methods like the renormalization group.

We believe therefore that the effect of correlations in inducing a metal-insulator transition is not yet satisfactorily understood. We thus wish to analyze in this paper the present level of understanding of the subject and to describe two recent approaches to it.

In the first one the Hubbard model is transformed into an effective Hamiltonian which is expressed in terms of physically relevant quantities (such as spin and charge fields) while maintaining its symmetry properties(6). This has allowed to clarify the mechanism of the transitions in different regions of the phase diagram of the system. This phase diagram is however still imperfectly known. One may wish therefore to obtain further information about it by computer simulation. The second approach(7) allows for a Monte Carlo simulation of a condensed matter system with fermionic degrees of freedom. This approach is a generalization of a method due to Fucito et al.(8) which has been successfully applied to problems in elementary particle physics.

## 2. - PHENOMENOLOGICAL THEORIES -

In three dimensions an antiferromagnetic insulating phase (AFI) should be present at low temperature for any value of U/W at least for a simple cubic lattice (W is the band width proportional to t). At large values of U/W the Hubbard model is equivalent to the Heisenberg model with an antiferromagnetic exchange coupling $\sim t^2/U$. In this limit the ground state is an ordered array of singly occupied sites alternating with spin up and down ($|\uparrow\rangle$, $|\downarrow\rangle$: magnetic sites).

By increasing the temperature a second order phase transition occurs from the AFI to a paramagnetic system of localized electrons (PI). By decreasing the ratio U/W a certain number of doubly occupied or empty sites ($|0\rangle$, $|\uparrow\downarrow\rangle$: non-magnetic sites) are generated by the hopping term. At sufficiently low values of U/W, by increasing the temperature a transition from the AFI to the paramagnetic metal (PM) takes place. This transition is expected to be of first order.

A direct PM - PI transition possibly appears at finite temperature. It could extend down to zero temperature if magnetic order were not present.

All these expectations are summarized in the following phase diagram:

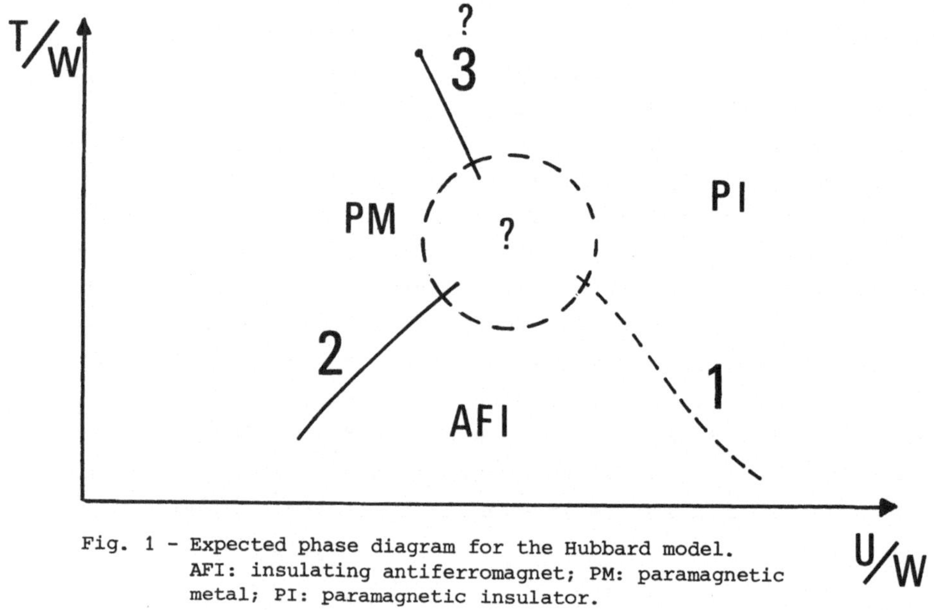


Fig. 1 - Expected phase diagram for the Hubbard model. AFI: insulating antiferromagnet; PM: paramagnetic metal; PI: paramagnetic insulator.

Let us now discuss how sound are the arguments behind them.

The original Hubbard decoupling procedure of the single particle Green's function equations of motion provides an interpolating formula between the two extreme cases: the localized one electron per atom and the free electron system. Magnetic effects are not taken into account. At T = 0 in the limit $U/W \to \infty$, it gives rise to a gap (equal to U) between the two bands for removing or adding one electron to the ground state. The possibility of propagation of the empty or doubly occupied sites lowers the value of the gap from U to U-2zt. The critical condition that the gap vanishes agrees with the physical idea that the system switches from an insulator to a metal when the delocalizing effect of the hopping term balances the localizing repulsive coupling U.

This is essentially the contents of all known approaches. The effort of deriving any more precise statement encounters serious difficulties.

Starting from the ground state associated to the large U/W limit at low temperature, the gap is decreased by increasing the number of "excited" electrons, i.e. the number of doubly occupied and empty sites or non-magnetic sites, given by:

$$x = \langle n_{i\uparrow}\, n_{i\downarrow} \rangle + \langle (1 - n_{i\uparrow})(1 - n_{i\downarrow}) \rangle = 2 \langle n_{i\uparrow}\, n_{i\downarrow} \rangle \quad . \tag{2.1}$$

The value of x is increased either by increasing the hopping term or by thermal excitation.

The introduction of x as the parameter driving the metal-insulator transition has been considered in several phenomenological approaches[9,10,11]. All of them are self-consistent mean field type approximations. They present however serious shortcomings and cannot be considered as a good starting point for a more appropriated theory.

In Ref. (9) the magnetic order is absent and a variational ground state with x as a parameter leads at zero temperature to a transition from a metallic phase to an insulator with x equal to zero.

In Ref. (10) a first order transition between a phase rich in magnetic sites (insulator) with $x < \frac{1}{2}$ and a phase rich in non-magnetic sites (metal) with $x > \frac{1}{2}$ is derived by thermodynamic considerations. A balance between the energy related to an effective potential for the electrons induced by the randomly placed holes and the entropy of the holes is considered. The crudeness of the simplification does not take into account magnetic order and moreover the obtained metallic phase is out of the physical region since x cannot exceed the value $\frac{1}{2}$ assumed by the free electron system for which

$$x = 2 < n_{i\uparrow} > \; < n_{i\downarrow} > \; = 1/2. \qquad (2.2)$$

In Ref. (11) the gap is assumed to depend on the mean sublattice magnetization s. The free energy for the coupled system of localized and excited free electrons is then minimized with respect to x and s. The transition becomes of first order when the final phase is a PI with a sufficiently small intrinsic gap or is PM.

Along the same lines Economou and White[12] have developped a model which strongly relies on the concept of local magnetic moments $m_i = \langle n_{i\uparrow} - n_{i\downarrow} \rangle$ in the framework of an alloy analogy.

They actually linearize the on site Coulomb interaction term

$$U\, n_{i\uparrow}\, n_{i\downarrow} \rightarrow \varepsilon_{i\sigma}\, n_{i\sigma} \; ; \quad \varepsilon_{i\sigma} = U/2 \pm m_i\, U/2. \qquad (2.3)$$

Assuming $m_i = \pm$ m they divide the lattice in sites A and sites B according to an up or down local moment with

$$\begin{aligned} \varepsilon^A_\uparrow &= \varepsilon^B_\downarrow = (1 - m)\, U/2 \;; \\ \varepsilon^A_\downarrow &= \varepsilon^B_\uparrow = (1 + m)\, U/2 \;. \end{aligned} \qquad (2.4)$$

Magnetic effects are introduced by an ad hoc probability P that a site is of type A if its nearest neighbours are of type B. This probability is written via an Ising interaction. P and m are the two variational parameters. The dynamics of the local moments is added by changing the Ising interaction into an Heisenberg one. The connection with the original Hubbard model is at this point rather loose and one expects that this approach works only in the limit of large U/W, where the magnetic sites dominate.

All the previous approaches fail to consider simultaneously the effects of the delocalization of the electrons and the band broadening due to the hopping term both at zero and finite temperature, the variation with the temperature of the number of thermally excited electrons and the magnetic order. Clearly all these effects are simultaneously important in the most interesting region where U is of the same order of W, when the direct metal insulator transition should take place and the magnetic transition switches from second to first order.

## 3. - FUNCTIONAL INTEGRAL METHODS -

A direct evaluation of the thermodynamic properties of the Hubbard model has been attempted[13] by means of the functional formulation of statistical mechanics.

In this approach the grand partition function of the system is expressed in terms of a functional integral over classical fields. First the operatorial identity[14]

$$\exp(A^2/2) = (2\pi)^{-\frac{1}{2}} \int d\phi_A \; e^{-\phi_A^2/2 - \phi_A A} , \qquad (3.1)$$

valid for any bounded operator A, is used to transform the two body interaction term in (1.1) in favour of single particle couplings with classical fields. Then the trace over the fermion degrees of freedom is carried out leading to a formal definition of the functional Lagrangian of the classical fields.

Moreover, correlation functions for the classical fields are related to correlation functions for operators of the quantum system.

At large values of U/W only magnetic sites ($|\uparrow\rangle$, $|\downarrow\rangle$) are important. At lower the value of U/W the non magnetic sites ($|0\rangle$, $|\uparrow\downarrow\rangle$) play an essential role. This means that both spin and charge operators

$$S_i^z = n_{i\uparrow} - n_{i\downarrow} \quad ; \quad \rho_i^z = n_{i\uparrow} + n_{i\downarrow} - 1 , \qquad (3.2)$$

are important when $U \sim W$.

Because of the rotational invariances of the Hubbard model, one has to introduce[6,15] the x and y components of the spin and charge operators which can be written in terms of raising and lowering operators

$$\begin{aligned} S_i^+ &= c_{i\uparrow}^+ c_{i\downarrow} , & \rho_i^+ &= c_{i\uparrow}^+ c_{i\downarrow}^+ , \\ S_i^- &= c_{i\downarrow}^+ c_{i\uparrow} , & \rho_i^- &= c_{i\downarrow} c_{i\uparrow} . \end{aligned} \qquad (3.3)$$

The spin operators act on magnetic sites, while the charge operators act on non-magnetic sites.

The fermion character of the creation and annihilation operators implies relations among the components of $\vec{S}$ and $\vec{\rho}$, as it can be easily verified by using the anticommutation rules of the c-operators.

$$\begin{aligned}
&(S_i^\nu)^2 + (\rho_i^\mu)^2 = 1 \quad ; \quad S_i^\nu \, \rho_i^\mu = 0 \quad ; \\
&(S_i^\nu)^{2m} = (S_i^\mu)^{2m} = (S_i^\nu)^2 = (S_i^\mu)^2 \; ; \; (\rho_i^\nu)^{2m} = (\rho_i^\mu)^{2m} = (\rho_i^\nu)^2 = (\rho_i^\mu)^2 \quad ; \\
&(S_i^\nu)^{2m} \, S_i^\mu = S_i^\mu \quad ; \quad (\rho_i^\nu)^{2m} \, \rho_i^\mu = \rho_i^\mu \quad ;
\end{aligned} \tag{3.4}$$

$\nu, \mu = x, y, z$; $m =$ integer

The interaction term of the Hubbard hamiltonian (1.1) can be reduced to a sum of squares of the local spin and charge operators in various ways[15] in order to apply (3.1).

We have in general

$$\begin{aligned}
n_{i\uparrow} n_{i\downarrow} = C + \alpha_0 S_i^z + \alpha_1 (S_i^x)^2 + \alpha_2 (S_i^y)^2 + \alpha_3 (S_i^z)^2 \; + \\
+ \beta_0 \rho_i^z + \beta_1 (\rho_i^x)^2 + \beta_2 (\rho_i^y)^2 + \beta_3 (\rho_i^z)^2 \; ,
\end{aligned} \tag{3.5}$$

provided

$$\alpha_0 = 0 \; ; \; \beta_0 = \frac{1}{2} \; ; \quad \alpha_1 + \alpha_2 + \alpha_3 + C = 0; \quad \beta_1 + \beta_2 + \beta_3 + C = \frac{1}{2} \; . \tag{3.6}$$

The two most used transformations are known under the names of Schrieffer[16]

$$n_{i\uparrow} n_{i\downarrow} = \frac{1}{2} + \frac{1}{2} \rho_i^z - \frac{1}{2} (S_i^z)^2 \quad , \tag{3.7a}$$

and Hamann[17]

$$n_{i\uparrow} n_{i\downarrow} = \frac{1}{4} + \frac{1}{2} \rho_i^z + \frac{1}{4} (\rho_i^z)^2 - \frac{1}{4} (S_i^z)^2 \quad . \tag{3.7b}$$

Transformation (3.7a) leads to a Lagrangian depending only on the one-component spin field $\phi_s$, whereas the Lagrangian corresponding to

(3.7b) depends on both one-component spin $\phi_s$ and charge $\phi_\rho$ fields. Transformation (3.5) in general leads to a Lagrangian depending on multicomponent fields. All of them are of course equivalent as long as the problem is treated exactly.

Transformation (3.7b) is usually preferred [13,18] in the literature. The grand partition function for the Hubbard Hamiltonian reads in this case

$$Z = \mathrm{tr}\, e^{-\beta H} = \int D\phi_s\, D\phi_\rho \; e^{-L_0[\phi_s,\phi_\rho] \; -L_I[\phi_s,\phi_\rho]} \quad , \tag{3.8}$$

where

$$L_0\,[\phi_s,\, \phi_\rho] = \frac{1}{2} \int_0^\beta d\tau \sum_i \left[\phi_{si}^2(\tau) + \phi_{\rho i}^2(\tau)\right] \quad , \tag{3.8a}$$

$$L_I\,[\phi_s,\, \phi_\rho] = -\ell n\, \mathrm{tr}\, \{ T_\tau \; e^{-\beta H_0} \exp\, [- \int_0^\beta d\tau \sum_i \; [U\,(\tfrac{1}{4} + \tfrac{1}{2}\,\rho_i^z(\tau)) \tag{3.8b}$$

$$+ \sqrt{U/2}\;\; \phi_{si}\,(\tau)\; s_i^z\,(\tau) + \sqrt{-\,U/2}\;\; \phi_{\rho i}\,(\tau)\;\; \rho_i^z\,(\tau)]]\} \quad ;$$

and the time-ordering operator $T_\tau$ has been introduced to take into account the non commutativity of the two terms of the Hubbard Hamiltonian.

The expectation values of the fields $\phi_s$, $\phi_\rho$ are proportional to the expectation values of the corresponding operators

$$\langle\phi_{si}\rangle = -\sqrt{U/2}\;\langle s_i^z\rangle \quad ; \quad \langle\phi_{\rho i}\rangle = -\sqrt{-\,U/2}\;\langle\rho_i^z\rangle \quad . \tag{3.9}$$

The following approximations are constantly used in the literature:

1) The imaginary time dependence of the fields is neglected (static approximation).
2) Charge fluctuations are completely neglected. This means that $\phi_\rho$ is assumed to be equal to its average value $\phi_\rho = \langle\phi_\rho\rangle = \langle\rho^z\rangle = 0$.
3) A power expansion of the Lagrangian in terms of the fields is often considered [18].

The first two approximations are introduced for simplicity. They are based on purely phenomenological considerations and are expected to be valid only in the case of localized electrons for the large values of U/W. They are however commonly used over the whole range of U/W.

Within these approximations an analogy with magnetic alloys can be derived[(13)].

One should however be doubtful about the static approximation when evaluating in general the single particle Green's functions. The time independence of the fields introduces some features of the Anderson problem for disordered systems which could be totally misleading in the understanding of the dynamical behaviour of the Hubbard model.

Moreover, when we neglect the charge fluctuations, we reduce the scheme (3.7b) to scheme (3.7a) with a different value of the coupling $\tilde{U} = U/2$. We thus obtain the original Hubbard system with a wrong value of the coupling. It is then clear that, if we introduce the charge field we must treat it together with the spin field.

In fact, the charge and the spin operators are not independent since they are related by Eqs. (3.4) which specify the fermion character of the original system. Eqs.(3.4), when imposed on the functional formulation, lead to constraints for the correlation functions (Ward Identities) which have to be satisfied by the approximation procedure one is going to use if one wants to keep the symmetry properties of the Hubbard Hamiltonian[(15)]. In the static approximation they imply a nonpolynomial form of the Lagrangian. Moreover any polynomial truncation of (3.8b) would lead to a thermodynamical instability.

Although not reliable for a quantitative description, one could hope that a polynomial expansion of the Lagrangian in powers of the fields could be used[(18)] in the framework of renormalization group methods in order to find out the critical properties of the model.

The expansion gives rise to an effective Landau-Ginzburg Lagrangian when higher order powers of the fields are neglected according to the standard criterion of irrelevant couplings.

If we assume however to work either in the scheme (3.7a) or (3.7b) a more generally, in what we call the Ising type scheme derived by (3.5) with $\alpha_1 = \alpha_2 = \beta_1 = \beta_2 = 0$, we obtain

$$n_{i\uparrow}\, n_{i\downarrow} = -\alpha_3 + \frac{1}{2}\, \rho_i^z + \alpha_3\, (S_i^z)^2 + (\frac{1}{2} + \alpha_3)\, (\rho_i^z)^2 \quad . \tag{3.10}$$

Apart from problems of thermodynamic instability[15] for $-\frac{1}{2} < \alpha_3 < 0$, if a polynomial form of the resulting Lagrangian in terms of the spin field is obtained, then in the large U/W limit the magnetic transition would show a critical temperature depending on $\alpha_3$ and would belong to the class of universality n = 1, n being the number of components of the order parameter.

On the contrary, a Heisenberg-type transformation can be derived from (3.5) with $\alpha_1 = \alpha_2 = \alpha_3 = \alpha$ and $\beta_1 = \beta_2 = \beta_3$

$$n_{i\uparrow} n_{i\downarrow} = -3\alpha + \frac{1}{2} \rho_i^z + \alpha \vec{S}_i \cdot \vec{S}_i + (\frac{1}{6} + \alpha) \vec{\rho}_i \cdot \vec{\rho}_i \quad . \tag{3.11}$$

In the present scheme, under the same conditions of the previous case (3.10), the magnetic transition would belong to the class of universality n = 3 . The n = 2 case could also be easily obtained.

One has therefore to take seriously into account the symmetry properties of the original Hubbard model and in particular its local constraints (3.4) which imply the arbitrariness in (3.5).

They reflect essentially the Pauli principle which is expected to be the relevant mechanism in determining the behaviour of the model when competition between the localizing on site interaction and the delocalizing hopping takes place.

In the next chapter we shall only use the invariance properties of the Hubbard model to introduce an effective Hamiltonian expressed in terms of the spin and charge operators (3.2) and (3.3). This allows for a more direct physical interpretation at least as far as the magnetic transitions of Fig. 1 are concerned.

## 4. - AN EFFECTIVE MODEL -

Following the general idea behind the renormalization group approach the Hubbard Hamiltonian was trasformed into an effective model retaining the symmetry properties and the quantum nature of the original system. In this way some information about the critical behaviour of the Hubbard model were obtained[6].

We consider the two unitary transformations

$$U_s = \prod_i e^{i\alpha_i \vec{k}_i \cdot \vec{S}_i} \quad , \tag{4.1}$$

$$U_\rho = \prod_i e^{i\gamma_i \vec{q}_i \cdot \vec{\rho}_i} \quad , \tag{4.2}$$

in order to derive the invariance properties of the Hubbard Hamiltonian (1.1).

The first transformation is a local change of the spin quantization axis. The second one is a generalization of the particle-hole transformation $c_{i\sigma} \to \sigma c^+_{i-\sigma}$, $c^+_{i\sigma} \to \sigma c_{i-\sigma}$ which is obtained for $\vec{q}_i = (0,1,0)$ and $\gamma_i = \frac{\pi}{2}$. Under particle-hole transformations H is invariant except for a trivial phase change in the Wannier representation if the system has one particle per site ($\mu = U/2$). Within the same condition H is invariant under both $U_s$ and $U_\rho$ if $\vec{k}_i = \vec{k}$, $\alpha_i = \alpha$, $\vec{q}_i = \vec{q}$ for any lattice site, and, with $\vec{q}$ in the x-y plane, $\gamma_i$ assumes the opposite values $\pm\ \gamma$ on the two equivalent sublattices in which we divide the original lattice. Each one of the two sublattices is made of the nearest neighbour sites of the other one. When $\vec{q}$ is oriented in the z direction $\gamma_i$ has to be equal for all lattice points.

If we define an effective Hamiltonian by performing a partial trace over the degrees of freedom of one of the two sublattices

$$e^{-\beta H_{eff}} = tr'\, e^{-\beta H} \quad , \tag{4.3}$$

then $H_{eff}$ must be globally invariant under (4.1) and (4.2).

Confining ourselves to nearest neighbour interactions $H_{eff}$ reads

$$\begin{aligned} H_{eff} = &- J \sum_{<ij>} \vec{S}_i \cdot \vec{S}_j - k \sum_{<ij>} S_i^2\, S_j^2 + \Delta \sum_i S_i^2 \; + \\ &- I \sum_{<ij>} \vec{\rho}_i \cdot \vec{\rho}_j + D \sum_{<ij>} \sum_\sigma (c^+_{i\sigma} c_{j\sigma} + c^+_{j\sigma} c_{i\sigma})\,(1 - n_{i-\sigma} - n_{j-\sigma}) \quad . \end{aligned} \tag{4.4}$$

The first four terms are all the invariants that can be written with the spin and charge operators. In fact, due to the constraints (3.4) no other independent invariant exists. The D-term is the most general linear combination of products of odd numbers of c-operators on each single site invariant under (4.1) and (4.2).

While the exact form of the various terms of (4.4) is known from symmetry properties, the coefficients have to be evaluated by performing the decimation procedure, which can be carried out in a finite lattice approximation[19].

Without being confined to limited regions of the parameters U/W, T/W, the model Hamiltonian (4.4) generalizes the effective Heisenberg Hamiltonian that can be derived from (1.1) by second order perturbation theory in the limit $U/W \to \infty$ , and reduces to it in that limit.

Apart from the D-term, (4.4) explicitly shows the effective spin and charge couplings which the original local interation induces between different sites via the hopping term. Both the spin-spin interaction term leading to the magnetic order and the interactions related to the charge fluctuations appear in (4.4). The effect of the charge fluctuations is not confined to the I-term. Through the constraints (3.4) they are related also to the K- and Δ-terms.

Just as the J-term leads to magnetic ordering for U positive, the I-term introduces a charge ordering at U negative. In fact, if we consider, e.g.,the transformation $c_\downarrow^+ \to c_\downarrow$, $c_\downarrow \to c_\downarrow^+$ followed by a phase change in the Wannier representation, the Hubbard Hamiltonian with $\mu = U/2$ is invariant (up to an additive constant term) provided $U \to -U$. At the same time in the effective Hamiltonian the $\vec{S}$ operators are transformed into the charge operators $\vec{\rho}$ and viceversa. The I and J terms interchange their roles. For positive U the I-term plays no special role and can therefore be neglected in studying critical properties.

If also the term associated with D were absent the model Hamiltonian would correspond, except for minor details, to the Blume-Emery-Griffiths model[20] which was introduced to study the $He^3$ - $He^4$ mixture.

This model has been extensively studied both in mean field approximation[20] and by renormalization group approach[21].

Depending on the relative values of the parameters of the Hamiltonian (4.4), the system undergoes:

a) A second order phase transition from a magnetically ordered phase to a normal system rich in magnetic sites (singly occupied sites) which should correspond to the AFI-PI transition of Fig. 1.

b) A first order phase transition from a magnetically ordered system to a disordered system with a jump in both the magnetization and the concentration of non magnetic sites (2.1) which in the present formulation is given by

$$x = 1 - \langle (S_i^z)^2 \rangle \quad .$$

This is the continuation of the second order phase transition line at lower values of U/W, when the number of non-magnetic sites becomes relevant.

c) A direct first order phase separation ending to a critical point between two non-magnetic phases with a jump in x. It would be natural to identify this last transition due to the direct interplay (K- and Δ-terms in (4.4)) between singly occupied sites and doubly occupied or empty sites with the direct PI-PM transition of Fig.1.

The direct phase separation occurs however between two phases rich in magnetic and non-magnetic sites respectively, with a critical point which is expected to be characterized by $x_c = \frac{1}{2}$. As we have seen in chapter 2, the phase rich in non-magnetic sites where x is larger than $\frac{1}{2}$ is out of the physical region when $U > 0$. We have at this stage the same trouble which appears in the phenomenological theories. We have not taken properly into account the dynamic nature of the interplay between magnetic and non-magnetic sites. Actually the D-term interchanges a magnetic site with a non-magnetic site allowing for a relative motion of the two components.

It would therefore introduce important physical features in the model as far as the MIT is concerned.

The D-term, being expressed in terms of the original c-operators, reintroduces most of the difficulties of the original Hubbard model. An attempt to study its effect in a simplified model of spinless interacting fermions by means of the Migdal-Kadanoff decimation approach has not led to any definite conclusion[22].

The introduction of the effective model (4.4) has clarified the mechanism leading to the magnetic transition, showing that the Hubbard model contains the ingredients for a first order (magnetic) transition. It has however left the problem of the direct PM-PI transition open, confirming the central role of the dynamics in the metal insulator transition.

It is therefore much required to have more detailed information about this transition. Since it is hard to identify in the experiments those features which can be directly related to the Hubbard model, it is highly desirable to have numerical simulations.

The way to approach this problem is described in the next section.

## 5. - MONTE CARLO APPROACH -

The main difficulty one faces in setting up a simulation scheme in our problem lies in the fermion character of the operators. Fucito, Marinari, Parisi and Rebbi[8] have recently introduced a method which reducing the integration over fermionic degrees of freedom to an integration over ordinary (commuting) variables allows for explicite evaluation with reasonable computing time. This method has been successfully applied in elementary particle physics[23] and is now being extended to condensed matter physics[7]. The approach agrees with the general philosophy of the functional integral formulation, with the difference that one attempts to evaluate the integrals over a finite sample and with a discrete imaginary time variable. We expect that different transformations (3.5) lead to equivalent results, since our only approximation (the discretization) does not violate their equivalence nor the local constraints (3.4). One has to take into account that the ordinary expressions of the functional integrals like (3.8) are marred with ambiguities, and it is hard to discriminate between licit and illicit manipulations without recurring to perturbative theory. We choose therefore to introduce an algebra of anticommuting variables in order to define a well established discrete form of these integrals.

We consider a Hubbard Hamiltonian (1.1) over a simple cubic lattice of N sites in d dimensions with periodic boundary conditions.

By standard manipulations[24] the partition function of (1.1) may be written as a functional integral over anticommuting variables $\eta$, $\bar{\eta}$, which are functions of the imaginary time argument $\tau$ :

$$Z = \mathrm{tr}\,\exp(-\beta H) = \lim_{M\to\infty} Z_M \quad ; \tag{5.1}$$

$$Z_M = \int \prod_{\nu=1}^{M} \prod_{i=1}^{N} \prod_{\sigma} d\bar{\eta}_{i\sigma\nu} \quad \exp L\,[\eta, \bar{\eta}] \;, \tag{5.2}$$

$$\begin{aligned} L\,[\eta, \bar{\eta}] = &\sum_{\nu i\sigma} \left[-\bar{\eta}_{i\sigma\nu}\,(\eta_{i\sigma\nu} - \eta_{i\sigma\nu-1}) + (\beta\mu/M)\;\bar{\eta}_{i\sigma\nu}\;\eta_{i\sigma\nu-1}\right] \\ &- (\beta t/M) \sum_{<ij>} \sum_{\sigma\nu} \bar{\eta}_{i\sigma\nu}\,\eta_{j\sigma\nu-1} \\ &- (\beta U/M) \sum_{\nu i} \bar{\eta}_{i\uparrow\nu}\,\eta_{i\uparrow\nu-1}\,\bar{\eta}_{i\downarrow\nu}\,\eta_{i\downarrow\nu-1} \quad . \end{aligned} \tag{5.3}$$

The anticommuting fields $\bar{\eta}$, $\eta$ satisfy periodic boundary conditions on the space label i and antiperiodic ones on the imaginary time label $\nu$, which spans the M intervals into which the imaginary time interval $[0,\beta]$ has been divided. The rule of integration over anticommuting variables[24] gives:

$$\int \prod d\bar{\eta} \;\; d\eta \; \exp\,(-\,\bar{\eta}A\eta) = \det\,(A) \quad . \tag{5.4}$$

In order to apply equation (5.4) one needs a Lagrangian quadratic in the anticommuting fields: one obtains this by the standard Stratonovich-Hubbard[14] transformation (3.1) which introduces a boson field $\phi_{i\nu}$ satisfying periodic boundary conditions:

$$Z_M = \int \prod_{\nu i} d\phi_{i\nu} \prod_{\nu i\sigma} d\bar{\eta}_{i\sigma\nu}\, d\eta_{i\sigma\nu} \;\exp L\,[\bar{\eta},\eta,\phi] \quad , \tag{5.5}$$

$$L\,[\bar{\eta},\eta,\phi] = -\frac{1}{2} \sum_{\nu i} \phi_{i\nu}^2 - \sum_{i\mu} \sum_{j\nu} \bar{\eta}_{i\uparrow\mu}\, A_{i\mu,j\nu}\,[\phi]\, \eta_{j\uparrow\nu} - \sum_{i\mu} \sum_{j\nu} \bar{\eta}_{i\downarrow\mu}\, A_{i\mu,j\nu}\,[-\phi]\, \eta_{j\downarrow\nu} \; . \tag{5.6}$$

$$A_{i\mu,j\nu}\,[\phi] = \delta_{ij}\, (\delta_{\mu\nu} - \delta_{\mu\nu+1}) + \delta_{\mu\nu+1} \left[ (\beta t/M)\, \varepsilon_{ij} - (\beta\mu/M - (\beta U/M)^{\frac{1}{2}}\, \phi_{i\mu})\, \delta_{ij} \right] . \tag{5.7}$$

where $\varepsilon_{ij} = 1$, if i,j are nearest neighbours and = 0 otherwise.

The integration upon the $\eta$ fields can be now performed according to Eq. (5.4). One thus obtains

$$Z_M = \int \prod_{\nu i} d\phi_{i\nu} \;\exp\Big(-\frac{1}{2} \sum_{i\nu} \phi_{i\nu}^2\Big)\, \zeta\,[\phi]\, \zeta\,[-\phi] \quad , \tag{5.8}$$

where

$$\zeta\,[\phi] = \det\,(A\,[\phi]) \quad , \tag{5.9}$$

Although Eq.(5.9) is explicit, it would take too long computer time to evaluate the determinant directly. It is therefore necessary to evaluate it by other means[8].

It is important to remark that $\zeta[\phi]$ is expected to be positive definite in the limit of large M. If we take the continuum limit it is easy to show in fact that imaginary time reversal $\tau \to -\tau$ leads, in the Fourier transform, to the change of A into $A^+$ on the one hand, but should amount on the other hand to a reshuffling of the matrix A. The set of eigenvalues of A coincides therefore with the set of their complex conjugates.

Both in the small $\phi$-limit and in the small t-limit all eigenvalues are complex and the determinant is positive definite.

If that is true, then

$$\zeta\,[\phi] = \det\,(A\,[\phi]) = \det^{\frac{1}{2}}\,(A\,[\phi]\,A^{+}\,[\phi]) = \det^{\frac{1}{2}}\,(\Delta\,[\phi])\;, \tag{5.10}$$

In order to compute this last expression, one introduces an ordinary (commuting) field $\psi$ and applies the usual Gaussian integration formula:

$$\zeta\,[\phi] = \det^{\frac{1}{2}}\,(\Delta[\phi]) = \left[\,N \int D\,\psi\,\exp\,(-\frac{1}{2}\,\psi\,\Delta\,[\phi]\,\psi\,)\right]^{-1}\;, \tag{5.11}$$

where $N$ is a suitable normalization constant.

One can thus compute averages over the Gibbs distribution by a "nested" Monte Carlo scheme: in order to update the configuration of the boson field $\phi$ at each Monte Carlo step one needs to evaluate

$$\exp\left[-\frac{1}{2}\,\Sigma\,(\phi+\delta\phi)^{2}\right]\,\zeta\,[\phi+\delta\phi]\,\zeta\,[-(\phi+\delta\phi)]\cdot \tag{5.12}$$

$$\cdot\,\{\exp\,(-\frac{1}{2}\,\Sigma\,\phi^{2})\,\zeta\,[\phi]\,\zeta\,[-\phi]\}^{-1}.$$

For small $\delta\phi$ this may be expressed in terms of $\phi$ and of correlation functions of $\psi$ fields which may be computed via the usual Monte Carlo scheme (with say p steps per site) at fixed $\phi$. Experience on the Schwinger model(23) shows that the number p need not be too large (up to about 120) even near a transition. We hope that this method will provide useful information on the phase diagram of the Hubbard model and will be useful as a starting point for more sophisticated approaches.

- REFERENCES -

(1) F.Wegner, this Conference and references quoted there.

(2) T.Lubensky, this Conference and references quoted there.

(3) B.L.Altshuler, A.G.Aronov and P.A.Lee, Phys. Rev. Lett. 44, 1288 (1980).

R.Oppermann, this Conference.

(4) See e.g. N.F.Mott, Metal Insulator Transitions, Taylor and Francis London (1974).

(5) J.Hubbard, Proc. Roy. Soc. A276, 238 (1963).

(6) C.Castellani, C.Di Castro, D.Feinberg and J.Ranninger, Phys. Rev. Lett. 43, 1957 (1979).

(7) C.Castellani, C.Di Castro, F.Fucito, E.Marinari, G.Parisi and L. Peliti, Poster presented in this Conference.

(8) F.Fucito, E.Marinari, G.Parisi and C.Rebbi, to appear in Nucl. Phys. B (F.S.) (1981).

(9) W.F.Brinkman and T.M.Rice, Phys. Rev. B2, 4302 (1970).

(10) S.Doniach, Adv. Phys. 18, 819 (1969).

(11) L.N.Bulaevskii and D.I.Khomskii, Sov. Phys. Solid State 9, 2422, (1968).

(12) E.N.Economou and C.T.White, Phys. Rev. Lett. 38, 289 (1977).

R.De Marco, E.N.Economou and D.C.Licciardello, Solid State Commun. 21, 687 (1977).

(13) J.M.Cyrot, J. Phys. (Paris) 33, 125 (1972).

(14) R.L.Stratonovich, Sov. Phys. 2, 416 (1958).

J.Hubbard, Phys. Rev. Lett. 3, 77 (1959).

(15) C.Castellani and C.Di Castro, Phys. Lett. 70A, 37 (1979).

D.Leuratti, Dissertation, Università de L'Aquila (1980).

(16) S.Q.Wang, W.E.Evenson and J.R.Schrieffer, Phys. Rev. Lett. 23, 92 (1969).

(17) D.R.Hamann, Phys. Rev. Lett. 23, 95 (1969).

(18) A.A.Gomes and P.Lederer, J.Phys. (Paris) 38, 231 (1977).

F.Brouers and C.M.Chaves, J.Phys. F7 , L233 (1977).

(19) C.Castellani, C.Di Castro, D.Feinberg, J.Ranninger, unpublished.

H.Takano and M.Suzuki, Preprint.

(20) M.Blume, V.J.Emery and R.Griffiths, Phys. Rev. A4, 1071 (1971).

(21) A.N.Berker and M.Wortis, Phys. Rev. B14, 4946 (1976).

(22) C.Castellani, C.Di Castro, J.Ranninger, to appear in Nucl. Phys. B(F.S.) (1981).

(23) E.Marinari, G.Parisi and C.Rebbi, CERN Preprint (May 1981).

(24) See e.g. C.Itzykson and J.-B.Zuber, Quantum Field Theory, McGraw--Hill (1980) p. 439.

# On Effects of Electron-Electron Interactions in Disordered Electronic Systems

R. Oppermann

Institut für Theoretische Physik, Universität Heidelberg, FRG

Abstract:

The effect of long-range electron-electron interactions on the existence of a mobility edge and on the characteristics of critical localization behaviour is studied for disordered systems by means of a 1/n-expansion in the finite temperature technique of many body theory. The lower critical dimension turns out to be two as in ensembles with interaction-free hamiltonians, and a subsequent d-2 expansion applies. In the case of time reversal invariance and in O(1/n), cancellations of correlation contributions leave the conductivity behaviour unchanged when the Fermi energy approaches the still existing mobility edge(continuous transition). Many body effects however introduce criticality into one-particle properties and the density of states $\rho(E_F)$ vanishes with the critical exponent $\beta=1/(d-2)+O((d-2)^0)$ when $E_F-E_c$ goes to zero on the metallic side of the transition. In the case of broken time reversal invariance the O(1/n) approximation gives rise to speculations on a first order transition, but $O(1/n^2)$ calculations are indispensable for a reliable conclusion.

## 1. Introduction and results

In contrast to the correlation-induced Mott transition the Anderson localization is found in a variety of models, loosely classified as Anderson models or Wegner's gauge invariant tight binding models, which neglect any many body effect. Altshuler, Aronov, Lee et al have been first to reveal that electron interactions cause anomalous behaviour in the entire noncritical metallic region of disordered systems |1-3|. They employed a Hartree Fock approximation in exactly two dimensions (2D) and in 3D.

Here I present results which cover both noncritical and critical region thus dealing simultaneously with interactions and critical localization. In section 2 I introduce two local gauge invariant models, one for the case of time reversal invariance the other for broken t.r.i. Both include nonrandom long range (bare Coulomb) electron-electron interactions. Section 3 contains comments on the exact solution of the large n limit as basis of a 1/n expansion which forms an exact microscopic expansion. In the noncritical region the results agree with those of the HF theory of Lee et al |3|, although the present models use a localized basis in contrast to the nearly free electrons(NFE) of |3|. A d-2 expansion of the 1/n coefficients for d>2 organizes critical contributions and thus provides insight into the critical region. The derivation of the following results is commented upon in sections 4 and 5. All indeed highly singular correlation contributions to the dc conductivity cancel in O(1/n) such that the interaction-free terms prevail:

$$\sigma_{dc} = \sigma_\infty \left( 1 - E_o^4(16\pi A(E_o^2-E_F^2))^{-1} \frac{1}{n(d-2)} + O(n^{-1}(d-2)^o, n^{-2}) \right), \qquad (1)$$

where A and $E_o$ are model parameters, given in secs.2 and 3, with $\pm E_o$ playing the role of the band edges of the impurity band. Eq.(1) refers to the orthogonal case (time reversal invariance), while the phase-invariant model (broken t.r.i.) lacks the $1/(n(d-2))$ term as had been deduced for the corresponding interactionless models |4|. The power law interpretation of the d-2 expansion was confirmed by renormalized field theory |5|, whence we again interprete (1) as

$$\sigma_{dc} \sim |E_F-E_c|^s \qquad \text{with} \quad s = 1 + O(d-2) \qquad (2)$$

and the mobility edge $E_c$ is located by

$$E_o^2-E_c^2 = E_o^4/(16\pi An(d-2)) \quad . \qquad (3)$$

Clearly any higher order in 1/n could upset this interpretation, if cancellations in interaction graphs would not occur as expected from O(1/n). Thus the all order valid mechanism and the symmetry behind it has to be found.

The density of states, which turned out to be an unconventially behaving order parameter with no sign of criticality at the localization transition of 'interactionless' ensembles, becomes dramatically affected by long range electron interactions developing $n^{-k}(d-2)^{-2k}$ singularities. In O(1/n) one finds for both t.r.i. and broken t.r.i.

$$\rho(E_F) = \rho_\infty(E_F)\{ 1 - E_o^4(16\pi A(E_o^2-E_F^2))^{-1}n^{-1}(d-2)^{-2} + O(n^{-2}(d-2)^{-4})\}. (4)$$

Together with Eq.(3) (t.r.i.) this yields

$$\rho(E_F) \sim |E_F-E_c|^\beta \qquad \text{with} \quad \beta = \frac{1}{d-2} + O((d-2)^o) \quad , \qquad (5)$$

which states that $\rho(E_F)$ goes to zero continuously when the Fermi energy approaches the mobility edge from the metallic side and is expected to stay zero in the localized region |6|. Familiar features of the order parameter are thus restored by interactions. Note that $\beta$ agrees with the correlation length exponent $\nu$ in lowest order. The fact that $\sigma_{dc}$ and $\rho(E_F)$ cannot obey power laws with a common zero in O(1/n) for the phase invariant model suggests a first order transition there, but again $O(1/n^2)$ might change the situation.

## 2. Model

Consider the full second quantized interaction hamiltonian of an electron-ion system with Coulomb interactions. Neglecting phonons the ionic fields are integrated out as nondynamic variables. This transfers effects of atomic randomness into parameters of the resulting electronic model. The transformation into occupation number space is achieved by expanding the electron field operators with respect to the basis of the n orbital wavefunctions $\phi_{r\alpha}(x)$ per site r, $\alpha=1...n$ by

$$\hat{\psi}(x) = \sum_{r\alpha} \phi_{r\alpha}(x)\, a_{r\alpha} \quad . \tag{6}$$

In contrast to the NFE Ansatz ($\phi_k(x)$ plane waves, k some wave number) the orbital wavefunctions are localized eigenfunctions of some atomic hamiltonian. They bring forth the idea of local gauge invariance, which was introduced by Wegner as the idealized form of rapid phase coherence decay in interactionless disordered systems, and also proves meaningful in the presence of interactions.

The grand canonical hamiltonian then becomes

$$K = -\mu N + \frac{1}{\sqrt{n}} \sum_{\substack{rr' \\ \alpha\beta}} t_{r\alpha r'\beta} a^{\dagger}_{r\alpha} a_{r'\beta} + \frac{1}{2n} \sum_{\substack{rr' \\ \alpha\beta}} U^{b}_{r-r'} \hat{n}_{r\alpha} \hat{n}_{r'\beta} + H_{ext} \quad , \tag{7}$$

where the $t_{r\alpha,r'\beta}$ are independent, gaussian distributed, random matrix elements with short ranged correlation. Local gauge invariance under

$$\phi_{r\alpha}(x) \rightarrow \begin{cases} \exp(iT_{r\alpha}(x))\phi_{r\alpha}(x) & \text{phase invariant ensemble} \\ \pm\phi_{r\alpha}(x) & \text{real matrix (orthogonal) ensemble} \end{cases} \tag{8}$$

results in the ensemble average

$$\langle t_{\underline{\alpha}\underline{\beta}} t_{\underline{\gamma}\underline{\delta}} \rangle_{Ens} = \begin{cases} M_{r-r'} \delta_{\underline{\alpha}\underline{\delta}} \delta_{\underline{\beta}\underline{\gamma}} & \text{phase invariant} \\ M_{r-r'} \{\delta_{\underline{\alpha}\underline{\gamma}} \delta_{\underline{\beta}\underline{\delta}} + \delta_{\underline{\alpha}\underline{\delta}} \delta_{\underline{\beta}\underline{\gamma}}\} & \text{orthogonal} \end{cases} \quad \underline{\alpha} = (r,\alpha) \tag{9}$$

with short ranged moments $M_{r-r'}$, such that for small q

$$M(q) = M(0) - Aq^2 \quad . \tag{10}$$

In the graph theory for ensemble-averaged quantities, M(q) appears as a time-independent 'effective interaction', whereas

$$U_b(q) = 2^{d-1} \pi^{(d-1)/2} \Gamma\left(\frac{d-1}{2}\right) e^2 q^{1-d} \tag{11}$$

represents the instantaneous bare Coulomb interaction in d dimensions.

## 3. Method

The 1/n expansion applied to the gauge invariant models generates a systematic graph theory and as such forms a conserving approximation. The conservation law for the total particle number is correctly represented in the perturbation theory.

Expanding in 1/n implies that only $-\mu N$, which moreover is a c-number due to particle number conservation, can be viewed as the free hamiltonian at the beginning. The rest is treated in a many body theory in finite temperature technique (allowing for the $T\rightarrow 0$ limit though) with an S-matrix

$$S(\beta) = T_\tau \exp\{-\int_0^\beta d\tau'(K(\tau')+\mu N)\} \quad , \quad \beta=1/T \ . \tag{12}$$

The interpretation of the simple starting-point is as follows: all orbital wavefunctions used to define the hamiltonian K become degenerate with respect to the free hamiltonian trivially. The levels are then occupied by chance. This however does not even cause a fictitious problem, if one uses retarded or imaginary-time ordered Green's functions. The oversimplified free hamiltonian mainly serves to provide the exact solubility of the nontrivial large n limit as the true starting-point.

The ensemble-averaged imaginary-time ordered Green's function

$$G(x_1-x_2,\tau) = -<<T_\tau\{\hat\psi(x_1,\tau)\hat\psi^\dagger(x_2,0)S(\beta)\}>_o/<S(\beta)>_o>_{Ens} \tag{13}$$

becomes local due to local gauge invariance. In the large n limit the totally free Green's function $G_o(z_n)=(iz_n+\mu)^{-1}$ is renormalized to

$$G_\infty(z_n) = 2E_o^{-2}(iz_n+\bar\mu-\{(iz_n+\bar\mu)^2-E_o^2\}^{1/2}) \ , \ z_n=(2n+1)\pi T, \ E_o^2=4M(0) \ , \tag{14}$$

where the renormalized chemical potential $\bar\mu$ has absorbed some irrelevant constants in addition to $\mu$. Bare Hartree- and other graphs with $U_b(0)$ lines are subtracted due to overall electroneutrality (infinite shift in $\mu$), whence $G_\infty$ is U-independent. Also unchanged remains the double line propagator $g(q;z_n,z_n-\omega_m)$, which is proportional to the T-matrix for impurity scattering and develops a diffusion pole when $z_n(z_n-\omega_m)<0$. The dynamically screened Coulomb interaction becomes

$$U^R_{scr}(q,\omega) = U_b(q)/\{1+U_b(q)\Pi(q,\omega)\} = U_b(1)q^{1-d}\ \frac{\omega+iDq^2}{\omega+iDq^2+i\kappa q^{3-d}} \tag{15}$$

in the large n limit with $\kappa=U_b(1)\rho_\infty D$, $D=2\pi A\rho_\infty$, and $\rho_\infty$ the density of states at $E_F$. $\Pi(q,\omega)\sim q^2$, for $q\to 0$ due to particle number conservation, is the density polarization part. Both scaling regions $\omega\sim q^2$ and $\omega\sim q^{3-d}$ in Eq.(15) are relevant. Note however that 3D is a special case.

## 4. Conductivity

The conductivity is derived by a linear response theory with respect to the external perturbation $H_{ext}=\int_x j(x,t)A(t)$ with $A(t)=A_o\exp(-izt)$, $Im\{z\}>0$. A many body generalization of the Kubo Greenwood formula yields

$$Re\ \sigma(\omega+io) = \frac{ne^2}{dV}\,\omega\frac{\partial}{\partial q^2}\ Im\ D_R(q,\omega)\Big|_{q=0} \ , \tag{16}$$

where $D_R$ denotes the retarded density correlation function. In fig.1 the U-dependent diagrams of $O(1/n)$ are shown. Individually they contain $(d-2)^{-2}$ and $(d-2)^{-1}$ singularities but their sum gives a nonsingular contribution to $\sigma_{dc}$, yet existing terms like $(\omega^{d-2}-\omega^{d/2-1})/(d-2)$ in $\sigma(\omega)$

change over into a ln $\omega$-singularity in 2D in agreement with the result of Lee et al. In fact this term derives from the HF subset in O(1/n).

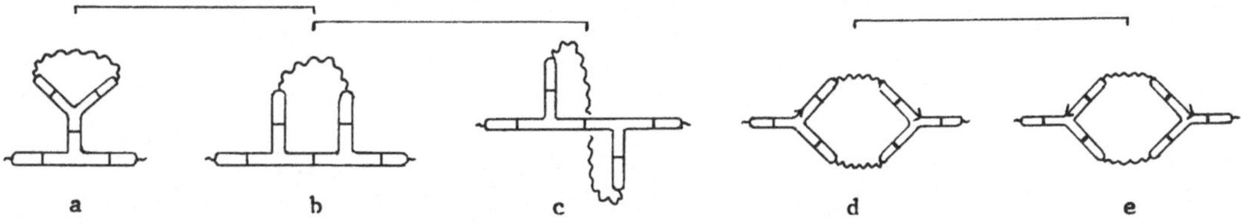


$$— = G\ ,\quad ═ = g(q;z_n,z_n-\omega_m)\ ,\quad \sim\sim = U_{scr}$$

Fig.1: Coulomb corrections to the density correlation in O(1/n)

Cancellations of singularities due to correlation are indicated by brackets in fig.1. Details can only be given elsewhere. Finally the U-free graphs |4| give the sole singular contribution and one obtains Eq.(1).

## 5. Density of states

The leading O(1/n) contribution to the density of states $\rho(E) = -\pi^{-1}\mathrm{Im}\, G_R(E)$ is contained in the one particle insertion in figure 1a. There is no cancellation and the leading correction is given by

$$\frac{\delta\rho(E)}{\rho_\infty(E)} = \frac{2}{n}(E_o^2-E^2)^{-1}\mathrm{Re}\int_q \frac{1}{2\pi i}\int_{-E_o}^{E-E_F} d\omega\ g_c^2(q,\omega)\ U_{scr}^R(q,\omega) \tag{17}$$

where the diffusion pole carrying propagator $g_c$ can be inferred from Ref.4. The evaluation of (17) results in Eq.(4). The result holds for both ensembles, since diagram 1a does not contain a parallel double-line propagator, which must be zero in the phase invariant case. In 2D I finally compare the O(1/n) approximation for the present gauge invariant models with the HF theory for the NFE model |3| by

$$\frac{\delta\rho(E)}{\rho_\infty(E)} = \begin{cases} -E_o^4(64\pi nA(E_o^2-E^2))^{-1}\ \ln^2(|E-E_F|/E_o) & \text{gauge invariant} \\ -(2\pi E_F\tau)^{-1}\ln^2(|E-E_F|\tau) & \text{NFE model} \end{cases} \tag{18}$$

which after replacement of some unimportant model parameters agree well.

## References

|1| B.L. Altshuler and A.G. Aronov, Solid State Comm.36, 115 (1979)
|2| B.L. Altshuler, D.Khmelnitskii, A.I. Larkin, and P.A. Lee, Phys.Rev.B22, 5142 (1980)
|3| B.L. Altshuler, A.G. Aronov, and P.A. Lee, Phys.Rev.Lett.44, 1288 (1980)
|4| R. Oppermann and F. Wegner, Z.f.Phys. B34, 327 (1979)
|5| L. Schäfer and F. Wegner, Z.f.Phys. B38, 113 (1980)
|6| A.L. Efros, J.Phys. C9, 2021 (1976)

# FINITE SIZE SCALING APPROACH TO ANDERSON LOCALISATION

J.L. Pichard and G. Sarma
Service de Physique du Solide et de Résonance Magnétique
CEN-SACLAY, 91191 Gif-sur-Yvette, Cedex, FRANCE

Abstract : The Anderson localisation problem is studied by a new scaling theory of finite systems which has been proved to be powerful for different phase transition problems [1].

In three dimensions, there is a transition at a critical value $W_c$ of the disorder parameter from a region of exponentially localized states to a region of *extended* states. When $W$ decreases to $W_c$, we find a value equal to 0.66 for the critical exponent $\nu$ relative to the divergence of the localisation length.

In two dimensions, a critical value $W_c$ of the disorder parameter separates a region of exponentially localized states ($W > W_c$) from a region of *"quasi extended" states* which are *non square summable* and fall off as $R^{-\eta(W)}$. We give the variation of the exponent $\eta(W)$ in the whole "quasi extended" region. On the other hand, we show that the divergence of the localisation length when W decreases to $W_c$ is now controlled by an *essential singularity*.

The behaviour of the dimensionless conductance is given in all the cases. In particular, in the two dimensional weak disorder phase, it is shown to have a *power law decay* $(\frac{L}{\lambda})^{-2\eta(W)}$ versus the "size" L.

## I INTRODUCTION

In this short communication, we can only present our essential results and rapidly sketch their derivation. For full details, we refer the reader to two previous publications [2], [3].

The basis of our method lies in the study of topologically one dimensional systems, taken infinite in one direction, as a function of the finite value $\ell$ of the (d-1) remaining transverse directions. Such an approach *never provides a direct simulation* of the d-dimensional infinite lattice, but allows to predict its behaviour thanks to a *scaling theory* whose parameter is precisely the finite transverse length $\ell$.

With the help of *Oseledec's theorem* on random matrix products, we defined and calculated in [2] the *relevant* localisation length $L_{max}$ for "strips" and "bars", leading respectively to two or three dimensional systems. These calculations were made for a tight binding model at zero energy (band centre) with "diagonal" disorder characterized by a rectangular distribution of width W of the random potential.

## II THE LOCALIZED REGION

II.a Three dimensional systems. One can find in [2] the variation of $L_{max}$ versus the side $\ell$ of a bar for different values of W. For $W > W_c$, $L_{max}$ converges for increasing $\ell$ towards a finite limit which is the *localisation length* $\xi_\infty$ of the infinite lattice. Our analysis of this region is based upon the fundamental *scaling ansatz :*

$$\frac{L_{max}(W,\ell)}{\xi_\infty(W)} = f\left(\frac{\ell}{\xi_\infty(W)}\right) \tag{1}$$

At the transition, where $\xi_\infty(W)$ becomes infinite, the ansatz (1) obviously implies a linear growth of $L_{max}$ versus $\ell$, which gives the critical value of the disorder parameter $W_c \sim 19 \pm 0.5$. Our method for the determination of the critical behaviour of

$\xi_\infty$ consists in the research of which postulated divergence will actually give the scaling function f defined by (1). We have looked for a divergence of the form :

$\xi_\infty \sim (W-W_c)^{-\nu}$, so as to map all the ratios $L_{max}(\ell,W)/\xi_\infty(W)$ of the localized region into the single function f(x). This requirement is satisfied with $\nu \sim 0.66 \pm 0.02$ in agreement with the result of Stein and Krey [4].

II.b Two dimensional systems. We refer to [2] for the variation of $L_{max}$ versus the width $\ell$ of a strip for different values of W. For $W > W_c$ ($W_c \sim 5.95 \pm 0.05$), $L_{max}$ again converges towards a finite limit for increasing $\ell$, corresponding to *exponentially* localized states. But here, we failed in all our trials for finding a scaling function f(x) with a power law divergence of $\xi_\infty$. This led us to look for an *essential singularity* of the form $\xi_\infty \sim \exp(W-W_c)^{-\vartheta}$. In such a case, a new scaling assumption has to be postulated [5]. This turns out to be successful and gives a value of $\vartheta \sim 0.5 \pm 0.1$.

## III THE EXTENDED (or "quasi extended") REGION

For $W < W_c$, $L_{max}$ diverges versus increasing $\ell$, which implies that we cannot define a finite localisation length for the infinite lattice. The problem is to relate this to the actual asymptotic behaviour of the wave function in the infinite lattice.

In the whole region $W < W_c$, we find that : $L_{max} \sim a(W)\,\ell$ and $L_{max} \sim b(W)\,\ell^2$ respectively for strips and bars. If we assume that $|\psi(R,W)|$ in the infinite lattice falls off isotropically, it was shown in [3] that

$$-\frac{1}{a(W)} = \pi R \frac{d}{dR} \log |\psi(R)| \quad \text{in two dimensions} \qquad (2)$$

$$-\frac{1}{b(w)} = 2\pi R^2 \frac{d}{dR} \log |\psi(R)| \quad \text{in three dimensions} \qquad (3)$$

III.a The three dimensional system. Integrating (3) gives the following *asymptotic* behaviour :

$$|\psi(R,W)| \sim |\psi_\infty(W)| \exp \frac{1}{2\pi\, b(W)\, R}$$

This shows that in three dimensions, the disorder for $W < W_c$ is *insufficient* for localising the states, $|\psi(R,W)|$ *decreasing* towards a *finite limit* $|\psi_\infty(W)|$ when $|\psi(R,W)|$ is constrained to be equal to unity in some finite region of the system. The limit is given by : $|\psi_\infty(W)| = \exp - \frac{c}{b(W)}$ where c is some constant. A non zero value of $|\psi_\infty|$ is characteristic of *extended states*. The figure 1 shows as it should be that $|\psi_\infty|$ goes to one in the limit of vanishing disorder and to zero when W reaches $W_c$. So $|\psi_\infty(W)|$ appears to us as a natural "order parameter" in the region of extended states. To our knowledge, such a concept has not been proposed till now.

III.b Two dimension systems. Integrating (2) gives

$$|\psi(R,W)| \sim R^{-\eta(W)} \quad \text{with} \quad \eta(W) = \frac{1}{\pi\, a(W)}$$

This result proves a *power law decay* of the wave functions in the *whole "quasi extended" region* ($W \leq W_c$). The knowledge of the numbers a(W) allows us to plot $\eta(W)$ versus W (Figure 2). At the transition, we find $\eta(W_c) = \frac{1}{2}$. In the whole region $W < W_c$, the wave functions are *non square summable*, which led us to call then "quasi-extended".

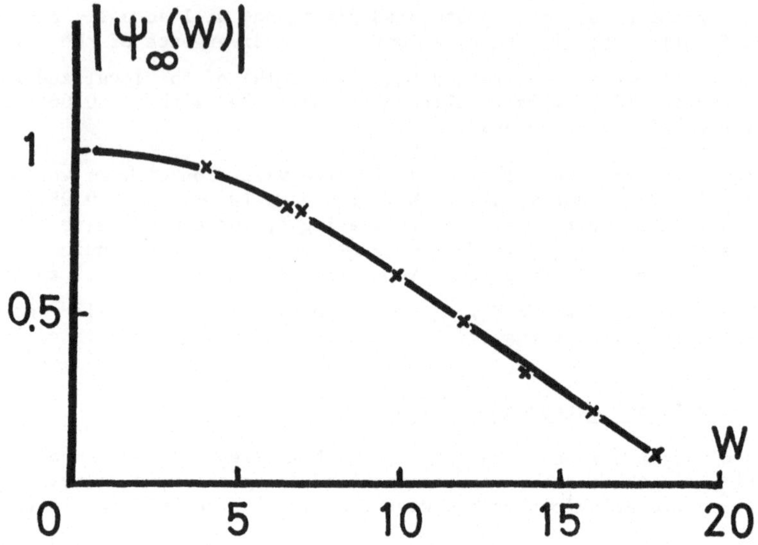


Fig. 1 : $d = 3$, $W < W_c$

$|\psi_\infty(W)|$ versus the disorder parameter W.

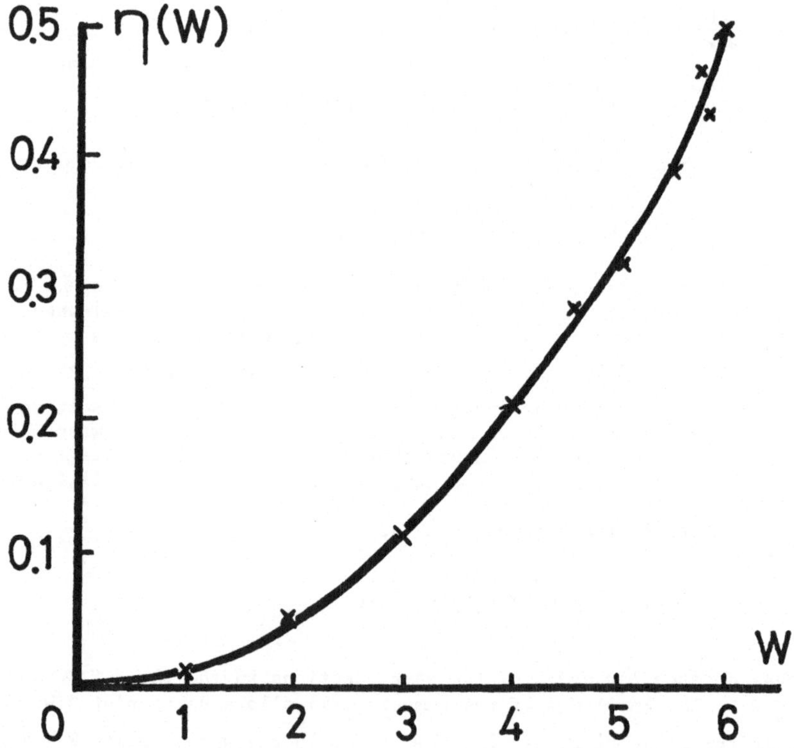


Fig. 2 : $d = 2$, $W \leq W_c$

the exponent $\eta(W)$ of the *"quasi-extended"* states versus W. Note that $\eta(W) \sim W^2$ up to $W \sim 0.8\ W_c$.

## IV RESULTS FOR THE CONDUCTIVITY

For any topologically one dimensional system, it was shown in [2] that the wave functions decay as exp $-\gamma_\alpha|Z|$ in $\ell$ independent "Oseledec subspaces", thus giving $\ell$ transmission coefficients $T_\alpha(Z) = \exp -2\gamma_\alpha|Z|$. These Oseledec subspaces correspond to the "channel" concept of Anderson et al. [6]. The generalisation of Landauer's formula [7] gives the dimensionless conductance

$$g(Z) = \sum_\alpha \frac{T_\alpha(Z)}{1-T_\alpha(Z)} \tag{4}$$

It is obvious that for large Z and for any *finite* value of $\ell$, this reduces to the single term corresponding to $\gamma_{min} = (L_{max})^{-1}$

IV.a The localized region. the infinite lattice limit obviously gives $g(Z) = \exp - \frac{2Z}{\xi_\infty}$ in any dimension.

IV.b The three dimensional extended region. $g(Z,\ell) = \left[\exp\left(\frac{2|Z|}{b(w)\,\ell^2}\right) - 1\right]^{-1}$. Taking the limit $Z = k\ell$, $k \gg 1$, $\ell \to \infty$, one gets $g(Z,\ell) = \frac{b(W)\ell^2}{2Z}$, whence the conductivity :

$$\sigma(W) = \frac{e^2}{\pi\hbar}\,\frac{b(W)}{2} \quad \text{(figure 3)}$$

For a definite conclusion about the minimum metallic conductivity, we have to look more carefully to the behaviour of b(W) close to $W_c$.

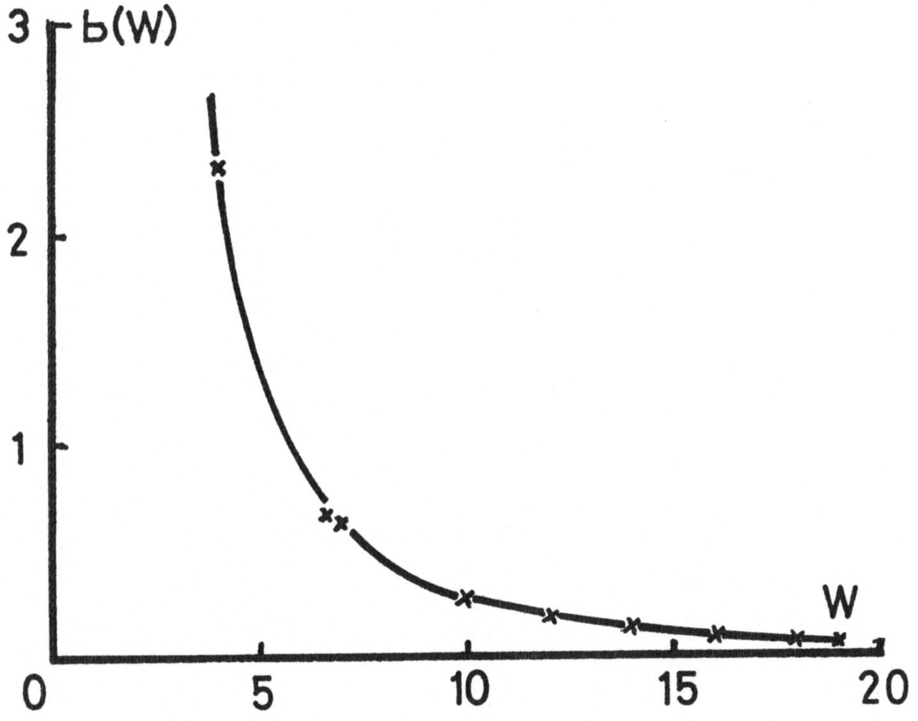


Fig. 3 : $d = 3$, $W < W_c$
The conductivity multiplied by $\frac{2\pi\hbar}{e^2}$ versus W.

## IV.c The two dimensional "quasi-extended" region

For the strip, $g(Z,\ell) \sim \exp - \frac{2|Z|}{a(W)\ell}$. In the same way as for the wave function's decay, this implies the scaling law $g(Z,\ell) \sim \ell^{-2\eta} f(\frac{Z}{\ell})$ which gives :

$$g(Z) \sim Z^{-2\eta(W)}$$

since $\ell$ must disappear in the infinite lattice limit. This result, which gives a *vanishing* conductance for $Z \to \infty$, has been recently derived in a different way by Mott and Kaveh [8], and a first order expansion in powers of $\eta(W)$ gives the well known "logarithmic correction" approximation [9].

## REFERENCES


1. B. Derrida, L. de Seze and J. Vannimenus, This conference
2. J.L. Pichard and G. Sarma, J. Phys. C 14,L127 (1981)
3. J.L. Pichard and G. Sarma, Submitted to J. Phys. C
4. J. Stein and U. Krey, Zeit. für Phys. B 34, 287 (1979)
5. C.J. Hamer and M.N. Barber, J. Phys. A 14, 259 (1980)
6. P.W. Anderson, D.J. Thouless, E. Abrahams and D.S. Fisher, Phys. Rev. B 22 8, 3529 (1981)
7. R. Landauer, Philos. Mag. 21, 863 (1970)
8. N.F. Mott and M. Kaveh, Preprint 1981
9. E. Abrahams, P.W. Anderson, D.C. Licciardello and T.V. Ramakrishnan, Phys. Rev. Lett. 42, 673 (1979)

# SCALING THEORY OF ANDERSON LOCALIZATION: A RENORMALIZATION GROUP APPROACH

by

Eytan Domany
Department of Electronics
The Weizmann Institute of Science
Rehovot, Israel

and

Sanjoy Sarker*
Laboratory of Atomic and Solid State Physics
Cornell University
Ithaca, NY 14853

## ABSTRACT

A position space renormalization-group method, suitable for studying the localization properties of electrons in a disordered system, was developed. Two different approximations to a well defined exact procedure were used. The first method is a perturbative treatment to lowest order in the inter-cell couplings. This yields a localization edge in three dimensions, while in two dimensions no fixed point is found, indicating localization even for small randomness. The second method is an application of the finite lattice approximation, in which the inter-cell hopping between two (or more) cells is treated to infinite order in perturbation theory. This method was used in two dimensions only, yielding results that are in agreement with those of the lowest order approximation.

I. INTRODUCTION

The Anderson model describes non-interacting electrons on a lattice in terms of the Hamiltonian

$$H = \sum_{\vec{r}} \varepsilon_{\vec{r}} |\vec{r}><\vec{r}| + \frac{1}{2} \sum_{\vec{r}\vec{r}'} V_{\vec{r}\vec{r}'} (|\vec{r}><\vec{r}'| + |\vec{r}'><\vec{r}|) \tag{1.1}$$

where the state $|\vec{r}>$ corresponds to a single atomic orbital of energy $\varepsilon_{\vec{r}}$ localized at the site $\vec{r}$. $V_{\vec{r}\vec{r}'}$ is the overlap between different orbitals, enabling electrons to hop from site to site, and is taken to be non-zero only when $\vec{r},\vec{r}'$ are nearest neighbors. Disorder can be introduced by taking either or both $\varepsilon_{\vec{r}}$ and $V_{\vec{r}\vec{r}'}$ to be random variables. The case that received the most attention[1-5] is the problem of diagonal disorder in which the hopping elements $V_{\vec{r}\vec{r}'}$ are assumed to be constant and the site energies $\varepsilon_{\vec{r}}$ are chosen independently from a rectangular distribution of width W.

The dimensionless quantity $\sigma = W/V$ is a measure of the degree of randomness in the system. The limiting cases $\sigma = 0$ and $\sigma = \infty$ can be trivially solved. For $\sigma = 0$ the states are infinitely extended plane waves, whereas for $\sigma = \infty$ the eigenstates are given by the orbitals $|\vec{r}>$ and therefore are completely localized. Between these two limits there must be a transition from extended to localized states. This transition has been named the Anderson transition and is reminiscent of phase transitions in magnetic systems. The manner in which such phase changes take place usually depends on the spatial dimensionality of the system. An eigenstate of the Hamiltonian (1.1) corresponding to eigenvalue E can be written as

$$|\psi>_E = \sum_{\vec{r}} a_{\vec{r}E} |\vec{r}> \tag{1.2}$$

By exponential localization we mean that the magnitude $|a_{\vec{r}E}|$ falls off exponentially away from a "center of localization" $\vec{r}_o$ as $|a_{rE}| \sim \exp[-\frac{r-r_o}{L}]$. The localization length L, averaged over the ensemble, is a function of energy and the degree of disorder, $\sigma$.

We expect that localized and extended states will not co-exist in energy, since the slightest perturbation will cause the former to delocalize by mixing with the corresponding extended state of the same energy. We can then talk about regions of localized and extended states. Mott[6] argued that there exists a sharp boundary $E_c(\sigma)$ that separates these two regions.

The localization length L increases as the "mobility edge" $E_c$ is approached from the localized regime and is expected to diverge at $E_c$ as

$$L \sim |E-E_c|^{-\nu'} \tag{1.3}$$

where the "critical exponent" $\nu'$ depends on the dimensionality d. Alternatively, if $\sigma_c(E)$ is the randomness at which a given state of energy E becomes extended, then as $\sigma_c$ is approached from above the localization length is expected to diverge as

$$L \sim (\sigma-\sigma_c)^{-\nu} \tag{1.4}$$

where the exponent $\nu$ is of the same order as $\nu'$, but the exact values may be unequal.

Due to the lack of any exact information, numerous approximate methods have been devised to analyze the Anderson model. There are some excellent reviews[2-5] on the subject.

We have developed a scaling theory of localization[7-9] in the spirit of the position space RG ideas of Niemeijer and van Leeuwen.[10] The basic idea is to study the localization length L characteristic of a state vector of energy E, in two and three dimensions. To this end we have performed a sequence of calculations, which can be viewed as a set of systematically improving approximations to an exact treatment contained in the general formalism. In addition to diagonal randomness, we have also considered off-diagonal randomness of the type in which the signs of the hopping elements $V_{\vec{r}\vec{r}'}$ are random. Our procedure is based on a two-parameter RG, which are $\sigma$ and E.

The results can be summarized as follows: In three dimensions, (i) we find a localization edge. The "critical behavior", characterized by $1.25 < \nu < 1.75$, is governed by a fixed point of our RG transformation located at $E = 0$, $\sigma_c \simeq 7.0$. (ii) This fixed point is stable, implying that the localization length L diverges with the same exponents along the entire localization edge $E_c(\sigma)$. (iii) On the basis of a simple scaling argument, we predict a parabolic "phase boundary" $E^2_c(\sigma) \propto (\sigma_c - \sigma)$. (iv) In two dimensions, we have not found a fixed point, which implies that all states are localized, in agreement with Abrahams et. al.[11]

## II. RENORMALIZATION GROUP PROCEDURE

To construct the renormalization-group transformation for the Anderson problem, defined by the Hamiltonian

$$H = \sum_{\vec{r}} \varepsilon_{\vec{r}} |\vec{r}><\vec{r}| + \frac{1}{2} \sum_{<\vec{r}\vec{r}'>} [|\vec{r}><\vec{r}'| + |\vec{r}'><\vec{r}|]$$

$$<\varepsilon_{\vec{r}}> = 0; \qquad <\varepsilon_{\vec{r}}\, \varepsilon_{\vec{r}'}> = \sigma^2 \delta_{\vec{r}\vec{r}'} \tag{2.1}$$

we use nearly degenerate perturbation theory.[12] The Hamiltonian is written as $H = H_o + H_1$. The eigenvalues $E_n$ and the normalized eigenvectors of $H_o$ are assumed to be known

$$H_o|\phi_n> = E_n|\phi_n> ; \qquad <\phi_n|\phi_m> = \delta_{nm} \tag{2.2}$$

The Hilbert space $\Omega$ in which H operates can be spanned by the vectors $|\phi_n>$. For the Anderson model the number of such states is equal to the number of sites N of the lattice. The RG mapping we are seeking to establish reduces N to $N' = N/b^d$. To achieve this reduction we divide the Hilbert space $\Omega$ into two subspaces D and $\overline{D}$, containing N' and N - N' states, respectively.

If $|\psi>$ is an eigenstate of H, with energy E, denote by $|\psi_D>$ the projection of $|\psi>$ onto D. Then $|\psi_D>$ satisfies the equation

$$H_D|\psi_D> = E|\psi_D> ; \qquad H_D = H_o + V \tag{2.3}$$

and V solves the operator equation

$$V = H_1 + H_1 \frac{1-P_D}{E-H_o} V \tag{2.4}$$

We have thus reduced the problem from that of diagonalizing an N x N matrix to that of an N' x N' one, provided we can calculate the matrix elements of V in the subspace D. In general this is non-trivial since (i) V depends explicitly on the exact energy, E, which can only be obtained by solving the original secular equation and (ii) one has to solve an operator equation (2.4). The first difficulty can be overcome by noting that for a very large lattice the spectrum of H forms a continuum, so that for an E within the band there is a solution. Therefore the energy E can be used as a paramter in Eq. (2.4).

So far the analysis has been exact. To overcome the second difficulty, i.e., the calculation of V, we resort to perturbation theory, and expand V in powers of $H_1$

$$V = H_1 + H_1GH_1 + H_1GH_1GH_1 + \ldots \quad , \tag{2.5}$$

where

$$G = \frac{1 - P_D}{E - H_o} \tag{2.6}$$

We now break up the lattice into small cells of volume $b^d$, and include in $H_o$ all the elements of (2.1) which connect sites in the same cell. Hopping elements that connect sites of neighboring cells $\vec{R}$ are included in $H_1$. $H_o$ is thus a sum of uncoupled cell Hamiltonians $h_R$. Let $|\vec{R}i\rangle$ be the eigenvectors and $e_{\vec{R}i}$ the corresponding eigenvalues of the cell Hamiltonian $h_{\vec{R}}$ (i runs from 1 to $b^d$). Now we construct the model subspace D by keeping only one of the $b^d$ states in each cell. For best convergence, choose that state from the cell $\vec{R}$, for which $|E - e_{\vec{R}i}|$ is smallest. Denoting this state by $|\vec{R}\rangle$, the Hamiltonian $H_D$ takes the form

$$H_D = \sum_R (e_{\vec{R}} + V_{\vec{R},\vec{R}})|\vec{R}\rangle\langle\vec{R}| + \frac{1}{2}\sum_{\vec{R}\neq\vec{R}'} (V_{\vec{R},\vec{R}'} \vec{R}\rangle\langle\vec{R}'| + h.c.) \tag{2.7}$$

Note that so far the formalism is exact. However, since the operator equation (2.4) cannot be solved exactly, we must resort to various approximations.

## A. First-order Approximation[7]

To first order in $H_1$ we can write

$$H_D^{\text{approx.}} = \sum_{\vec{R}} e_{\vec{R}} |\vec{R}\rangle\langle\vec{R}| + \frac{1}{2}\sum_{\vec{R}\vec{R}'} V'_{\vec{R}\vec{R}'} (|\vec{R}\rangle\langle\vec{R}'| + h.c.) \tag{2.8}$$

where $V'_{\vec{R}\vec{R}} = \langle\vec{R}|H_1|\vec{R}'\rangle$. This Hamiltonian looks very much the same as the Anderson Hamiltonian (1.1). The "site" energies $e_{\vec{R}}$, being functions of different sets of random variables, are themselves random and independent of each other. However, the hopping elements $V_{\vec{R}\vec{R}'}$ are no longer uniform. We proceed by replacing the actual $V'_{\vec{R}\vec{R}}$ distribution that is generated by some fixed distribution. Various choices were tried;[7,9] (a) $V_{\vec{R}\vec{R}'} \to V_{eff}$, where $V_{eff} = \langle|V_{RR'}|\rangle$ or $\langle V_{RR'}{}^2\rangle^{1/2}$; (b) $V_{\vec{R}\vec{R}'} = V_{eff}\,\eta_{RR'}$ where $\eta_{RR'} = \pm 1$ (randomly assigned). Also, since for $E \neq 0$ one has $\langle e_R\rangle \neq 0$, in contrast to (2.1), we subtract the diagonal operator $\langle e_R\rangle |R\rangle\langle R|$ from $H_D^{\text{approx.}}$. After division by $V_{eff}$ we obtain recursion relations for the eigenvalue E and the second moment of the diagonal element distribution;

$$E' = (E - \langle e_{\vec{R}}\rangle)/V_{eff} = g(\sigma,E) \tag{2.9}$$

$$\sigma' = \langle(e_{\vec{R}} - \langle e_{\vec{R}}\rangle)^2\rangle/V_{eff}{}^2 = f(\sigma,E) \tag{2.10}$$

The numerical implementation of our procedure is described in detail elsewhere;[7,9] results will be summarized below, for both two and three dimensions.

B. Finite Lattice Approximation[9]

To our knowledge, the finite lattice approximation has not been applied previously to quantum systems. In the language of a perturbative approach to the exact equation for the effective cell Hamiltonian, e.g. eqn. (4.9-10), this approximation is equivalent to summation to infinite order of a selected subset of the operators that constitute the perturbation $H_1$. To be more specific, in order to calculate the effective hopping between two neighboring cells, $V_{RR'}$, the part of $H_1$ that connects these two cells is summed to infinite order. To see how this is done, note that (2.4) can be written as an algebraic equation for the matrix elements

$$V_{\vec{R}\vec{R}'} = H_{1\vec{R}\vec{R}'} + \sum_{\vec{R}''\alpha''} \frac{H_{1\vec{R},\vec{R}''\alpha''} \; V_{R''\alpha'',\vec{R}'}}{E-e_{\vec{R}''\alpha''}} \tag{2.11}$$

where $|R'''a''\rangle$ is an eigenstate of cell $\vec{R}''$ with eigenvalue $e_{\vec{R}'',\alpha''}$, that was not assigned to D, (i.e. $|\vec{R}''\alpha\rangle \varepsilon \bar{D}$) and $H_{1R,R'\alpha'} = \langle\vec{R}|H_1|\vec{R}'\alpha'\rangle$. Thus, the evaluation of $V_{RR'}$, an element in D, requires the knowledge of all matrix elements of $H_1$ between any state in D and any state in $\bar{D}$. The equation for such elements, as obtained from (2.4) is given by

$$V_{R\alpha,R'} = H_{1R\alpha,R'} + \sum_{\vec{R}''\alpha''} \frac{H_{1R\alpha,R''\alpha''} \; V_{R''\alpha'',R'}}{E-e_{R''\alpha''}} \tag{2.12}$$

Note that matrix elements $V_{\vec{R}\alpha,\vec{R}'}$ of the same kind (i.e. that connect $\bar{D}$ to D) appear on both sides of this equation, and therefore, $V_{\vec{R}\alpha,\vec{R}'}$ can be calculated by solving the set of algebraic equations

$$\sum_{\vec{R}''\alpha''} M_{\vec{R}\alpha,\vec{R}''\alpha''} \; V_{\vec{R}''\alpha'',\vec{R}'} = H_{1\vec{R}\alpha,\vec{R}'} \tag{2.13}$$

where

$$M_{\vec{R}\alpha,\vec{R}''\alpha''} = \delta_{\vec{R}\vec{R}'',\alpha\alpha''} - \frac{H_{1\vec{R}\alpha,\vec{R}''\alpha''}}{E-e_{\vec{R}''\alpha''}} \tag{2.14}$$

Once $V_{\vec{R}''\alpha'',\vec{R}}$ are known, substitution into (2.11) immediately yields $V_{R,R'}$. Thus diagonal and off-diagonal elements of the effective cell Hamiltonian are obtained to infinite order in the unrenormalized hopping elements that connect the cells in our finite cluster.

This procedure was carried out in d=2, using finite clusters that contain two cells only; again, the distribution of the off-diagonal elements was replaced by $V_{eff} = \langle|V_{RR'}|\rangle$ and recursion relations of the form (2.9-2.10) were obtained for

E and $\sigma$.

## III. Results

We have obtained two kinds of results, that are given in detail elsewhere.[7-9]
1. On the basis of very general assumptions and symmetry considerations, we find[8,9] that if a fixed point of recursion relations such as eq. (2.9-10) exists at $(E=0, \sigma_c)$, and it is stable in the E direction, the divergence of the localization length L is given by

$$L \sim (\sigma - \sigma_c + \beta E^2)^{-\nu} \tag{3.1}$$

This scaling form may explain[9] the variation of numerical results for the exponent $\nu'$. The result (3.1) is obtained independently of our numerical work and approximations.
2. Numerical evaluation of the function $\sigma' = f(E=0,\sigma)$, based on first order perturbation theory,[7] indicates the existence of a fixed point at $\sigma_c \approx 7$ in d=3 dimensions, e.g. the existence of a localization edge. The estimate of the exponent $\nu$ is quite inaccurate because of statistical error; we find $1.25 \leq \nu \leq 1.75$. This fixed point was found[9] to be stable against E perturbations, which means that the localization length diverges with the same exponent $\nu$ along the entire edge $E_c(\sigma)$.

In two dimensions both methods used indicate[7,9] that no fixed point exists: we always get $\sigma' > \sigma$, and, therefore, the system flows to the $\sigma=\infty$ regime, and all states are localized.

This research was supported by the US-Israel Binational Science Foundation, Jerusalem, Israel.

## REFERENCES

* Present Address: Department of Physics, Rutgers University, Piscataway, NJ 08854

1. P.W. Anderson, Phys. Rev. 109, 1492 (1958).

2. N.F. Mott and E.A. Davis, Electronic Processes in Non-Crystalline Materials, Clarendon Press, Oxford, 1971.

3. D.J. Thouless, Physics Reports 13, 93 (1974).

4. R.J. Elliott, J.A. Krumhansl and P.L. Leath, Rev. Mod. Phys. 46, 465 (1974).

5. D.J. Thouless, in The Metal Non-Metal Transition in Disordered Systems, ed. L.P. Friedman and D.P. Tunstall, Proc. 19th Scott. Universities Summer School, Scotland (1978).

6. N.F. Mott, Adv. Phys. 16, 49 (1967).

7. E. Domany and S. Sarker, Phys. Rev. B20, 4726 (1979).

8. S. Sarker and E. Domany, J. Phys. C13, L273 (1980).

9. S. Sarker and E. Domany, Phys. Rev. B (1981), in print.

10. Th. Niemeijer and J.M.J. van Leeuwen, Physica 71, 17 (1974); also in Phase Transitions and Critical Phenomena, vol. 6, ed. C. Domb and M.S. Green, Academic Press, 1976.

11. E. Abrahams, P.W. Anderson, D.C. Licciardello and T.V. Ramakrishnan, Phys. Rev. Lett. 42, 673 (1979), P.W. Anderson, E. Abrahams and T.V. Ramakrishnan, Phys. Rev. Lett., 43, 718 (1979).

12. B.H. Brandow, Rev. Mod. Phys. 39, 771 (1967).

# ELECTRON SPIN RESONANCE IN A FERMI GLASS

K.A. Müller
IBM Zurich Research Laboratory, 8803 Rüschlikon, Switzerland

T. Penney, M.W. Shafer and W.J. Fitzpatrick
IBM T.J. Watson Research Center, Yorktown Heights, N.Y. 10598

Abstract. A broad Lorentzian ESR line in H-doped $CaV_2O_6$ crystals has been observed between 300 K and 4 K and assigned to itinerant electrons in $d_{x^2-y^2}$ orbitals. The spin susceptibility is nearly temperature independent, whereas the conductivity varies strongly, consistent with Mott's $\exp-(1/T)^{1/4}$ variable range hopping law over three orders of magnitude. Our observations indicate Pauli paramagnetism in this random system.

Introduction. In disordered solids a conductivity variation as a function of temperature in the form:

$$\sigma = \sigma_0 \exp - [T_0/T]^{1/4} \tag{1}$$

is observed quite frequently which Mott[1] deduced for random local potentials. For such solids Fermi statistics will be obeyed even when the conductivities are very low and Anderson[2] has named them accordingly "Fermi Glasses". Therefore, these systems should show Pauli's temperature-independent susceptibility to zero order if electron-electron correlation energies and ferro- or antiferro-magnetic interactions are small compared to kT. We investigated protonated $CaV_2O_6$ crystals which follow Eq. (1) and do indeed show Pauli paramagnetism as measured by electron spin-resonance (ESR) intensity.

Experiments. Our crystals were grown by the Czochralski method from slightly $V_2O_5$-rich solutions.[3] By proper control of such parameters as the melt composition, its temperature profile and the rotation and pull rate, we were able to pull clear inclusion-free cystals over 2 cm in length and about 0.5 cm in diameter. The crystals selected for hydrogen doping were thin slices cleaved along the $(20\bar{1})$ plains, and were heated in pure $NH_3$ gas for various times and temperatures from $350^o$ to $500^oC$. The samples obtained in this way, can be classified by their ESR spectra: Two principally different signals were observed:

a) A *localized* $V^{4+}(3d^1)$ EPR signal with $g_{\|} = 1.930$ and $g_{\perp} = 1.975$ showing a main hyperfine interaction with one $^{51}V$, $I = 7/2$, nuclear spin as well as weak super-hyperfine interactions with a nearby localized proton ($I = 1/2$) and two other $V^{5+}$ ions. This spectrum is identical to the one reported previously by a Russian group.[4] Upon cooling, the signals grew $\propto 1/T$ as expected for localized particles.[5]

b) A new broad ESR line was observed. Its width $\Delta H_{pp}$ varied from sample to sample in narrow limits from 260 to 320 Gauss at 300 K. This broad line could be fitted to a Lorentzian shape and was found to be accurately Lorentzian over most of the temperature range. After the $NH_3$ treatment, three types of samples could be distinguished: Samples showing only the narrow $V^{4+}$ lines previously reported are called Type 1; samples which show in addition the new broad Lorentzian ESR line are Type 2. Finally, samples which show only the broad line are called Type 3. We do not know whether Type-2 samples are macroscopic mixtures of Types 1 and 3 or consist of microscopic inhomogeneities due to randomness of proton distributions.

Resistivity as a function of temperature was measured using the van der Paw technique. Contacts were made with In solder. Typical sample dimensions were 3 × 2 × 0.1 mm. Data were taken between 300 and 5 K. The data were plotted as log conductivity versus $T^f$ for $f = -1, -1/2, -1/4$ and $-1/8$. Reasonable fits were obtained for $f = -1/4$ and $-1/8$. A characteristic result of a Type-3 sample is shown in Fig. 1 obeying the -1/4 law over three orders in magnitude. These data are thus consistent with Mott's variable range hopping (- 1/4).

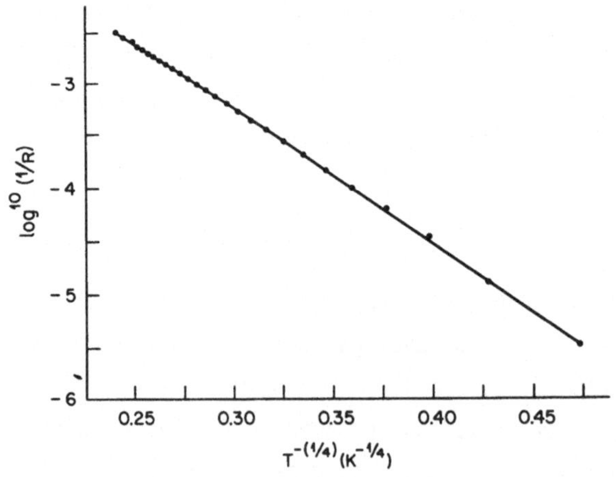


Fig. 1. The inverse resistivity plotted logarithmically as a function of $T^{1/4}$ for a $CaV_2O_6$ sample heated for 20 hours at 470°C. $\sigma_0 = 3.9\ \Omega^{-1}\ cm^{-1}$ and $T^{1/4} = 30\ K^{1/4}$ [see Eq. (1)]. ( © 1981 by The American Physical Society.)

In Fig. 2, the ESR line width $\Delta H_{pp}$ and the intensity are plotted as a function of temperature for the sample whose resistivity is shown in Fig. 1. The behaviors of other samples were similar. From the logarithmic temperature scale, one sees that the intensity $\propto \chi_s(T)$ is nearly temperature independent between 300 and 20 K, i.e., in the range where Mott's law (Fig. 1) holds. Thus, this behavior can be better looked at by a Pauli rather than a Curie law. Below 20 K, $\chi_s(T)$ drops by a factor of about three before rising again. Decreases of this sort in the region of 10 to 40 K were also observed in the half dozen other samples investigated although less pronounced.

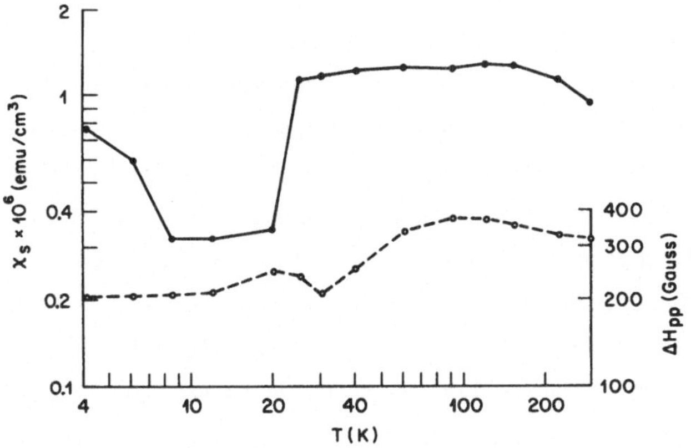


Fig. 2. CESR line susceptibility and width as a function of $\ell nT$ for the sample as in Fig. 1. Due to macroscopic inhomogeneities the scale of the spin susceptibility is only approximate. (© 1981 by The American Physical Society.)

<u>Interpretation</u>. The broad Lorentzian ESR lines could not be saturated. Furthermore, their widths were the same at 9 and 19 GHz. Thus, we can assume that they are homogeneous with $T_1 = T_2 = \tau = \hbar/\beta\Delta H_{pp}$. For a line 300 Gauss wide, this yields $\tau \sim 4 \times 10^{-10}$ sec. As the broad line always correlates with the strongly conducting crystals, we conclude that this broad line is due to "itinerant" charge carriers, i.e., not strongly localized.

Typical g-values measured for the broad lines were $g_{\parallel} = 1.955$; $g_{\perp} = 1.962$. Their average value, $\bar{g} = (1/3)(g_{\parallel} + 2g_{\perp}) = 1.959$, is very close to that of the average $\bar{g}$ of the localized $V^{4+}$ hyperfine split line with $\bar{g} = 1.962$. The latter has 0.999 $d_{x^2-y^2}$ character.[4] Therefore, the broad line stems from electrons in predominantly vanadium $d_{x^2-y^2}$ orbitals.

Analysis of the conductivity and line-width data gives more insight into our system: the percolative derivation of Eq. (1),[6] yields for the expression of the temperature $T_0$,

$$T_0 = 16\,\alpha^3 / N(E_F), \tag{2}$$

where $\alpha^{-1}$ is the extension of the quasi-localized wave function $\Psi \propto \exp - 2\alpha r$, and $N(E_F)$ is the density of electron states at the Fermi energy. For the half dozen samples we measured, $T_0^{1/4}$ varied between 30 and 10 $K^{1/4}$. The latter value occurs for higher doping levels.

There are two macroscopic models which we shall consider. In one, the protons are distributed completely at random. The result is the Mott-Anderson behavior treated by AHL. The Lorentzian ESR line then comes from the weakly localized electrons in the AHL-connected "regions". A possible objection to this model is that one might expect the ESR line width to be temperature dependent since the degree of localization is strongly $T$ dependent. The second model is that instead of a totally random distribution there is a tendency for protons to cluster. In this case, Eq. (1) still holds because the system is in the same universality class. The weak temperature dependence of the ESR line width follows because the relaxation is internal to the cluster.

For conduction ESR(CESR) and s-type orbitals, the well-known formula of Elliott for spin-lattice relaxation holds:[7]

$$\Delta\nu = 1/\tau = c(v_F/\ell)\,\Delta g^2\ , \tag{3}$$

where $\Delta g = g - g_{free}$ is proportional to the spin-orbit coupling, and $c$ is a factor of the order of unity. We can assume that Eq. (3) also holds qualitatively for d-orbitals.

If the velocity of the carriers in the clusters is $v_F$, then the Fermi velocity, for $E_F > kT$ is temperature independent. This is the case as long as correlation and other effects do not play a role. Thus the spin-relaxation time, $\tau$, is temperature independent if the scattering length $\ell$ inside the cluster does not change.

The small variation of the CESR line width's $\Delta H_{pp}$ was uncorrelated to that of the $T_0^{1/4}$'s and can be further discussed by using Eq. (3): It requires a fairly constant ratio of $v_F/\ell$ within the conducting clusters. This can be very tentatively done by assuming the existence of clustered protons with more or less homogeneous density within these clusters. From experimental evidence beyond this report cluster-surface scattering can be excluded.

We would like to thank B.I. Halperin, J. Hubbard, H. Thomas and S. Kirkpatrick for enlightening discussions. We benefited from expert experimental help by R.A. Figat, J.M. Rigotty and W. Berlinger.

## References

[1]N.F. Mott, J. Non-Cryst. Solids 1, 1 (1968).

[2]P.W. Anderson, Comments on Solid State Phys. II, 193 (1970).

[3]M.W. Shafer, T. Penney, K.A. Müller, and R. Figat, to be presented at the Intl. Conf. on Crystal Growth, San Diego, 1981.

[4]Yu. N. Belyaninov, V.S. Grunin, Z.N. Zonn, V.A. Ioffe, I.B. Patrina, and I.S. Yanchevskaya, Phys. Status Solidi (a) 27, 165 (1975).

[5]This was checked by comparing the intensity of the $V^{4+}$, $m_I = 7/2$ hyperfine line with those of impurity $Mn^{2+}$ ions present on $Ca^{2+}$ sites. Care was taken not to saturate either line.

[6]V. Ambegaokar, B.I. Halperin, and I.S. Langer, Phys. Rev. 4, 2612 (1971).

[7]R.J. Elliott, Phys. Rev. 96, 266 (1954).

# MODELS OF DISORDERED SYSTEMS

Scott Kirkpatrick
IBM Research
Yorktown Heights, N.Y. 10598 USA

Model calculations have been useful in the study of disordered systems for getting at questions of principle and underlying mechanism as well as for understanding the details of specific experiments and materials. I will describe several recent calculations which attempt to address such issues in the theories of electron localization and spin glasses. The questions addressed are:

1) Are there singular features in the one-electron density of states at the energy where the mobility vanishes?

2) What is the nature of the ordered phase in the simplest (infinite-ranged interactions) model of a spin glass?

3) How does the spin glass state in 3D differ from this?

4) Is frustration the essential microscopic mechanism for the group of phenomena we associate with spin glasses?

## Density of states at mobility edges.

As is customary, I treat the Anderson model Hamiltonian for a tight-binding band of non-interacting spinless electrons,

$$\mathcal{H} = \sum_{i} \varepsilon_i c_i^+ c_i + \sum_{<ij>} V\left(c_i^+ c_j + c_j^+ c_i\right) , \tag{1}$$

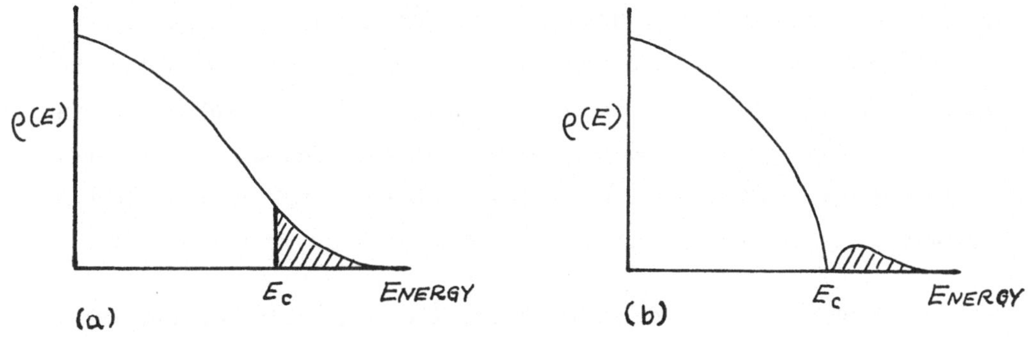


Fig. 1 (a) Conventional picture of the density of states in an energy band tail.
(b) Proposed behavior when $\rho(E)$ has a singularity at the mobility edge, $E_c$.

where the $\varepsilon_i$ are random energies distributed as

$$\rho(\varepsilon_i) \begin{array}{ll} = W^{-1} & -\frac{W}{2} \le \varepsilon_i \le \frac{W}{2}, \\ = 0 & \text{otherwise} . \end{array} \tag{2}$$

The conventional view (Mott and Davis, 1971) of the densities of extended and localized states in this model is expressed in Fig. 1a. The density of states is usually assumed to decrease smoothly into the band tail regions, crossing the mobility edge at $E_c$ without any perceptible change or structure at $E_c$.

An alternate view which has been advanced by several groups recently (Haydock, 1981a,b; Harris and Lubensky, 1981) and has been developed in this conference in the lectures of Prof. Lubensky, is given in Fig. 1b. In this latter view, the density of extended states vanishes at $E_c$,

$$\rho(E) \propto |E - E_c|^{\beta}, \tag{3}$$

with a characteristic exponent $\beta$ which need not take the conventional (or Van Hove) value of $\beta = d/2-1$. The localized states are described by a different function which vanishes with an essential singularity as E approaches $E_c$ from outside the mobility edge.

Lubensky and Harris's prediction is obtained using field theoretic techniques and an expansion valid only between 4 2/3 and 8 dimensions. A density of states like that of Fig. 1b has also been predicted for the 3D Anderson model by Haydock (1981b), who uses a continued fraction formalism for the single particle Green's function and obtains results which are claimed to be exact to lowest order in the scattering strength. Haydock finds $\beta = d/2-1$ and makes no specific prediction for the form of the density of localized states.

A theorem of Thouless and Edwards (Edwards and Thouless, 1971; Thouless, 1972) implies that Fig. 1b can occur only in the weak scattering limit, in which the mobility edges are close to the band edges and the band tails are small. They showed that for the distribution (2), essential singularities of the one-particle Green's function, if present, must lie within half the unperturbed band width of the energies $\pm W/2$. For $W >> zV$, which is the case in high dimensionalities when the two mobility edges come together and the Anderson transition occurs, the mobility edges lie outside these limits and cannot be accompanied by essential singularities. But for W of order a few times the band width or less, the band tails will be small and the position of the mobility edge can be estimated with reasonable accuracy to lie at the band edge position calculated in the coherent potential approximation (CPA) (Velický 1968). This gives

$$E_c \sim \pm \left| \frac{W}{2} + \frac{\alpha-1}{\alpha} zV \right| \tag{4}$$

where $\alpha$ is the value (in units of $(zV)^{-1}$) of the unperturbed Green's function at the original band edge. Since $\alpha$ is $< 1$ and decreases with increasing dimensionality, we find that for large d and $W \approx zV$ the CPA band edges lie just inside $\pm W/2$, within the band of energies to which the Thouless and Edwards theorem does not apply.

It can also be shown (Wegner, private communication) that $\rho(E)$ is strictly $> 0$ within the band for certain classes of disordered models, including (2). However, sharp structure in $\rho(E)$ which is some remnant of Fig. 1b can not be excluded by analytic arguments, so experimental results provide the definitive test.

Both the Haydock and Harris-Lubensky theories contain arguments which give cause for concern. In the field theoretic approach, the localized states are treated in an approximation which is not consistent with the treatment of the extended states, so the total number of states in the band is overestimated. In the continued fraction approach, it is not clear to me that one can identify a small parameter. The discussion below of the exponential growth of error in the closely related Lanczos procedure should make that difficulty apparent. But Fig. 1b, if true, has important consequences for transport, and should be tested. The clearest test, and the only one possible in the case of $4\ 2/3 < d \leq 8$, is computer experiment.

To determine the density of states in the band tails of some reasonable large samples of the model (1), I have used a Lanczos procedure related to the methods introduced by Licciardello and Thouless (1978) and used more recently by Stein and Krey (1980). This procedure transforms the original Hamiltonian into a tridiagonal matrix, for which it is relatively easy to extract eigenvalues. Using this method I was able to study band tails and band edges in samples of up to $8^5$ sites (in 5 dimensions).

The Lanczos procedure has very unsatisfactory numerical stability characteristics, so it is necessary to incorporate special precautions into the analysis. The procedure can be viewed as a transformation of the original problem, defined in a basis set of site orbitals, into a new basis set. One starts with an arbitrary normalized basis vector and generates each new vector by applying H to the previous basis vector and orthogonalizing it to the previous vectors in the new set. Formally, one can arrange things so that each vector need be explicitly orthogonalized only to the two preceding vectors, and orthogonality to the remaining basis vector follows automatically. The transformation at each stage has the form

$$\beta_{i+1} V_{i+1} = \mathcal{H} V_i - \alpha_i V_i - \beta V_{i-1} \tag{5}$$

where $||V_n|| = 1$. The choice of $\alpha_n$., $\beta_n$ which yields orthogonality also provides that (Edwards, 1980; Stein and Krey, 1980).

$$V_i^T \mathcal{H} V_i = \alpha_i$$

$$V_{i+1}^T \mathcal{H} V_i = \beta_{i+1}, \tag{6}$$

Thus we may interpret the matrix (6) of transformed coefficients $\alpha$ and $\beta$ as the Hamiltonian of a disordered 1 dimensional chain which has the spectrum of the original problem.

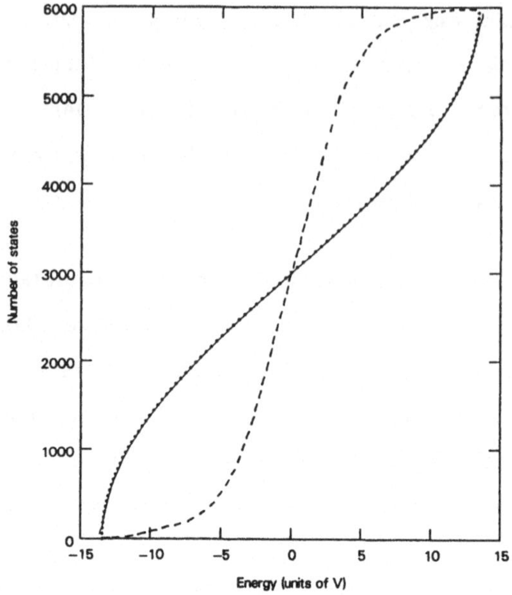


Fig. 2 Integrated density of states for a 5D Anderson model with W = 20 and $8^5$ sites after transforming to tridiagonal form using the Lanczos procedure. The dashes are the result expected for the 5D model, the solid line indicates the number of states found for a tridiagonal matrix of 6000 elements, and the dots are the result for a uniform chain with the same length and average off-diagonal matrix element.

It should be a bit surprising that a 1D chain can be constructed to represent a problem in as many as 5 dimensions, since the density of states of a 1D chain, even if disordered, doesn't look much like the density of states of the model problem. In fact, this equivalence can be made only after some interpretation. In Fig. 2, I have plotted the integrated density of states expected for a weakly disordered 5D energy band. By contrast, a 1D band has inverse square root singularities in $\rho(E)$ at its edges, giving rise to an integrated density of states like the dots in Fig. 2. In generating the dots I set the energy scale by making the hopping matrix element of a uniform 1D chain equal to the average of the elements $\beta_n$ generated in transforming a particular 5D model into tridiagonal form. Finally, in Fig. 2 the solid line indicates the integrated density of states actually obtained in the transformed tridiagonal matrix. Except in the furthest tails of the band, this is indistinguishable from the density of states of the uniform chain.

The reason the density of states comes out wrong is that the $V_n$ lose their formal orthogonality due to roundoff error in the computation. A small roundoff error is inevitable, and the resulting loss of orthogonality appears to increase (Thouless et. al.) exponentially with distance along the chain. As a result, some states are represented many times among the eigenfunctions of the 1D chain. The signal for this is that the associated eigenvalues are highly degenerate, with multiplicities of several hundred not uncommon near the band edges. Since each such eigenvalue arises from a single state of the original problem, the cure for this problem is to count degenerate eigenvalues only once. The tridiagonal matrix also has spurious eigenvalues, which are not eigenvalues of the original problem. There are ways to identify and reject these (Edwards, 1980; Cullum and Willoughby, 1980) but I took the simpler route of discarding all non-degenerate eigenvalues of the tridiagonal matrix. Following this procedure, it was possible to obtain the outermost several hundred states in the band tails after generating 6-10,000 Lanczos basis vectors.

3.

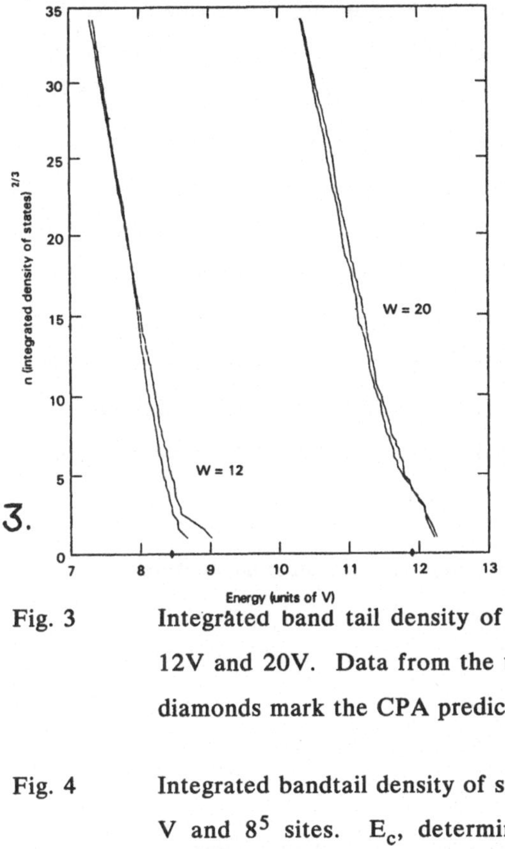


4.

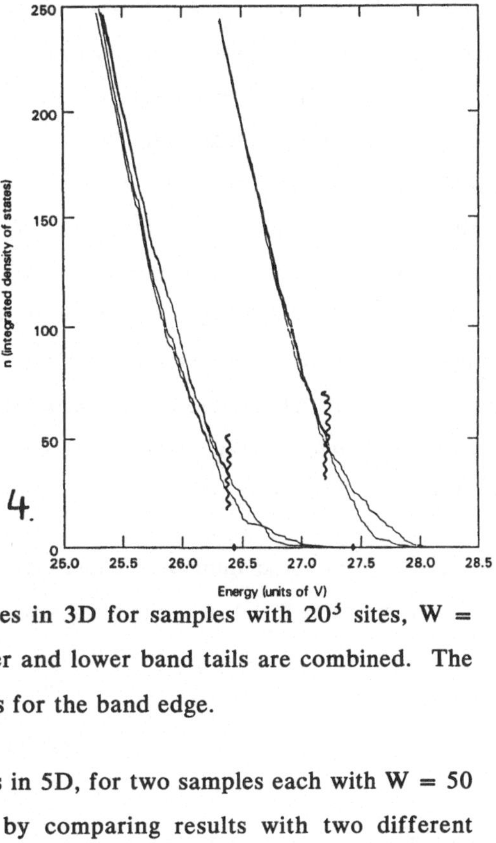


Fig. 3 Integrated band tail density of states in 3D for samples with $20^3$ sites, W = 12V and 20V. Data from the upper and lower band tails are combined. The diamonds mark the CPA predictions for the band edge.

Fig. 4 Integrated bandtail density of states in 5D, for two samples each with W = 50 V and $8^5$ sites. $E_c$, determined by comparing results with two different boundary conditions, is indicated with a wiggly line. In the second sample, all energies have been increased by V for the sake of clarity.

Some results of these experiments are presented in Figs. 3 and 4. In 3D the integrated density of states $\propto E^{3/2}$ at the band edges in the absence of disorder, so in Fig. 3 I have plotted the 2/3 power of the integrated densities of states found in two reasonably large 3D samples with W = 12V and 20V. (The Anderson transition is thought to occur at much larger values of W in 3D.) In each case, the density of states has the power law behavior expected at the band edge, except for the last few states which form a small band tail. The diamonds mark the band edges predicted by CPA in each case, and the Van Hove portion of the actual band edges extrapolates nicely to the CPA prediction in both samples. A gap or dip in the density of states would integrate into a flat spot in the curves shown in Fig. 3, but there is no evidence for such a feature.

In 5D, the largest samples which were treated had 8 sites on a side of the sample hypercube, and sample to sample variations in the characteristics of the band edge states were large. This should be expected, since the unperturbed density of states, which $\propto E^{3/2}$ in 5D, already resembles a band tail. Plots of integrated densities of states for disordered 5D models, raised to the 2/5 power, again look roughly like straight lines with the addition of small band tails. Data for two cases with W = 50V are plotted in Fig. 4. In addition to combining upper and lower bandtails in each case plotted, I have also plotted the results for two boundary conditions. Localized states can be distinguished from extended states by the fact that their energies are not affected by this change. There is a flat spot in one of the four band tails plotted in Fig. 4, but it occurs well outside the mobility edge identified by the boundary condition test. I believe that it is a statistical consequence of the small number of states in the tails, not a systematic effect.

I conclude from these numerical experiments that there are no singular features in the density of states at $E_c$, even in the weak scattering limit.

## Nature of the Ordering in the SK Model

The 1975 paper by Edwards and Anderson makes two very exciting suggestions. These are the ideas that (1) a random system may order in a random state, which exhibits no preferred direction in either real or Fourier space, and (2) that a new definition of an order parameter, calculable analytically with the help of replica methods of averaging, may nevertheless capture this type of ordering. Sherrington and Kirkpatrick (1975) introduced a model spin glass with infinite-ranged interactions (SK model) in the expectation that the model would prove soluble and would illuminate these two ideas. The SK Hamiltonian is

$$H = -\sum_{i>j} J_{ij} S_i S_i \tag{7}$$

where $<J_{ij}> = 0$ in the simplest case, $<J_{ij}^2> = (N-1)^{-1}$ for the proper scaling of all energies, and the $S_i$ are Ising spins.

The model has not been solved analytically, although it is known to have a phase transition at $T_c = 1$. There are several competing theories of the low temperature phase, which Parisi and Toulouse have described in their talks at this meeting. Peter Young and I have undertaken a numerical study of the properties of (7) to attempt to resolve these discrepancies. This work has been carried out by exact calculations on samples with N, the number of spins, finite. Properties calculated in this way must then be averaged over many different samples, and finally the extrapolation $N \to \infty$. must be understood and carried out. It is speculated that unusual analytic properties of the spin glass phase result because of the delicacy of this limit. Our calculations provide an opportunity to check this.

The Edwards-Anderson order parameter,

$$q(T) = <<S_i>_T^2>_J, \tag{8}$$

where $< >_T$ denotes thermal averaging and $< >_J$ denotes averaging over the random choice of bond values, will vanish for a finite sample unless some device, such as the imposition of a small external field, is used to single out an ordering direction. To avoid the ambiguity inherent in such devices, we work instead with a higher-order quantity,

$$q^{(2)} \equiv <<S_i S_j>_T^2>_J, \tag{9}$$

which is non-negative in the absence of applied fields. Note that $q^{(2)}$ also enters the internal energy, U, as evidence of frustration, since

$$<U(T)>_J = \left(1 - q^{(2)}(T)\right)/2T. \tag{10}$$

Above $T_c$, $q^{(2)}$ will vanish as $N \to \infty$. From (10) one sees that it must tend to 1 with corrections linear in T as $T \to 0$, in order for the ground state energy to be finite and non-zero.

For most of the numerical work, we specialize to the distribution in which $J_{ij} = \pm(N-1)^{-1/2}$, in order to reduce the task of evaluating the partition function at all temperatures and external fields to the simpler process of tabulating the number of

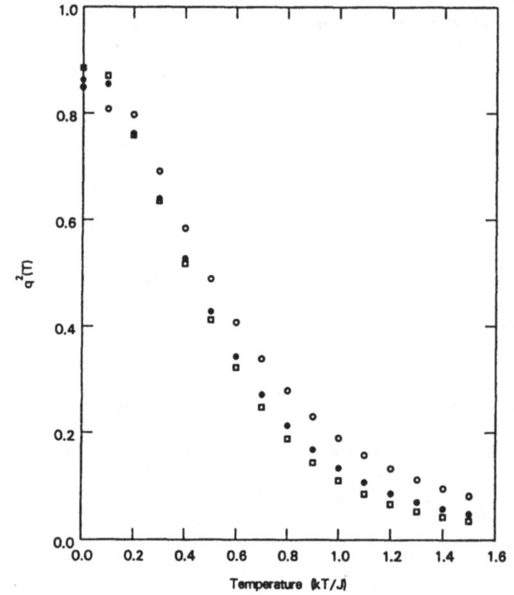


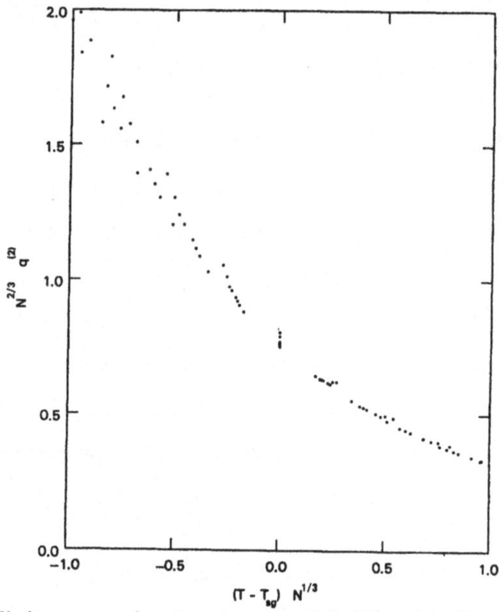


Fig. 5 Order parameter $q^{(2)}(T)$ for the infinite ranged spin glass model (7) with N spins. Here N = 20 (squares), 14 (dots), and 8 (circles).

Fig. 6 Data like that of Fig. 5 for samples with N = 5, 7, 9, 12, 14, 16, and 20 are rescaled as suggested in (13) to extract the scaling function.

configurations with a particular energy and total moment. Samples with as many as 24 spins could be treated in this way, and accurate evaluation of the $< >_J$ was possible.

Results for $q^{(2)}$ as a function of T for three sample sizes are shown in Fig. 5. At most temperatures, $q^{(2)}$ decreases with increasing N, more rapidly so at the higher temperatures, but at the lowest temperatures, $q^{(2)}$ increases with increasing N. This is in marked contrast with the results of Morgenstern and Binder (1979) on 2D and 3D Ising spin glass models, where the degree of correlation decreased with increasing sample size even at T = 0.

To interpret the results of Fig. 5, we need a theory of the size dependence of $q^{(2)}$. Above $T_c$, one expects that

$$q^{(2)}(t) \sim (Nt)^{-1}, \quad \text{where } t = \frac{T - T_c}{T_c}, \tag{11}$$

since $q^{(2)}$ is a susceptibility. At and below $T_c$, most of the analytic theories predict that

$$q^{(2)}(t) \sim t^2, \tag{12}$$

A scaling form fitted to these two limits is

$$q^{(2)}(N,t) \sim N^{-2/3} f\left(tN^{1/3}\right) \tag{13}$$

Replotting the data of Fig. 5 to extract the scaling function assumed in (13) we find in Fig. 6 a reasonably good fit to the assumed size and temperature dependence. Slight modifications to (13) can give even better fits, but they do not seem justified when there are higher-order corrections to the temperature dependence (12) which become important at the lower temperatures plotted.

In analyzing the behavior of $q^{(2)}$ at low temperatures we find that the size-dependence is $\propto N^{-1/2}$. Thus Fig. 7 shows that the overlap between two degenerate ground states (not differing by only a reversal of all the spins) tends to unity as $N^{-1/2}$. If different ground states were unrelated in direction, the overlap would instead tend to zero as $N^{-1/2}$, so Fig. 7 is evidence that there is indeed a unique ordered state in the SK model, with variations about it limited to a small number of spins.

The same $N^{-1/2}$ dependence is seen in other low-temperature properties and can in some cases be derived by analytic arguments. For example, the remanent moment at $T = 0(\lim_{h \to 0} m(h))$ can be shown to be (Young and Kirkpatrick, 1981)

$$N^{-1} << | \sum_i S_i | >_T >_J = (2/\pi N)^{1/2}. \tag{14}$$

Similar dependences are found for the overlap between ground states and their lowest-lying excited states. These differ not by the reversals of individual spins, but by changes in the orientation of small clusters of spins.

Finally, Figs. 8a-8c give some examples of the field dependance of the magnetization in samples with $N = 20$. Results for six samples are shown because the variation from sample to sample is so extreme. At the lowest temperature the magnetization simply passes through a series of steps as different configurations become the ground state when the field increases in strength. Increasing temperature smooths out these steps. The typical height and width of a step should be of order $N^{-1/2}$. This is the behavior conjectured by Parisi

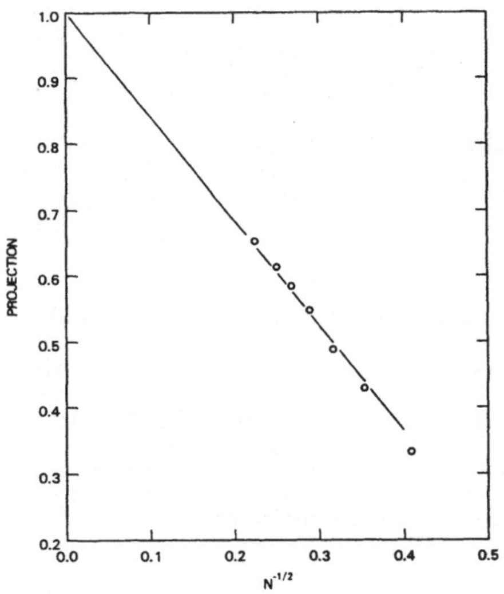


Fig. 7 Average overlap of two distinct ground states of the same SK model, for N = 6 to 20.

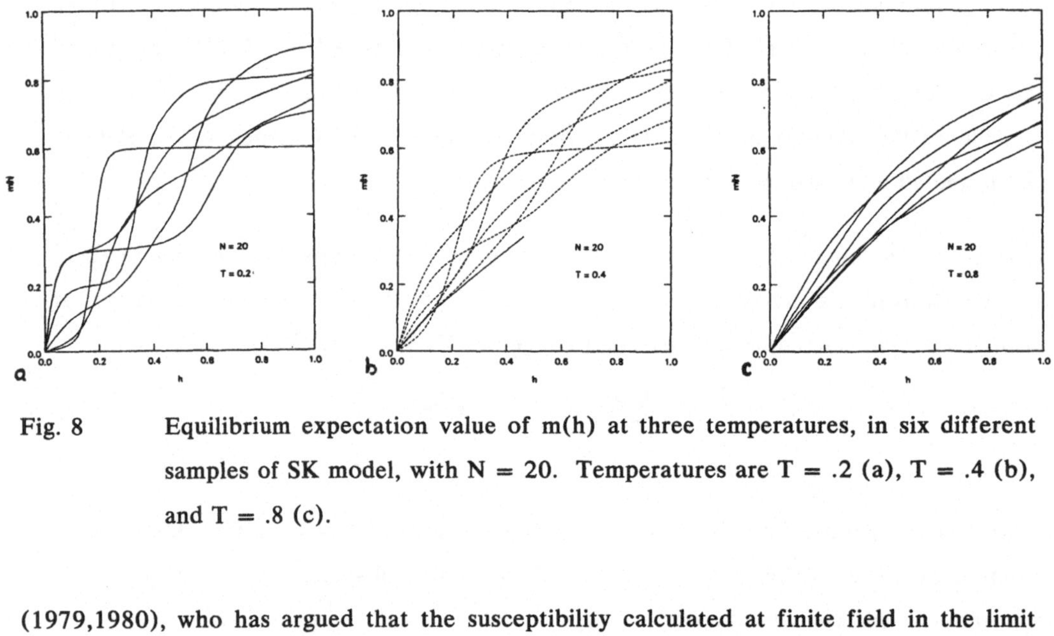


Fig. 8 Equilibrium expectation value of m(h) at three temperatures, in six different samples of SK model, with N = 20. Temperatures are T = .2 (a), T = .4 (b), and T = .8 (c).

(1979,1980), who has argued that the susceptibility calculated at finite field in the limit $N \to \infty$ is constant below $T_c$, while the usual zero-field limit of the susceptibility, calculated using the fluctuation-dissipation theorem, is zero at T = 0.

In the three cases shown in Fig. 8 the envelope of m(h) narrows with increasing temperature at finite field, but the center of that envelope does not shift appreciably, in rough agreement with Parisi's ideas. However, quantitative comparison of m(h) in small samples with the predictions of Parisi and those of Parisi and Toulouse (1980, see also Vannimenus,

1981) for larger fields (see Fig. 8b) will require further analysis of the size dependence of the averaged magnetization. This work is in progress.

## Pinning and Degeneracy in 3D Spin Glasses

Careful numerical evaluation (Morgenstern and Binder, 1979, Morgenstern and Horner, this conference) of correlation functions for small samples of Ising spin glasses with near neighbor interactions have made it clear that there is no long ranged order in 2D, even at T = 0. In 3D, the evidence is not as convincing but the calculated spin correlations again do not appear long ranged. To develop some qualitative ideas about why the 2D and 3D systems are so different from the SK model and the mean field theory introduced by Edwards and Anderson, I have studied the Ising model with random sign, uniform strength bonds on a simple cubic (3D) lattice. First I shall describe the effects of frustration in this model with no disorder, then introduce disorder by randomizing the location of the negative bonds and consider the modifications which this introduces.

An arrangement of ferro- and antiferro-magnetic interactions which produces a totally frustrated model with full cubic symmetry is shown in Fig. 9. (The f.c.c. Ising antiferromagnet is also totally frustrated, but gives rather different results than those I obtain for the simple cubic totally frustrated model. Villain in his contribution discusses the f.c.c. model.)

The ground state of the simple cubic model is highly degenerate, although the entropy still vanishes as T -> 0. The ferromagnetic spin arrangement is one ground state. From it, one can form other states of the same energy by reversing the signs of all spins in a linear chain extending across the system, as shown in Fig. 9. Since there are $L^2/4$ such chains in each direction for a system of $L^3$ sites, and each chain in a given direction can be reversed independently, there are more than $3 \times 2^{L^2/4}$ ground states available. Each ground state has quasi-1D excitations in which portions of these chains are reversed, by the generation of a kink-antikink pair. The energy of these excitations is independent of the separation between the kink and its partner. Therefore it seems reasonable to expect that the model has quasi-1D thermodynamic properties and possibly no transition at all.

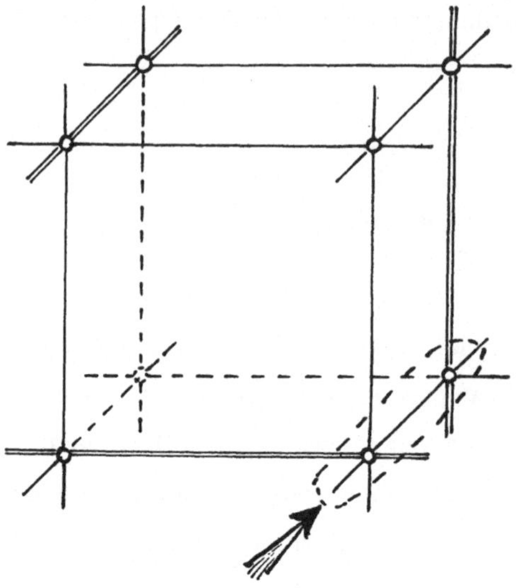

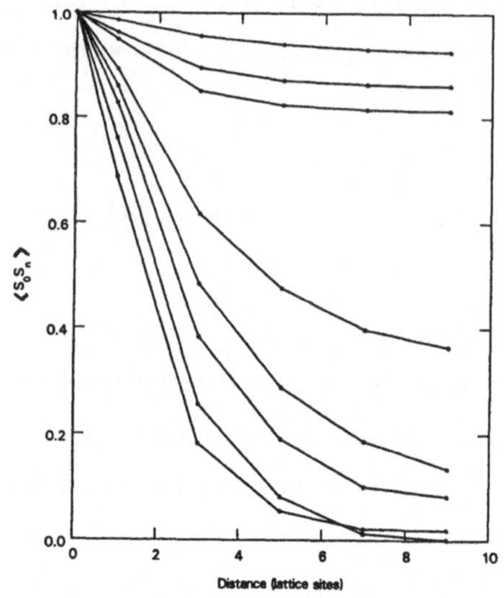


Fig. 9 Unit cell of an Ising model on a 3D simple cubic lattice which has all of its plaquettes frustrated. The double lines are antiferromagnetic interactions, and hidden lines are dashed. The arrow indicates a chain of spins which can be reversed without energy cost.

Fig. 10 Evidence for long-range correlations in the frustrated model of Fig. 9. $<S_0 S_n>$ is plotted for a $20^3$ site sample in which the spins in the surface layer 0 are held fixed in a ferromagnetic alignment.

However, the model has a rather well-defined freezing temperature at about T = 1.25 J. The spin correlations which freeze in are rather complicated because of the high degeneracy. One way to simplify them is to leave one face of a finite system aligned ferromagnetically, then observe the decay of the resulting alignment into the bulk. This decay is plotted in Fig. 10. Below T = 1.2 J it appears that the influence of the ferromagnetic surface becomes long-ranged.

Even stronger evidence for a continuous phase transition with a critical point comes from the specific heat observed in Monte Carlo calculations on a wide range of system sizes. Fig. 11 shows that the specific heat maximum increases by roughly equal amounts for each factor of 2 in the linear dimension of the system studied. This size dependence is as expected for a critical point with $\alpha \approx 0$. The critical behavior is removed by disorder,

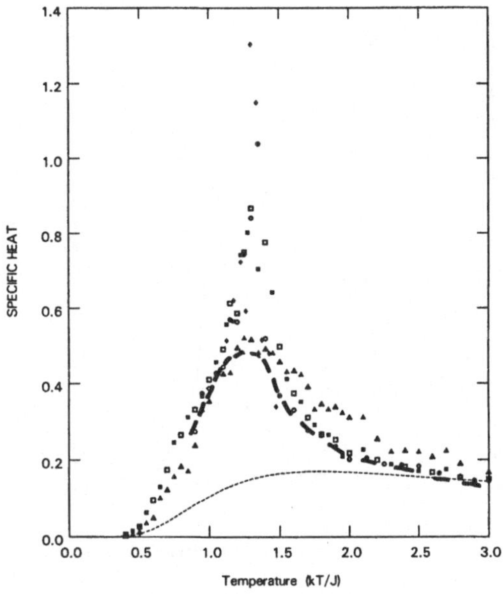


Fig. 11Specific heat for the 3D totally frustrated Ising model, plotted for several sample sizes. The sizes are, in order of increasing peak height, $2^3$, $4^3$, $8^3$, $20^3$, and $30^3$, indicated by a short-dashed line, triangles, squares, circles, and diamonds, respectively. The solid data points are dU(T)/dT; the open data points were obtained from the magnitude of energy fluctuations. The dashed line indicates C(T) for a disordered sample of $16^3$ spins, with 5 per cent of the bonds, selected at random, reversed in sign.

however. When a few per cent of the bonds are reversed in sign, as shown by the dashed line in Fig. 11, there is no singularity in C(T).

Although the order parameter is not known for this highly degenerate model, there is evidence in the diffraction which sheds some light on the type of order present. Fig. 12 shows S(q) for the totally frustrated model above and at its transition temperature. Scattering peaks which are much weaker than Bragg peaks from conventional ordering but still about 200 times the background intensity are found at each corner of the cubic Brillouin zone at $T_c$. There is no evidence for the formation of structure incommensurate with the lattice. Disorder again destroys this sharp and regular structure (Fig. 12b).

Notice that although the diffraction intensities showed cubic symmetry in the ordered model, the results for 10 per cent reversed bonds differ along different directions in the cube. This is evidence for a coarse domain structure in the disordered model. The domains occur because the effect of changing the sign of bonds and destroying the arrangement which led to perfect frustration is to pin together the phases of two or more of the infinite chains of spins which connect different ground states. From study of systems

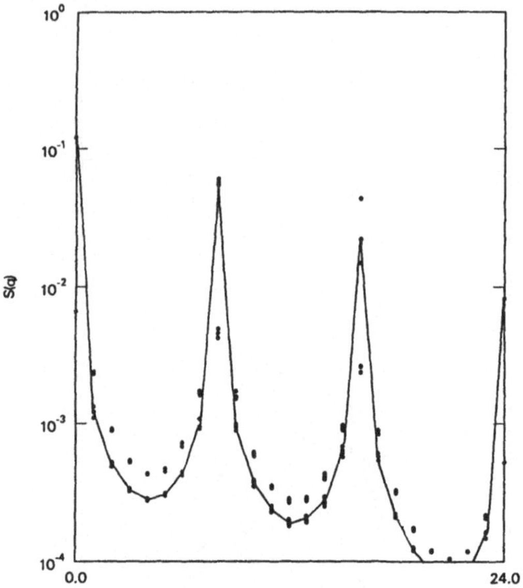


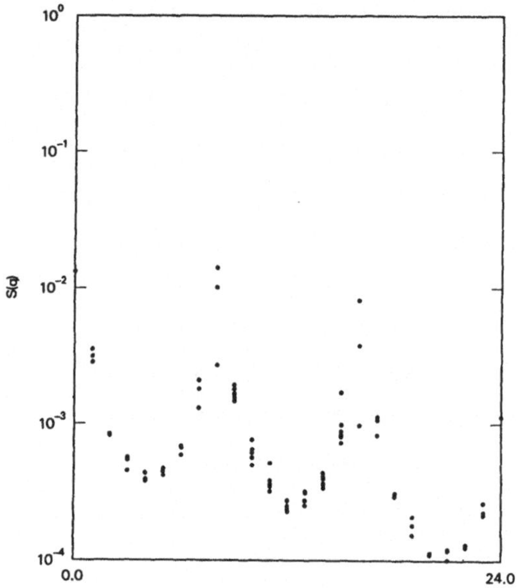


Fig. 12 Equal-time spin correlations, S(q), along the edges of the cubic Brillouin zone of the totally frustrated simple cubic Ising model, for a system with $16^3$ sites. Averages over 40,000 MCS were taken for each point plotted.

(a) Results for T = 1.25J are connected with solid lines, and those for T = 1.5 J are also shown.

(b) S(q) for a disordered model with x = 0.1, and T = 1.25J. Points were not connected because of the domain-induced scatter.

with quasi-1D charge density waves in the presence of pinning centers, we understand that perturbations which couple to objects which are infinite in spatial extent can change the nature of a critical point at arbitrarily small concentrations. That is what appears to be happening in this model as disorder is added.

The analogy to CDW systems with pinning has another consequence for spin glasses. Pinning centers have been shown to be like random external fields in their effect on CDW systems, and such fields will raise the lower critical dimensionality of a model, below which there can be no long ranged order. For a 3D Ising model with random external fields on every site, one presently expects either no long ranged order, or a transition at a finite temperature into a phase with power law decay of spin correlations (Pytte, Imry and Mukamel, 1981).

The analogy becomes less compelling at high concentrations of reversed sign bonds, when the chain-like excitations are probably broken up into the "clusters" of spins conventionally used to describe low energy excitations in spin glasses. However, the conclusion that disorder changes the lower critical dimension of a frustrated system would seem to retain its force beyond low concentrations, since it is difficult to imagine regaining some simpler sort of long range order through additional disorder. At the least, this analogy should serve as a warning of the danger in describing a 3D spin glass by mean field theory, or by appeals to results for the SK model.

## The Statistical Mechanics of a Travelling Salesman

The final model has many characteristics of a spin glass. In particular, it exhibits metastability below a freezing temperature which is not accompanied by any singularity in the specific heat or evidence of long ranged order. Although somewhat polymer-like, the model is drawn not from physics, but from optimization theory. I will describe the statistical mechanics of the Hamiltonian circuits passing through N fixed, randomly placed, points in a square region.

Finding the shortest such circuit for a given set of points is commonly called the "travelling salesman problem," with the points viewed as cities to be visited once each on the salesman's tour. This is an "NP-hard" problem (Aho, Hopcroft, and Ullman, 1974). As far as is known, one cannot obtain a provably minimal length tour connecting all N points without doing at least of order exp(N) computations. Similarly, one can show that finding the ground state of an Ising spin glass, that is, a Hamiltonian of the form

$$\mathcal{H} = -\sum_{<ij>} J_{ij} S_i S_j ,$$

where the $J_{ij}$ are generated at random, is NP hard in general, but restrictions on H may make easier solution possible.

Several authors have noted that questions about spin glasses may belong in the widely studied class of NP-complete combinatorial problems, and have conjectured that this may imply something about the ordered states of spin glasses (Palmer, 1980, Biéche, 1980,

Barahona, 1981). I will take a different tack here, and demonstrate that by considering all configurations of these optimization problems, not just the optimum solution, one obtains statistical models with the characteristics of spin glasses. We then observe that the common elements of NP-hard problems provide a generalization of the concept of microscopic frustration.

The objective function (quantity being minimized) in an optimization problem is analogous to the energy in statistical mechanics. In the case of the travelling salesman problem, the path length is the objective function, so a partition function for the problem is

$$Z = \mathrm{Tr} \exp(-\beta L), \tag{15}$$

where the trace is a sum over the N! permutations of the order in which the path passes through each site. One can show (Beardwood, 1959) that the minimum total path length is $\propto N^{1/2}$, so I shall express $L = \alpha N^{1/2}$, and all temperatures in units of $N^{-1/2}$, the average separation between nearest neighbor points.

A Monte Carlo program can be written to sample equilibrium configurations of a travelling salesman problem if we have some procedure for generating rearrangements of the path (the usual "moves" of a Monte Carlo program), and use the Metropolis rule (see Binder, 1979 for references) for accepting or rejecting each rearrangement. I have considered as moves all rearrangements made by cutting out a subsequence of sites along the path and inserting them between two other sites on the path, possibly reversing their sequence as well. These moves are the basis of Lin's deterministic (and zero-temperature) algorithm (Lin, 1965), which is usually successful in finding the shortest tours among up to $\approx$ 100 points, but fails for larger problems and becomes too time-consuming to be of practical value.

In Fig. 13a is shown a typical high-temperature configuration of the problem. Its length is about twice the optimal value, and equilibrium was quickly reached at this temperature. At half this temperature, the system begins to freeze, and equilibration times increase drastically. Such a configuration is shown in Fig. 13b. Finally, I show a nearly optimal solution in Fig. 13c, obtained by slowly cooling to a temperature well below the freezing temperature. There are many such near-optimal solutions, with different ones found on different cooling runs, just as different spin glass ground states are reached by different thermal histories.

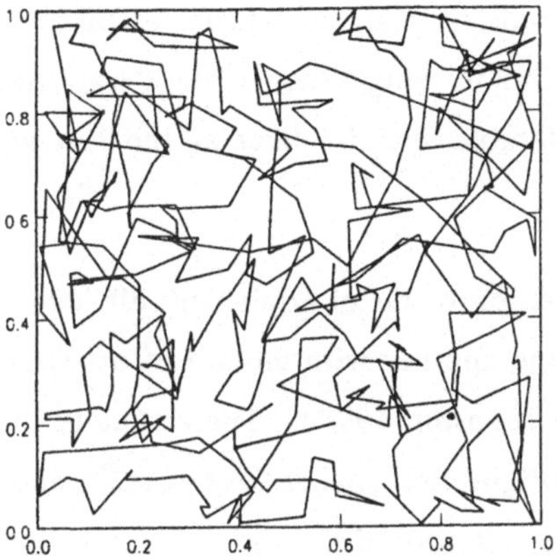


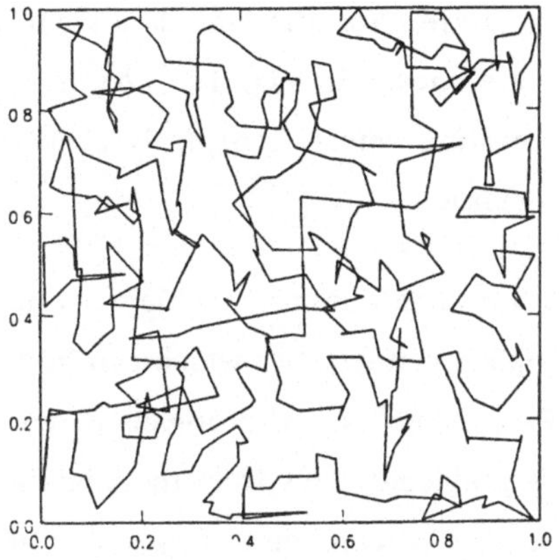


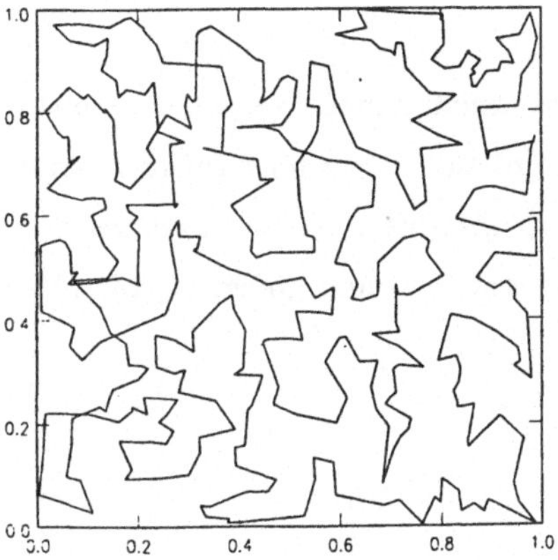


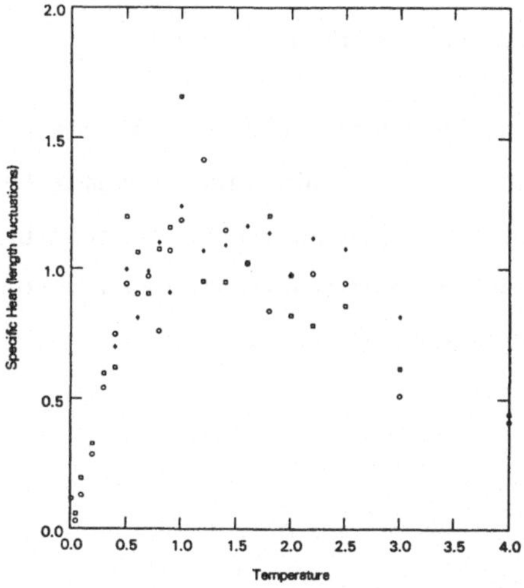


Fig. 13 A travelling salesman problem with 200 "cities" in a square region is pictured at three temperatures: (a) T = 1 and the length = 2.05 $N^{1/2}$; (b) T = .5 and length = 1.38 $N^{1/2}$; (c) T = .1 and length = 1.034 $N^{1/2}$.

Fig. 14 Specific heat determined from length fluctuations in equilibrium for two travelling salesman problems with 100 cities each (open data points) and one with 200 cities (diamonds).

One measure of the number of configurations lying just above the optimal solution is the specific heat, defined here as the derivative of the average length of equilibrium configurations with respect to temperature. By the usual fluctuation-dissipation theorem, it is given through the definition (15) in terms of the variance of the lengths observed at a given temperature. Specific heat data from the Monte Carlo runs on samples with N = 100 and 200 are shown in Fig. 14. As in spin glasses, we find no singularity in C(T) at the temperature where freezing is observed, and a considerable specific heat at the lowest temperatures.

The travelling salesman problem is one of several hundred problems of equivalent difficulty in applied mathematics which can be transformed into one another with modest effort (the "NP-complete" set of problems reviewed by Garey and Johnson, 1979). The difficulty of these problems stems from the combination of large configuration spaces (at least $2^N$, and sometimes N! configurations) and conflicting constraints, often on different scales. In the travelling salesman problem the conflict is between the short range requirement that each step of the path be as short as possible and the long range requirement that every point be visited once and the path be closed.

Frustration defined microscopically by local interactions which cannot be simultaneously satisfied by any configuration seems to be a special case of this. One physical realization of the more generalized constraints which may lead to glassiness in real systems is a conflict between local interactions and conservation laws, such as those governing the total moment along a particular axis.

**Acknowledgments**

The calculation of band tail states arose from a discussion with Dan Fisher.

Work on the infinite range model of an Ising spin glass has been carried out with Peter Young. Parts of it have been presented at the Nov. 1980 MMM conference, and appear in those proceedings. Further work on the problem is being submitted for publication.

For the suggestion of working on models with frustration, then adding disorder, I am indebted to Gerard Toulouse. Early parts of those calculations were carried out in collaboration with Kurt Binder. Discussions with C. Jayaprakash, Eduardo Fradkin, and David Mukamel are also gratefully acknowledged.

Finally, the Monte Carlo method for solving travelling salesman problems is one of several applications of the concepts and methods of statistical mechanics to optimization problems which I am presently exploring in collaboration with Dan Gelatt.

**References**

A. Aho, J. Hopcroft and R. Ullman 1974, "The Design and Analysis of Computer Algorithms", (Addison-Wesley, Reading, MA).

F. Barahona 1981, preprint.

J. Beardwood, J. H. Halton and J. M. Hammersley 1959, Proc. Canad. Phil. Soc. **55**, 299.

J. Biéche, R. Maynard, R. Rammal and J. P. Uhry 1980, J. Phys. **A13**, 2553.

K. Binder 1979, "The Monte Carlo Method - Statistical Mechanics", (Springer-Verlag, Hamburg).

J. Cullum and R. A. Willoughby 1979, Proc. IEEE Conference on Decision and Control, p. 45 (IEEE, New York).

J. T. Edwards, D. C. Licciardello and D. J. Thouless 1980, J. Inst. Math. Appl., to appear.

S. F. Edwards and P. W. Anderson 1975, J. Phys. **F5**, 965.

M. L. Garey and D. L. Johnson 1979, "Computers and Intractility, A Guide to the Theory of NP Completeness", (W. H. Freeman, San Francisco).

A. B. Harris and T. C. Lubensky 1981, preprints.

R. Haydock 1981a, J. Phys. C **14**, 229;
R. Haydock 1981b, Philosophical Magazine B, to appear.

D. C. Licciardello and D. J. Thouless 1978, J. Phys. **C11**, 925.

S. Lin 1965, Bell Syst. Tech. Journ., Dec. 1965, 2245.

I. Morgenstein and K. Binder 1979, Phys. Rev. Lett. **43**, 1615.

N. F. Mott and E. A. Davis 1971, "Electronic Processes in Non-Crystalline Materials", (Clarendon Press, Oxford).

R. G. Palmer 1980, talk at STATPHYS 14.

G. Parisi 1979, Phys. Rev. Lett. **43**, 1754.

G. Parisi 1980, J. Phys. **A13**, 1887.

G. Parisi and G. Toulouse 1980, J. Physique Lett. **41**, L-361.

E. Pytte, Y. Imry and D. Mukamel 1981, Phys. Rev. Lett. **46**, 1173.

D. Sherrington and S. Kirkpatrick 1975, Phys. Rev. Lett. **35**, 1792.

J. Stein and V. Krey 1980, Z. Phys. **B37**, 13.

J. Vannimenus, G. Toulouse and G. Parisi 1981, J. Physique Lett. **42**, 565.

B. Velicky, S. Kirkpatrick and H. Ehrenreich 1968, Phys. Rev. **175**, 747.

A. P. Young and S. Kirkpatrick 1981, in preparation.

## ADDRESS LIST OF PARTICIPANTS


- AHARONY A., Department of Physics & Astronomy - Tel Aviv University
  RAMAT-AVIV, TEL-AVIV (Israel)

- AOKI H., Cavendish Laboratory
  Madingley Road - CAMBRIDGE CB3 OHE (England)

- APEL W., Institut für Theoretische Physik - Universität Hannover
  Appelstrasse 2 - 3000 HANNOVER 1 (Germany)

- AUBRY S., Laboratoire Léon Brillouin - Orme des Mérisiers
  B.P. 2 - 91191 GIF-SUR-YVETTE (France)

- AUSLOOS M., Institut de Physique B5 - Université de Liège
  4000 LIEGE (Belgium)

- BAERISWYL D., Laboratoires RCA Ltd.
  Badenrstrasse 569 - 8048 ZUERICH (Switzerland)

- BARACCA A., Istituto di Fisica Teorica - Università degli Studi
  Largo E.Fermi, 2 - 50125 FIRENZE (Italy)

- BEAL-MONOD M.T., Laboratoire de Physique des Solides - Université de Paris-Sud
  Bat. 510 - 91405 ORSAY (France)

- BERNASCONI J., Brown Boveri Research Center
  5405 BADEN (Switzerland)

- BEVAART L., Neutron Physics Department - SCK/CEN
  2400 MOL (Belgium)

- BIANCONI A., Istituto di Fisica "G.Marconi" - Università degli Studi
  Piazzale Aldo Moro, 2 - 00185 ROMA (Italy)

- BRAY A.J., Department of Theoretical Physics - The University
  MANCHESTER M 13 9 PL (England)

- BROUERS F., Freie Universität Berlin
  FB 20, WE 5, Arnimallee 3 - 1000 BERLIN 33 (Germany)

- BOCCARA N., Service de Physique des Solides - Centre d'Etudes Nucléaires de
  Saclay - B.P. 2 - 91190 GIF-SUR-YVETTE (France)

- CAPIZZI M., Istituto di Fisica "G.Marconi" - Università degli Studi
  Piazzale Aldo Moro, 2 - 00185 ROMA (Italy)

- CARACCIOLO S., Laboratoire de Physique del'ENS
  24 rue Lhomond - 75231 PARIS Cedex 05 (France)

- CASTELLANI C., Istituto di Fisica "G.Marconi" - Università degli Studi
  Piazzale Aldo Moro, 2 - 00185 ROMA (Italy)

- CILIBERTO S., Istituto di Fisica - Università degli Studi
  Largo E.Fermi, 2 - 50125 FIRENZE (Italy)

- CLERC J.P., Université de Provence - Departement de Physique des Systèmes
  Désordonnés - Centre St. Jerôme
  13397 MARSEILLE Cedex 4 (France)

| | |
|---|---|
| - CLIPPE P., | Institut de Physique B5 - Université de Liège<br>4000 LIEGE (Belgium) |
| - CONIGLIO A., | Istituto di Fisica Teorica - Università degli Studi<br>Mostra d'Oltremare, Pad. 19 - 80125 NAPOLI (Italy) |
| - CORBELLI G., | Istituto di Fisica - Università degli Studi<br>Via Campi, 213/A - 41100 MODENA (Italy) |
| - DE PASQUALE F., | Istituto di Fisica - Facoltà di Ingegneria - Università<br>Piazzale Aldo Moro, 2 - 00185 ROMA (Italy) |
| - DERRIDA B., | DPhT CEN Saclay BP n.2<br>91190 GIF-SUR-YVETTE (France) |
| - DES CLOIZEAUX J., | DPhT CEN Saclay BP n.2<br>91190 GIF-SUR-YVETTE (France) |
| - DEUTSCHER G., | Department of Physics and Astronomy - Tel Aviv University<br>RAMAT-AVIV, TEL-AVIV (Israel) |
| - DEVORET M., | DPh/SRM CEN Saclay BP n.2<br>91190 GIF-SUR-YVETTE (France) |
| - DI CASTRO C., | Istituto di Fisica "G.Marconi" - Università degli Studi<br>Piazzale Aldo Moro, 2 - 00185 ROMA (Italy) |
| - DOMANY E., | Department of Electronics -Weizmann Institute of Science<br>REHOVOT (Israel) |
| - DUPLANTIER B., | DPhT CEN Saclay BP n.2<br>91190 GIF-SUR-YVETTE (France) |
| - ESTEVE D., | DPh/SRM CEN Saclay BP n.2<br>91190 GIF-SUR-YVETTE (France) |
| - FERT A., | Laboratoire de Physique des Solides, Université Paris-Sud<br>91405 ORSAY (France) |
| - FIORANI D., | Centre de Recherçhes sur les très basses Températures<br>25 Avenue des Martyrs - BP 166 X - 38042 GRENOBLE Cedex (France) |
| - FISCHER K.H., | Institut Laue-Langevin, 156<br>38042 GRENOBLE Cedex (France) |
| - FUCITO F., | Istituto di Fisica "G.Marconi" Università degli Studi<br>Piazzale Aldo Moro, 2 - 00185 ROMA (Italy) |
| - FONTANA M., | Istituto di Fisica - Università degli Studi<br>43100 PARMA (Italy) |
| - GABAY M., | Laboratoire de Physique des Solides, Université Paris-Sud<br>91405 ORSAY (France) |
| - GRIFFIN W.G., | Department of Theoretical Physics - University of Oxford<br>1, Keble Road - OXFORD OX1 3NP (England) |
| - GUNN M., | (T.C.M.) Cavendish Laboratory<br>Madingley Road - CAMBRIDGE (England) |

- HENTSCHEL G., Weizmann Institute of Science
REHOVOT (Israel)

- HERTZ J., Nordita
Blegdamsvej, 17 - 2100 COPENHAGEN Ø (Denmark)

- HIKAMI S., Research Institute for Fundamental Physics
Kyoto University - KYOTO 606 (Japan)

- HOCHLI U.T., IBM Research
8803 RÜSCHLIKON (Switzerland)

- HOUGHTON A., Physics Dept - Brown University
PROVIDENCE , Rhode Island 02912 (U.S.A.)

- HIPPERT F., Laboratoire de Physique des Solides - Université Paris-Sud
91405 ORSAY (France)

- JOHNSTON R., Laboratoire de Physique Théorique - E.P.F.L.
14, Avenue Eglise Anglaise - 1006 LAUSANNE (Switzerland)

- JONES K., Editorial Office of "Nuclear Physics" c/o Nordita
Blegdamsvej 17 - 2100 COPENHAGEN Ø (Denmark)

- KANEYOSHI T., Department of Physics - Nagoya University
NAGOYA 464 (Japan)

- KATSURA S., Department of Applied Physics - Tohoku University
SENDAI (Japan)

- KHURANA A., Nordita
17 Blegdamsvej - 2100 COPENHAGEN Ø (Denmark)

- KINZEL W., KFA-IFF
Postfach 1913 JULICH(Germany)

- KIRKPATRICK S., IBM Thomas Watson Research Center P.O.Box 218
YORKTOWN HEIGHTS, N.Y. 10598 (U.S.A.)

- KOUVEL J., Department of Physics - University of Illinois
CHICAGO, Illinois 60680 (U.S.A.)

- KREY U., Fachbereich Physik der Universität
8400 REGENSBURG (Germany)

- LAGE E.J., Dept. of Physics - Faculty of Science - University of Porto
4000 PORTO (Portugal)

- LE GUILLOU J.C., Université P. et M. Curie,Laboratoire de Physique Théorique et
Hautes Energies, Tour 16 - 75230 PARIS Cedex 05 (France)

- LEROUX HUGON P., Groupe de Physique des Solides de l'ENS - Université Paris VII
Tour 23 - 2, Place Jussieu - 75221 PARIS Cedex 05 (France)

- LIVI R., Istituto di Fisica Teorica c/o ICTP
Strada Costiera, 11 Miramare - 34100 TRIESTE (Italy)

- LONGA L.W., Instytut Fizyki UJ - Dept. of Statistical Physics
Reymonta 4 30-059 KRAKOW (Poland)

- LOVESEY S.W., Rutherford Laboratory
Chilton, DIDCOT OX11 OQX (Oxfordsh.) (England)

- LUBENSKY T., Dept. of Physics - University of Pennsylvania
PHILADELPHIA, PA 19104 (U.S.A.)

- MacKINNON A., Referat für Theoretische Physik - Physikalisch-Technische
Bundesanstalt, Bundesallee 100 - 3300 BRAUNSCHWEIG (Germany)

- MacLEAN W.L., Serin Physics Lab. Rutgers University
NEW BRUNSWICK, N.J. 08903 (U.S.A.)

- MARCHESONI F., Istituto di Fisica dell'Università
Piazza Torricelli, 2 - 56100 PISA (Italy)

- MARINARI E., Istituto di Fisica "G.Marconi" - Università degli Studi
Piazzale Aldo Moro, 2 - 00185 ROMA (Italy)

- METHFESSEL S., Ruhr - Universität
Postfach 102148 - 4630 BOCHUM (Germany)

- MICKLITZ H., Institut für Experimentalphysik Lehrstuhl IV
Postfach 102148 - 4630 BOCHUM (Germany)

- MONOD P., Laboratoire de Physique des Solides - Bat 510 - Université de
Paris-Sud - 91405 ORSAY (France)

- MOORE M.A., Dept. of Theoretical Physics - University of Manchester
MANCHESTER M13 9PL (England)

- MORANDI G., Istituto di Fisica - Università degli Studi
Via Campi 213/A - 41100 MODENA (Italy)

- MORGENSTEIN I., Theor. Physik
Philosophenweg 19 - 69 HEIDELBERG (Germany)

- MUELLER K.A., IBM Research Laboratory
Säumerst 4 - 8803 RUESCHLIKON (Switzerland)

- MYDOSH J.A., Kamerlingh Onnes Laboratorium der Rijksuniversiteit
Postbus 9506 - 2300 RA LEIDEN (Netherlands)

- NOBILE A., Istituto di Fisica Teorica c/o I.C.T.P.
Strada Costiera, 11 - 34100 TRIESTE (Italy)

- ONO I.O., Dept. of Physics - Tokyo Institute of Technology
Oh-okayama, Meguro-ku - TOKYO 152 (Japan)

- OPPERMANN R., Institut für Theoretische Physik
Philosophenweg 19 - 69 HEIDELBERG (Germany)

- PAQUET D., Centre National d'Etudes des Télécomunications - Dept. SPD
196 Rue de Paris - 92220 BAGNEUX (France)

- PARISI G., Istituto di Fisica - Facoltà di Ingegneria - Università
Piazzale Aldo Moro, 2 - 00185 ROMA (Italy)

- PASSARI L., Istituto di Fisica - Università di Ferrara
Via Paradiso, 12 - 44100 FERRARA (Italy)

- PEKALSKI A., Institute of Theoretical Physics - University of Wroclaw
Cybulskiego 36 - 50-205 WROCLAW (Poland)

- PELCOVITS R., Dept. of Physics - Brown University
Box 1843 - PROVIDENCE, R.I. 02912 (U.S.A.)

- PELITI L., Istituto di Fisica "G.Marconi" - Università degli Studi
Piazzale Aldo Moro, 2 - 00185 ROMA (Italy)

- PEPPER M., Cavendish Laboratory - University of Cambridge
CAMBRIDGE CB3 OHE (England)

- PICHARD J.L., DPh/SRM - CEN SACLAY
B.P. n. 2 - 91190 GIF-SUR-YVETTE (France)

- PIETRONERO L., Brown Boveri Research Center
5405 BADEN (Switzerland)

- PUOSKARI M., Nordita
Blegdamsvej 17 - 2100 COPENHAGEN O (Denmark)

- RANNINGER J., Groupe des Transitions des Phases - CNRS BP 166
GRENOBLE Cedex (France)

- REATTO L., Istituto di Fisica dell'Università
Via Celoria, 16 - 20133 MILANO (Italy)

- RICE T.M., Bell Laboratories
MURRAY HILL, N.J. 07974 (U.S.A.)

- ROUSSENQ J., Université de Provence - Department de Physique des Systèmes
Désordonnés - Centre St-Jerôme - 13397 MARSEILLE Cedex 4
(France)

- RUFFO S., Istituto di Fisica dell'Università
Piazza Torricelli, 2 - 56100 PISA (Italy)

- SARMA G., DPh/SRM - CEN SACLAY
B.P. n. 2 - 91190 GIF-SUR-YVETTE (France)

- SHAPIR Y., Dept. of Physics and Astronomy - Tel Aviv University
RAMAT AVIV (Israel)

- SHERRINGTON D., Physics Dept. - Imperial College
LONDON SW7 2BZ (England)

- SIGNORELLI G., Istituto di Fisica "G.Marconi" - Università degli Studi
Piazzale Aldo Moro, 2 - 00185 ROMA (Italy)

- SOUILLARD B., Centre de Physique Théorique - Ecole Polytechnique
91128 PALAISEAU (France)

- SOURLAS N., Lab. de Physique Théorique de l'Ecole Normale Supérieure
24, rue Lhomond - 75231 PARIS Cedex (France)

- SPARPAGLIONE M., Istituto di Fisica dell'Università
Piazza Torricelli, 2 - 56100 PISA (Italy)

- STANLEY E., Center for Polymer Studies - Boston University
111 Cummington Street - BOSTON, MA 02215 (U.S.A.)

- STAUFFER D., Institut für Theoretische Physik - Universität
5000 KOLN 41 (Germany)

- STEPHEN M.J., Physics Dept. - Rutgers University
PISCATAWAY, N.J. 08854 (U.S.A.)

- STINCHCOMBE R.B., Dept. of Theoretical Physics
1 Keble Road - OXFORD OX1 3NP (England)

- SUZUKI M., Dept. of Physics - Faculty of Science - University of Tokyo
Bunkyo-ku, TOKYO 113 (Japan)

- SZNAJD J., Inst. for Low Temperature and Structure Research -
Polish Academy of Science - 50-950 WROCLAW 2, P.O.Box 937
(Poland)

- TABET E., Laboratorio delle Radiazioni - Istituto Superiore di Sanità
Viale Regina Elena, 299 - 00161 ROMA (Italy)

- TERZI N., Istituto di Fisica dell'Università
Via Celoria, 16 - 20133 MILANO (Italy)

- TOULOUSE G., Laboratoire de Physique de l'ENS
24, Rue Lhomond - 75231 PARIS Cedex 05 (France)

- UENO Y., Institut de Physique Théorique - Université de Lausanne
1015 LAUSANNE (Switzerland)

- UZELAC K., Laboratoire de Physique des Solides - Université Paris-Sud
Bat. 510 - 91405 ORSAY (France)

- VAN HEMMEN J.L., Universität Heidelberg, SFB 123
Im Neuenheimer Feld 294 - 6900 HEIDELBERG (Germany)

- VANNIMENUS J., Physique des Solides - Ecole Normale Supérieure
24, rue Lhomond - 75231 PARIS Cedex 05 (France)

- VICSEK T., Dept. of Physics - King's College, Strand
LONDON WC2R 2LS (England)

- VERSTRAETEN G., Dept. of Physics - University of Antwerpen
ANTWERPEN (Belgium)

- VILLA M., Istituto di Fisica "A.Volta" - Università degli Studi
27100 PAVIA (Italy)

- VILLAIN J., DRF/DN/CENG/85x
38041 GRENOBLE Cedex (France)

- VITICOLI S., "Laboratorio Teoria e Struttura" - Area della Ricerca di Roma
Via Salaria km 29,500 - C.P. n.10 - 00016 MONTEROTONDO STAZIONE
(Roma - Italy)

- WEGNER F.J., Institut für Theoretische Physik - University of Heidelberg
Philosophenweg 12 - 6900 HEIDELBERG (Germany)

- ZANNETTI M., Istituto di Fisica - Università di Salerno
84100 SALERNO (Italy)

- ZIEGLER K., Institut für Theoretische Physik - University of Heidelberg
Philosophenweg 12 - 6900 HEIDELBERG (Germany)

# Lecture Notes in Physics

Vol. 114: Stellar Turbulence. Proceedings, 1979. Edited by D. F. Gray and J. L. Linsky. IX, 308 pages. 1980.

Vol. 115: Modern Trends in the Theory of Condensed Matter. Proceedings, 1979. Edited by A. Pekalski and J. A. Przystawa. IX, 597 pages. 1980.

Vol. 116: Mathematical Problems in Theoretical Physics. Proceedings, 1979. Edited by K. Osterwalder. VIII, 412 pages. 1980.

Vol. 117: Deep-Inelastic and Fusion Reactions with Heavy Ions. Proceedings, 1979. Edited by W. von Oertzen. XIII, 394 pages. 1980.

Vol. 118: Quantum Chromodynamics. Proceedings, 1979. Edited by J. L. Alonso and R. Tarrach. IX, 424 pages. 1980.

Vol. 119: Nuclear Spectroscopy. Proceedings, 1979. Edited by G. F. Bertsch and D. Kurath. VII, 250 pages. 1980.

Vol. 120: Nonlinear Evolution Equations and Dynamical Systems. Proceedings, 1979. Edited by M. Boiti, F. Pempinelli and G. Soliani. VI, 368 pages. 1980.

Vol. 121: F. W. Wiegel, Fluid Flow Through Porous Macromolecular Systems. V, 102 pages. 1980.

Vol. 122: New Developments in Semiconductor Physics. Proceedings, 1979. Edited by F. Beleznay et al. V, 276 pages. 1980.

Vol. 123: D. H. Mayer, The Ruelle-Araki Transfer Operator in Classical Statistical Mechanics. VIII, 154 pages. 1980.

Vol. 124: Gravitational Radiation, Collapsed Objects and Exact Solutions. Proceedings, 1979. Edited by C. Edwards. VI, 487 pages. 1980.

Vol. 125: Nonradial and Nonlinear Stellar Pulsation. Proceedings, 1980. Edited by H. A. Hill and W. A. Dziembowski. VIII, 497 pages. 1980.

Vol. 126: Complex Analysis, Microlocal Calculus and Relativistic Quantum Theory. Proceedings, 1979. Edited by D. Iagolnitzer. VIII, 502 pages. 1980.

Vol. 127: E. Sanchez-Palencia, Non-Homogeneous Media and Vibration Theory. IX, 398 pages. 1980.

Vol. 128: Neutron Spin Echo. Proceedings, 1979. Edited by F. Mezei. VI, 253 pages. 1980.

Vol. 129: Geometrical and Topological Methods in Gauge Theories. Proceedings, 1979. Edited by J. Harnad and S. Shnider. VIII, 155 pages. 1980.

Vol. 130: Mathematical Methods and Applications of Scattering Theory. Proceedings, 1979. Edited by J. A. DeSanto, A. W. Sáenz and W. W. Zachary. XIII, 331 pages. 1980.

Vol. 131: H. C. Fogedby, Theoretical Aspects of Mainly Low Dimensional Magnetic Systems. XI, 163 pages. 1980.

Vol. 132: Systems Far from Equilibrium. Proceedings, 1980. Edited by L. Garrido. XV, 403 pages. 1980.

Vol. 133: Narrow Gap Semiconductors Physics and Applications. Proceedings, 1979. Edited by W. Zawadzki. X, 572 pages. 1980.

Vol. 134: $\gamma\gamma$ Collisions. Proceedings, 1980. Edited by G. Cochard and P. Kessler. XIII, 400 pages. 1980.

Vol. 135: Group Theoretical Methods in Physics. Proceedings, 1980. Edited by K. B. Wolf. XXVI, 629 pages. 1980.

Vol. 136: The Role of Coherent Structures in Modelling Turbulence and Mixing. Proceedings 1980. Edited by J. Jimenez. XIII, 393 pages. 1981.

Vol. 137: From Collective States to Quarks in Nuclei. Edited by H. Arenhövel and A. M. Saruis. VII, 414 pages. 1981.

Vol. 138: The Many-Body Problem. Proceedings 1980. Edited by R. Guardiola and J. Ros. V, 374 pages. 1981.

Vol. 139: H. D. Doebner, Differential Geometric Methods in Mathematical Physics. Proceedings 1981. VII, 329 pages. 1981.

Vol. 140: P. Kramer, M. Saraceno, Geometry of the Time-Dependent Variational Principle in Quantum Mechanics. IV, 98 pages. 1981.

Vol. 141: Seventh International Conference on Numerical Methods in Fluid Dynamics. Proceedings. Edited by W. C. Reynolds and R. W. MacCormack. VIII, 485 pages. 1981.

Vol. 142: Recent Progress in Many-Body Theories. Proceedings. Edited by J. G. Zabolitzky, M. de Llano, M. Fortes and J. W. Clark. VIII, 479 pages. 1981.

Vol. 143: Present Status and Aims of Quantum Electrodynamics. Proceedings, 1980. Edited by G. Gräff, E. Klempt and G. Werth. VI, 302 pages. 1981.

Vol. 144: Topics in Nuclear Physics I. A Comprehensive Review of Recent Developments. Edited by T.T.S. Kuo and S.S.M. Wong. XX, 567 pages. 1981.

Vol. 145: Topics in Nuclear Physics II. A Comprehensive Review of Recent Developments. Proceedings 1980/81. Edited by T. T. S. Kuo and S. S. M. Wong. VIII, 571-1.082 pages. 1981.

Vol. 146: B. J. West, On the Simpler Aspects of Nonlinear Fluctuating. Deep Gravity Waves. VI, 341 pages. 1981.

Vol. 147: J. Messer, Temperature Dependent Thomas-Fermi Theory. IX, 131 pages. 1981.

Vol. 148: Advances in Fluid Mechanics. Proceedings, 1980. Edited by E. Krause. VII, 361 pages. 1981.

Vol. 149: Disordered Systems and Localization. Proceedings, 1981. Edited by C. Castellani, C. Castro, and L. Peliti. XII, 308 pages. 1981.

Selected Issues from

# Lecture Notes in Mathematics